Microbial Toxins

VOLUME VIII
FUNGAL TOXINS

Microbial Toxins

Editors: Samuel J. Ajl
Alex Ciegler
Solomon Kadis
Thomas C. Montie
George Weinbaum

Microbial Toxins

Edited by

Solomon Kadis

Research Laboratories
Albert Einstein Medical Center
Philadelphia, Pennsylvania

Alex Ciegler

United States Department of
   Agriculture
Agricultural Research Service
Northern Utilization Research
   and Development Division
Peoria, Illinois

Samuel J. Ajl

Research Laboratories
Albert Einstein Medical Center
Philadelphia, Pennsylvania

VOLUME VIII

FUNGAL TOXINS

1972

ACADEMIC PRESS • *NEW YORK AND LONDON*

ACADEMIC PRESS, INC.
111 Fifth Avenue, New York, New York 10003

United Kingdom Edition published by
ACADEMIC PRESS, INC. (LONDON) LTD.
24/28 Oval Road, London NW1 7DD

LIBRARY OF CONGRESS CATALOG CARD NUMBER: 78-84247

PRINTED IN THE UNITED STATES OF AMERICA

Contents

Section A

1. The Isolation and Identification of Toxic Coumarins

DONALD E. RICHARDS

2. The Biological Action of the Coumarins

LESTER D. SCHEEL

3. The Natural Occurrence and Uses of the Toxic Coumarins

VERNON B. PERONE

8. A Phytotoxin from *Didymella applanata* Cultures

F. Schuring and C. A. Salemink

9. Compounds Accumulating in Plants after Infection

Joseph Kuć

10. The Toxic Peptides of *Amanita* Species

Theodor Wieland and Otto Wieland

11. Mushroom Toxins Other than *Amanita*

Robert G. Benedict

12. Ergot

Detlef Gröger

List of Contributors

Numbers in parentheses indicate the pages on which the authors' contributions begin.

ROBERT G. BENEDICT (281), *College of Pharmacy, University of Washington, Seattle, Washington*

JOSEPH FORGACS (95), *Automated Biochemical Laboratories and Ramapo General Hospital, Spring Valley, New York*

V. E. GRACEN, JR. (131, 139),* *Department of Agronomy, University of Georgia, Athens, Georgia*

DETLEF GRÖGER (321), *German Academy of Sciences at Berlin, Institute for Biochemistry of Plants, Halle/Saale, German Democratic Republic*

JOSEPH KUĆ (211), *Departments of Biochemistry and Botany and Plant Pathology, Purdue University, Lafayette, Indiana*

H. H. LUKE (131, 139), *U.S. Department of Agriculture, Crops Research Division and Department of Plant Pathology, University of Florida, Gainesville, Florida*

VERNON B. PERONE (67), *Environmental Health Service, Environmental Control Administration, Bureau of Occupational Safety and Health, Cincinnati, Ohio*

DONALD E. RICHARDS (3), *U.S. Department of Health, Education, and Welfare, Public Health Service, National Institute for Occupational Safety and Health, Cincinnati, Ohio*

C. A. SALEMINK (193), *Organic Chemistry Laboratory, State University of Utrecht, Utrecht, Holland*

LESTER D. SCHEEL (47), *U.S. Department of Health, Education, and Welfare, Public Health Service, National Institute for Occupational Safety and Health, Cincinnati, Ohio*

*Present address: Department of Plant Breeding and Biometry, New York State College of Agriculture, Cornell University, Ithaca, New York.

F. SCHURING (*193*), *Keuringsdienst van Waren* (*Food Inspection Service*), *Utrecht, Holland*

G. E. TEMPLETON (*169*), *Department of Plant Pathology, University of Arkansas, Fayetteville, Arkansas*

OTTO WIELAND (*249*), *Klinisch-Chemisches Institut, Städt. Krankenhaus München-Schwabing, Munich, Germany*

THEODOR WIELAND (*249*), *Max-Planck-Institut für medizinische Forschung, Abteilung Chemie, Heidelberg, Germany*

Preface

Ergot toxicity has been known to man for much of his recorded history; man, in addition, probably learned early to distinguish between the toxic and edible mushrooms. Yet, in spite of these clues and the fact that molds have been associated with illness as early as the nineteenth century, mycotoxicoses remained the "neglected" diseases until 1961 when the aflatoxins were discovered and found to be the cause of the "turkey X" syndrome. These findings resulted in a renaissance in mycotoxin research followed by a rapid accumulation of a massive body of literature. The mycotoxin problem was soon found to be worldwide, although it was most severe in the nonindustrialized areas of Africa and Asia. Epidemiological studies indicated that some of the mycotoxins played an etiological role in cancer, particularly in southern Africa. Yet the role as causative agents of disease of many of the mycotoxins discovered in the past decade remains to be determined.

In the three volumes devoted to the algal and fungal toxins (VI, VII, VIII) of this multivolume treatise on microbial toxins, the authors have attempted to comprehensively and critically review what has been accomplished as well as indicate future lines of research. These volumes are intended for scientists and graduate students in the various scientific disciplines for a multidisciplinary approach is required to resolve and find solutions to the many problems presented.

The extensiveness of the mycotoxin literature forced the editors to divide the originally planned single volume on algal and fungal toxins into three: VI, VII, and VIII. However, the divisions are arbitrary with no implication intended that the toxins in any given group are related with respect to structure, function, mode of action, or biosynthesis. Volume VI includes the toxins produced by the Aspergilli and Penicillia, Volume VII, the algal toxins and those mycotoxins produced by species in the genera *Fusarium, Rhizoctonia,* and *Pithomyces,* and this volume, the toxins produced by the fungal phytopathogens, the mushrooms, and those toxins synthesized in plants in response to fungal invasion or other injury.

We are grateful to the contributors and the staff of Academic Press for their fine cooperation.

SOLOMON KADIS
ALEX CIEGLER
SAMUEL J. AJL

Contents of Other Volumes

Section A

CHAPTER 1

The Isolation and Identification of Toxic Coumarins

DONALD E. RICHARDS

I. Introduction

Among the fungal toxins associated with animal and human food materials are included several derivatives of coumarin that evoke varied pharmacological and physiological responses in the animal body. One such derivative, 3,3'-methylenebis(4-hydroxycoumarin), has been identified as the toxicant responsible for the fatal hemorrhagic disorder of domestic animals, principally cattle and sheep, which feed on moldy sweet clover. Schofield (1924) isolated an *Aspergillus*, a *Penicillium*, and a *Mucor* from a toxic sweet clover hay and cultured these fungi separately on nontoxic sweet clover stalks. By feeding the artificially spoiled sweet clover to rabbits, this investigator induced the disease symptoms in the experimental animals. This work of Schofield was the first in which a toxicosis was reproduced experimentally from toxigenic fungi that had been isolated from a mold-infested food material involved in a naturally occurring outbreak of mycotoxicosis. Two furocoumarins, 8-methoxypsoralen and 4,5',8-trimethylpsoralen, that possess strong phototoxic properties have been isolated from celery infected with the filamentous fungus *Sclerotinia sclerotiorum* (Scheel *et al.*, 1963). The

3

aflatoxins consist of a group of structurally similar, highly toxic compounds, related to coumarin, which are metabolites of certain strains of the mold *Aspergillus flavus*. These mycotoxins are potent hepatotoxins possessing carcinogenic activity. In 1960, the "turkey X" disease that killed an estimated 100,000 turkey poults in England was caused by the use of a diet containing groundnut meal contaminated with aflatoxins. Disease and death of other domestic animals including ducklings, chickens, cattle, and swine also resulted from consumption of such contaminated rations. The toxin-producing fungus isolated from whole unprocessed groundnuts was identified as *Aspergillus flavus* Link ex Fries (Sargeant *et al.*, 1961).

This review will present a detailed summary of the isolation and identification of toxic coumarin derivatives obtained as fungal metabolites from food materials. The aflatoxins, however, will be omitted from consideration, since these compounds have been treated thoroughly in Volume VI of this treatise. Coumarin itself is a fungal metabolite, but as yet has been obtained only in laboratory culture. The reaction sequence for the biosynthesis of coumarin in higher plants is reviewed for comparison with the biogenic route, so far as it is known, by which coumarin is formed in the fungus *Phytophthora infestans* and with the biochemical pathway in fungi leading to 4-hydroxycoumarin, the immediate precursor of 3,3'-methylenebis(4-hydroxycoumarin). The toxicities, expressed as LD_{50} values or as minimum effective concentrations in the case of the phototoxins, are given for the mycotoxins considered.

II. Coumarin

Coumarin, $C_9H_6O_2$, is a condensed heterocyclic compound. Synonyms for this compound include cumarin, 2H-1-benzopyran-2-one, 1,2-benzopyrone, 2-oxo-1H-benzopyrone, and o-coumarinic acid lactone. The system of numbering this heterocycle, which is in accordance with the rules approved by the International Union of Pure and Applied Chemistry, is shown in Fig. 1. Coumarin and its biogenic precursors are widely distributed in higher plants and coumarin itself has been isolated as a fungal metabolite.

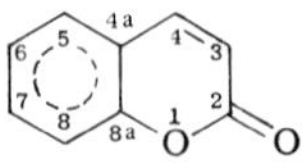

FIG. 1. Coumarin.

A. BIOSYNTHESIS OF COUMARIN BY HIGHER PLANTS

The biochemical pathway by which coumarin is synthesized in higher plants has been slowly elucidated through the research efforts of numerous investigators. Early researchers isolated and assayed coumarin in the extracts of plant tissues and erroneously concluded that the coumarin obtained represented to a very close approximation the actual coumarin content in the normal intact plant. With increased experimentation, it became evident that a glucosidic material, called "bound coumarin," also existed in plant tissue and was converted to coumarin under hydrolytic conditions used in most isolated procedures. The careful isolation and identification of this conjugate as coumarinic acid β-glucoside and subsequent acidic or enzymatic hydrolysis to coumarin established this glucoside as an intermediate in the coumarin biosynthesis. More recently, chemical feeding experiments with ^{14}C-labeled compounds have demonstrated that a wide variety of plant constituents are coumarin precursors, and as a result of these experiments and others, a logical development of the reaction sequence leading from L-phenylalanine to coumarin has been advanced. The biogenesis has been studied, for the most part, in species of the genus *Melilotus*.

It was demonstrated by Guérin and Goris (1920) that leaves of *Melittis melissophyllum* L. contained a substance, suggested as a glucoside, that following emulsin digestion yielded coumarin. Proof of the existence in plant tissues of three *Melilotus* species and *Asperula odorata* L. of such a glucosidic principle that could be hydrolyzed to coumarin and D-glucose was given by Bourquelot and Hérissey (1920). These researchers first extracted leaves and green branches of *Melilotus officinalis* with boiling water and removed existing traces of coumarin from this extract by distillation and ether extraction of the distillate. Portions of this coumarin-free extract released coumarin upon subsequent treatment with either sulfuric acid or a powder obtained from *M. officinalis* by extracting the fresh plant tissue with 95% ethanol, removing the ethanol, washing the plant tissue with ethanol, and finally drying the residual plant tissue at 34-35°C. Hydrolysis with the latter material showed the presence of an enzyme in sweet clover capable of catalyzing the release of coumarin. In another experiment, the fresh plant in bloom was first extracted with boiling ethanol. After removal of the ethanol by distillation, the residue was dissolved in water. Following treatment with emulsin, a change in optical rotation of the aqueous solution occurred that was indicative of the presence of D-glucose. By the same method, aqueous solutions were prepared from ethanolic extracts of leaves and stems of *Melilotus arvensis* Wallr. and dry seeds of *Melilotus leucantha* Koch. Emulsin digestion of these extracts also yielded D-glucose identified by optical rotation shifts. An extract of freshly harvested plants of *Asperula odorata* L. after treatment with emulsin formed coumarin and D-glucose. The D-glucose was shown to be present by a change in the optical rotation of the solution. The coumarin was removed from the enzyme-hydrolyzed ethanol extract by steam distillation. Ether extraction of the distillate followed by evaporation of the ether gave coumarin in crystalline form. Although this research established the presence of a glucosidic material in plant tissue that under the influence of emulsin is lysed to form coumarin and D-glucose, the chemical constitution of this glucoside was not established.

Further insight into the structural nature of the glucoside was given by Charaux (1925) with the isolation from the dried blossoms of *Melilotus altissima* Thuil. and *Melilotus arvensis* Wallr. of a compound identified as coumaric acid glucoside. This compound crystallized from water as the monohydrate, which upon drying at 105-110°C lost the water of hydration. Hydrolysis of this compound with emulsin or sulfuric acid produced coumaric acid and D-glucose in equimolar quantities. The D-glucose was identified by specific optical rotation. The coumaric acid was extracted into ether, and upon evaporation of the ether, the compound crystallized. Identification was made by the melting point, mixed melting point, fluorescence characteristics, and phenol color reactions.

By allowing ground seed tissue of common sweet clover (*Melilotus*) to autolyze at 40°C for 30 minutes, Roberts and Link (1937b) showed that the coumarin content increased from an initial 0.02% to 1.96%. A similar assay showed intact green leaf tissue of sweet clover to contain 0.63% coumarin; after autolysis, however, 1.43% coumarin was found. These increases in coumarin content were attributed to an enzymatic conversion to coumarin of a material contained in the plant tissue. It was suggested that this material, termed "bound coumarin," was a glycoside. These findings were in accordance with those of Bourquelot and Hérissey (1920) that tissue of *M. officinalis* contained an enzyme that catalyzed the formation of coumarin from a glucoside. Bound coumarin was later suggested to be coumarinic acid β-glucoside (Rudorf and Schwarze, 1958). This proposed identity was deduced from a knowledge of the hydrolysis products of the combined form. When the bound coumarin, extracted from leaves of *Melilotus alba,* was subjected to hydrolytic procedures, coumarin was identified as a hydrolyzate of emulsin digestion and both coumarin and glucose were found to be products of hydrochloric acid-catalyzed hydrolysis. Consequently, it was reasoned that bound coumarin must be a glucoside that upon enzymatic or non-enzymatic hydrolysis gives glucose and coumarinic acid, the cis form of *o*-hydroxycinnamic acid, which spontaneously lactonizes to form coumarin.

The identity of bound coumarin with coumarinic acid β-glucoside was confirmed experimentally by Kosuge (1961). This researcher showed that the quantity of coumarinic acid β-glucoside isolated from an extract of sweet clover tissue accounted for nearly all of the bound coumarin in this extract as determined by the colorimetric method of Roberts and Link (1937b). Fresh shoots of *M. alba* Desr. were frozen and ground, and the resulting powder was extracted with a hot 5% ethanol solution. After the plant residue was filtered off, the filtrate was acidified and extracted with ether to remove the coumarin. This coumarin-free filtrate then was neutralized and allowed to stand for 30 minutes at 37°C for the enzymatic hydrolysis of the bound coumarin. The quantity of coumarin formed during this period, as determined by the colorimetric method, was equivalent to 1.64 mg of coumarinic acid β-glucoside/gm of fresh tissue. Coumarinic acid β-glucoside was isolated from another portion of the plant extract, not subjected to conditions of hydrolysis, by paper chromatography. This method, shown by a control experiment to have a recovery efficiency of 88%, gave 1.39 mg of coumarinic acid β-glucoside/gm of fresh tissue. By correcting this value for the recovery efficiency, the quantity of coumarinic acid β-glucoside in the plant tissue was calculated to be 1.58 mg/gm of fresh tissue. The similarity of the quantities of bound coumarin and coumarinic acid β-glucoside found in the plant extract indicated that the two materials were identical.

Further substantiation of the identity of bound coumarin was given by Stoker and Bellis (1962b). These researchers isolated bound coumarin in pure form from the seeds of *M. alba* using paper chromatography and freeze-drying techniques. By chemical analysis, melting point determination, ultraviolet absorption characteristics, and paper electrophoretic and chromatographic behavior, the isolated material was shown to be identical to synthetically prepared coumarinic acid β-glucoside.

Even before the identity of bound coumarin had been established, intensive studies on the biosynthesis of coumarin by higher plants had begun. Weygand and Wendt (1959) using isolated root tissue cultures of *M. officinalis* studied the biogenic formation of phenylpropanoid derivatives that serve as coumarin precursors. Coumarin was isolated from the root tissues grown in nutrient solutions containing 3% sugar to which sodium acetate-1-^{14}C, uniformly labeled L-phenylalanine-^{14}C, or glucose-1-^{14}C had been added. The ^{14}C of both L-phenylalanine-^{14}C and glucose-1-^{14}C was incorporated into coumarin, but sodium acetate-1-^{14}C was not utilized. The coumarin obtained from root tissues grown on the medium containing isotopically labeled L-phenylalanine was converted by alkali fusion to salicylic acid which contained 73% of the specific radioactivity of the original coumarin. Had the intact 9-carbon skeleton of L-phenyl-alanine been transferred into coumarin and then undergone fission to form the 7-carbon skeleton of salicylic acid, 78% ($\frac{7}{9}$) of the ^{14}C of the original coumarin would have remained in the salicylic acid. It was thereby shown that little decomposition and reformation of the phenyl-propanoid structure took place in the biosynthetic conversion of phenyl-alanine to coumarin. After incorporation of glucose-1-^{14}C, the isolated coumarin was subjected to extensive chemical degradation studies to determine the distribution of radioactivity within the carbon skeleton. The ^{14}C in the benzene nucleus was localized to the extent of 87% in carbon atoms 5 and 7 (determined together) and 8a (see Fig. 1). If it is assumed that the radioactivity contribution of carbon atom 7 is negligible, the ^{14}C-rich positions become 5 and 8a, which are ortho to the propanoid side chain. Such a labeled-carbon distribution was similar to that which Srinivasan *et al.* (1956) found in shikimic acid isolated from the microorganism *Escherichia coli* mutant 83-24 after feeding on glucose-1-^{14}C. The radioactivity in the shikimic acid was concentrated in carbon atoms 2 and 6, adjacent to the carbonyl side chain, to the extent of 40 and 25%, respectively (Fig. 2). It was thus suggested that the benzene ring of coumarin is synthesized from carbohydrates by the shikimic acid pathway. Furthermore, the distribution of the radioactivity in the 3-carbon side chain that becomes attached to the ring during the conversion of shikimic acid to prephenic acid indicated that this 3-carbon unit is formed by the glycolysis of hexose.

The pathway by which coumarin is synthesized in *M. alba* Desr. was studied extensively by Kosuge and Conn (1959). Twenty-four hours after a solution of uniformly labeled shikimic acid-^{14}C, uniformly labeled glucose, DL-phenylalanine-3-^{14}C, *trans*-cinnamic acid (ring and 3-^{14}C), or sodium acetate-2-^{14}C and 28 hours after a solution of *o*-coumaric acid-2-^{14}C was administered to *M. alba* shoots, these plant tissues were assayed for radioactive coumarin. With the exception of *o*-coumaric acid, these compounds were not incorporated significantly into coumarin, and even the ^{14}C of *o*-coumaric acid underwent a dilution (S. A. Brown and

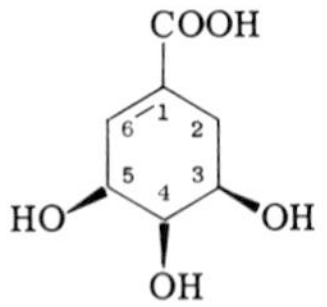

FIG. 2. Shikimic acid.

Neish, 1955) of 360 in the metabolic conversion. The *o*-coumaric acid was metabolized more efficiently, however, to its β-glucoside; a ^{14}C dilution of 12 was obtained for this transformation. *trans*-Cinnamic acid, DL-phenylalanine, and shikimic acid with radioactive-carbon dilutions of 29, 45, and 142, respectively, were effective precursors of *o*-coumaric acid which was isolated as the glucoside. A meaningful quantity of ^{14}C from labeled glucose also was found in *o*-coumaric acid, but sodium acetate was not utilized effectively. The incorporation of radioactivity from labeled shikimic acid and glucose into *o*-coumaric acid indicated that the aromatic ring of *o*-coumaric acid is synthesized by the shikimic acid pathway, and since *o*-coumaric acid is a precursor of coumarin, it was inferred that coumarin also is formed via the shikimic acid pathway.

The metabolism of coumarin in *M. alba* also was studied by Kosuge and Conn (1959). A solution of coumarin-3-^{14}C was administered to excised shoots of *M. alba*, and following a metabolic period of 28 hours, radioactive melilotic acid and melilotic acid β-glucoside, along with small amounts of *o*-coumaric acid β-glucoside and two unidentified compounds, were isolated from the sweet clover shoots. The melilotic acid and the melilotic acid β-glucoside accounted for 35 and 12%, respectively, of the total radioactivity given. Only 1% of the activity was incorporated into *o*-coumaric acid β-glucoside, whereas the two unidentified compounds acquired 1.5% of the total activity administered. It was suggested that this metabolism accounted for the low levels of ^{14}C found in coumarin following the administration of radioactive precursors of *o*-coumaric acid to *M. alba* shoots. In a later study conducted by Kosuge and Conn (1961), the metabolites of coumarin in *M. alba* were determined after a 24-hour exposure to coumarin-3-^{14}C. The major metabolites isolated from the plant tissue, melilotic acid and melilotic acid β-glucoside, acquired 12 and 59%, respectively, of the radioactivity from the isotopically labeled coumarin, whereas the coumarin isolated retained only 0.5% of the original activity.

The metabolism of coumarin to melilotic acid and melilotic acid β-glucoside in *M. alba* was confirmed by Stoker and Bellis (1962a). Radioactive coumarin, formed as a metabolite of *trans*-cinnamic acid-3-^{14}C administered to excised *M. alba* shoots, was partially metabolized further to melilotic acid and to a lesser extent to melilotic acid β-glucoside 48 hours but not 24 hours following the addition of radioactive *trans*-cinnamic acid to the shoots.

S. A. Brown *et al.* (1960) studied the biosynthesis of coumarin in the perennial sweet grass *Hierochloë odorata* Beauv. Ten carbon-labeled compounds were administered in sodium carbonate solution individually to excised shoots of *H. odorata* for evaluation as coumarin precursors. After a metabolic period of 7.5 hours following complete absorption of these radioactive compounds, coumarin was isolated from the shoots. This isolation was accomplished by first homogenizing the plant tissue in hot water, steam distilling the homogenate under reduced pressure, ether extracting the coumarin from the distillate, and finally evaporating the ether. Cinnamic acid-3-^{14}C, *o*-coumaric acid-1-^{14}C, randomly ^{14}C-labeled shikimic acid, and randomly ^{14}C-labeled L-phenylalanine with respective ^{14}C dilutions of 1200, 1200, 2900, and 4500 were most efficiently incorporated into coumarin. Acetate from acetic acid-1-^{14}C or acetic acid-2-^{14}C was utilized to a much lesser degree. It was concluded from these results that the shikimic acid pathway is the principal one operative in the coumarin biosynthesis—a conclusion reached earlier by Weygand and Wendt (1959) and Kosuge and Conn (1959) studying *Melilotus* species.

It was shown also that cinnamic acid served as an efficient precursor of *o*-coumaric acid in *H. odorata*. The tissue-containing residue from the steam distillation of homogenized *H. odorata* shoots to which cinnamic acid-3-^{14}C had been administered was extracted with ethanol, and the resulting slurry was acidified and freed from plant tissue and pigments prior to continuous ether extraction. Further purification was accomplished by treating the ether extract with 5% sodium hydrogen carbonate solution, acidifying the aqueous phase, and reextracting this aqueous layer with ether. The radioactive carbon in the *o*-coumaric acid, which was isolated from the ether extract by paper chromatography, underwent a dilution of only 3.2 in the conversion from cinnamic acid. Under similar conditions it was shown that *o*-coumaric acid-1-^{14}C was converted to radioactive *o*-coumaric acid β-glucoside by excised shoots of *H. odorata* with a ^{14}C dilution of 5.9.

Since cinnamic acid and *o*-coumaric acid were utilized with relative efficiency in the biogenic formation of coumarin, it was suggested that one intermediate step in the coumarin biosynthesis is the direct ortho hydroxylation of cinnamic acid. The efficient transformation of cinnamic acid to *o*-coumaric acid also supports this conclusion.

S. A. Brown and his co-workers (1960) also used chemical degradation of radioactive coumarin obtained from *H. odorata* shoots to which *o*-coumaric acid-1-^{14}C, *trans*-cinnamic acid-1-^{14}C, or *trans*-cinnamic acid-3-^{14}C was administered to show that the intact carbon skeleton of these phenylpropanoid acids is incorporated into coumarin. The coumarin formed from each of the phenylpropanoid acids was subjected to two separate degradation procedures. In the first of these, the radioactive coumarin was converted directly to salicylic acid by alkali fusion. Only the coumarin biosynthesized from *trans*-cinnamic acid-3-^{14}C gave rise to radioactive salicylic acid; this degradation product contained nearly 95% of the theoretical ^{14}C. In the second degradation procedure, an alkaline solution of the radioactive coumarin was treated with mercuric oxide to form labeled *o*-coumaric acid which then was thermally decarboxylated. The liberated carbon dioxide obtained from coumarin formed from the carboxyl-labeled *o*-coumaric and *trans*-cinnamic acids was radioactive. Of the ^{14}C administered as *o*-coumaric acid-1-^{14}C, 82% was recovered as carbon dioxide-^{14}C, and of that originating from *trans*-cinnamic acid-1-^{14}C, 94% was found in carbon dioxide-^{14}C.

Studies by S. A. Brown (1960, 1962) showed more conclusively that L-phenylalanine is a precursor of coumarin in higher plants. In the first of these, randomly ^{14}C-labeled L-phenylalanine was administered to *H. odorata* shoots under conditions similar to those used by S. A. Brown *et al.* (1960). In this work, however, the metabolic period was increased to 24 hours, and a ^{14}C dilution of 2240 was obtained. In the second study, L-phenylalanine was shown to be a precursor of coumarin in lavender. Six to seven days after the administration of a solution of generally ^{14}C-labeled L-phenylalanine through the roots of hydroponically grown *Lavandula officinalis* Chaix plants, radioactive coumarin was isolated by homogenization of the plant tissue in hot ethanol, removal of the ethanol under vacuum, and ether extraction of the aqueous residue. The dilution of radioactive carbon during this transformation was 201.

S. A. Brown (1960) suggested that *trans*-cinnamic acid is formed from L-phenylalanine and that this conversion could be enzyme mediated. The presence of such an enzyme, L-phenylalanine deaminase, in various higher plants including *M. alba* that catalyzes the direct deamination of L-phenylalanine to *trans*-cinnamic acid was demonstrated by Koukol and Conn (1961). Heating an aqueous solution, buffered at pH 8.8, containing the enzyme, isolated from barley shoots, and L-phenylalanine for 1 hour at 40°C resulted in the formation of *trans*-cinnamic acid. This compound was isolated and identified by its absorption spectrum, chromatographic properties, and melting point. The reaction, which appears to be irreversible, is specific for L-phenylalanine and forms only the trans isomer.

The rapid conversion of *trans*-cinnamic acid to *o*-coumaric acid β-glucoside in *M. alba* was reported by Stoker and Bellis (1962a). Radioactive *o*-coumaric acid β-glucoside was isolated as the major product 24, 48, and 72 hours after the administration of *trans*-cinnamic acid-3-^{14}C to excised sweet clover shoots. Smaller quantities of labeled coumarinic acid β-glucoside and radioactive coumarin also were isolated from the shoots. The specific activity of *o*-coumaric acid β-glucoside measured after each period of metabolism was found to be highest after 24 hours and then decreased progressively with time. Concurrently, the specific activities of the coumarinic acid β-glucoside and coumarin increased with the longer metabolic periods. These results indicated that once o-coumaric acid β-glucoside had been formed it slowly underwent rearrangement to its cis isomer, coumarinic acid β-glucoside, which then was metabolized further to coumarin. Such a trans-cis isomerization in the biosynthetic reaction sequence leading to coumarin had been suggested previously by Kosuge and Conn (1961). These researchers showed that in white sweet clover o-coumaric acid β-glucoside was partially metabolized to coumarin. Twenty-four hours after ^{14}C-labeled o-coumaric acid β-glucoside was given to excised shoots of *M. alba* Desr., 0.6% of the administered radioactivity was found in coumarin which was isolated from the shoots. This isolation was accomplished by treating the plant material, which was ground in the frozen state, with 50% ethanol at 60°C for 30 minutes, removing the plant residue by filtration, extracting the filtrate with ether, and finally chromatographing the concentrated ether extract on paper. The coumarin, located by its ultraviolet-light quenching property, was eluted with 20% ethanol. It was inferred from this study that the metabolism of o-coumaric acid β-glucoside to coumarin in the white sweet clover plant must proceed by way of the cis isomer as an intermediate.

That this biosynthetic trans–cis isomerization might be photoactivated was suggested by Haskins and Gorz (1961). The ultraviolet-mediated transformation of o-coumaric acid and its β-glucoside to coumarin and coumarinic acid β-glucoside, respectively, had been reported previously. Stoermer (1909) showed that o-coumaric acid, dissolved in benzene or in methanol, was changed to coumarin in approximately 75% yield after exposure for several days to ultraviolet radiation from a mercury lamp. Later, Lutzmann (1940) demonstrated that in neutral aqueous solution, the sodium salt of o-coumaric acid β-glucoside underwent an 80 to 85% conversion to the cis isomer when subjected to ultraviolet irradiation. Haskins and Gorz (1961) prepared aqueous extracts of leaflets of *M. alba* Desr. by autoclaving excised intact leaflets submerged in water for 15 minutes at approximately 15 psig. Upon exposure of these aqueous plant extracts to standard cool white fluorescent light or ultraviolet radiation, a partial conversion of o-coumaric acid β-glucoside to coumarinic acid β-glucoside was effected. These investigators also found that subjection of excised intact leaflets of white sweet clover to sunshine or ultraviolet irradiation resulted in a decrease of o-coumaric acid β-glucoside accompanied by a corresponding increase of the cis isomer. This ultraviolet-mediated isomerization was confirmed by Kahnt (1962) who showed by ultraviolet absorption spectra that o-coumaric acid β-glucoside extracted from leaves of *M. alba* was converted to the cis compound in methanolic solution after only a 10-minute exposure to sunlight.

Further evidence was given by Gorz and Haskins (1964) that the trans-cis transformation in higher plants is a photochemical reaction. These investigators reported that extracts of leaves from fifteen species of *Melilotus* and seven species of *Trigonella* plants, grown from seed in growth chambers at 80°F and 50% relative humidity under constant cool white fluorescent light, contained on the average greater than 90% of the o-hydroxycinnamic acid β-glucoside in the trans form. After these same plants were transferred to a greenhouse illuminated by natural daylight supplemented with incandescent light to give an 18-hour photoperiod and were allowed to grow an additional week, the trans compound in leaf extracts was reduced to an average 35% of the total. Although somewhat inconsistent with the conclusions drawn earlier by these researchers (Haskins and Gorz, 1961) that cool white fluorescent light induced the conversion of o-coumaric acid β-glucoside to the cis isomer, it was inferred from this more recent work that, whereas natural daylight supplemented with incandescent light effected the trans to cis isomerization, the fluorescent light was relatively ineffective in producing this conversion. The apparent contradiction in the results of these two investigations was resolved later by Haskins et al. (1964) who showed that the spectral emission characteristics of the General Electric Power Groove lamps used in the study reported in 1964 differed from those of Amplex cool white fluorescent lamps used in the earlier work. The radiation of the General Electric lamps was completely ineffective in converting an aqueous solution of authentic o-coumaric acid β-glucoside to coumarinic acid β-glucoside, but that of the Amplex lamps was indeed capable of catalyzing the isomerization reaction. Subsequently, it was demonstrated that only radiation having wavelengths shorter than 360 nm induced the trans to cis conversion. This higher wavelength limit was determined by subjecting detached leaflets of *M. alba* Desr. to sunlight illumination, filtered to give various transmission characteristics, for 0.5, 1, and 2 hours and then analyzing the irradiated leaflets for o-coumaric acid and total o-hydroxycinnamic acid.

Haskins et al. (1964) provided indirect evidence that the conversion of o-coumaric acid β-glucoside to coumarinic acid β-glucoside in *M. alba* Desr. was promoted entirely by photocatalysis and that no photosensitive isomerase system was involved in the reaction. These researchers showed that ultraviolet irradiation of detached leaflets placed on ice cubes and thus at a temperature only slightly above 0°C for 2 hours produced essentially the same amount of cis isomer (49 ± 2%, by difference) as that formed in leaflets irradiated at 29°C for a like period (43 ± 2%, by difference). Such a lack of temperature sensitivity would not be expected if a photosensitive isomerase system were active. In leaflets maintained in darkness, 100 ± 2% of the o-hydroxycinnamic acid was in the trans form. Radioisotope studies conducted by Edwards and Stoker (1967) further showed that no enzyme is present in M. officinalis which catalyzes the isomerization reaction in the absence of light. These workers administered o-coumaric acid-2-^{14}C dissolved in 5% aqueous sodium hydrogen carbonate to two groups of excised shoots. One group then was subjected to artificial illumination for 24 hours while the other was kept in darkness during this time. Following this metabolic period, coumarin was isolated from the shoots comprising each group by extraction and chromatographic techniques. This isolation involved an initial grinding of the plant tissue with sand in distilled water followed by treatment of the resulting mixture with β-glucosidase for 1 hour at 25°C. After the addition of acetone and acidification, the mixture was shaken mechanically for 1 hour and filtered. The filtrate then was diluted with water and extracted repeatedly with ether. Following evaporation of the ether from the bulked extracts, the coumarin was isolated and purified by chromatography on thin-layer plates and paper. The coumarin isolated from the shoots exposed to light contained a considerable quantity of ^{14}C. A small amount of radioactivity was measured also in the coumarin obtained from the shoots maintained in the dark. It had been observed, however, that solutions of o-coumaric acid-2-^{14}C in 5% aqueous sodium hydrogen carbonate at pH 8.5, after acidification, yielded ether extracts which when chromatographed gave radioactive spots corresponding in position to coumarin. Further experimentation revealed that o-coumaric acid undergoes a spontaneous slow progressive isomerization in the dark. Therefore, a control experiment was conducted to determine the amount of o-coumaric acid isomerized in the dark in the presence of plant tissue in which all enzymes had been heat inactivated. Fresh plant tissue was ground, heated to 100°C, and allowed to cool. A solution of o-coumaric acid-2-^{14}C then was added, and after a 24-hour period in darkness, coumarin was isolated and measured for radioactivity. The amount of radioactive coumarin formed in this control experiment was not significantly different from that produced in the feeding experiment conducted in the dark. It was thereby concluded that the isomerization reaction in the biosynthesis of coumarin in M. officinalis is not enzyme-mediated but is induced by photocatalysis. These results confirmed the earlier findings of Haskins et al. (1964).

The coumarinic acid β-glucoside thus formed in the plant tissue undergoes enzymatic hydrolysis to the unstable coumarinic acid which spontaneously lactonizes to form coumarin. Bourquelot and Hérissey (1920) showed that *M. officinalis* contained an enzyme capable of liberating coumarin from bound coumarin (coumarinic acid β-glucoside). These researchers prepared a powder that possessed β-glucosidase activity from fresh tissue of *M. officinalis* Willd. by extracting this plant tissue with 95% ethanol, removing the ethanol, and drying the residual plant tissue. An aqueous solution containing coumarinic acid β-glucoside then was obtained from *M. officinalis* by extracting leaves and green branches with boiling water, an extraction process which inactivated the endogenous enzyme. This solution was freed of coumarin by distillation and ether extraction of the distillate. Addition of the crude enzyme powder to the distillate containing the coumarinic acid β-glucoside and allowing the enzyme-substrate system to stand at room temperature resulted in the liberation of coumarin. After 48 hours, this compound was isolated in crystalline form from the heterogeneous mixture by distillation, extraction of the distillate with ether, and evaporation of the organic solvent. Roberts and Link (1937b) showed quantitatively that autolysis of both ground seed tissue and intact green-leaf tissue of sweet clover (*Melilotus*) at 40°C for 30 minutes in aqueous media brings about an increase in the coumarin content of these materials and ascribed this coumarin increase to an enzymatic hydrolysis of bound coumarin. Schaeffer et al. (1960) isolated the enzyme β-glucosidase from *M. alba* plants by grinding the plant tissue in cold water, centrifuging the resulting suspension, and fractionally precipitating the supernate with cold acetone. Reaction of the most active fraction, which precipitated over an acetone concentration range representing 40 to 57% of the total volume, with a preparation of bound coumarin, obtained from a hot water extract of *M. alba* leaves and partially purified by paper chromatography, at 30°C resulted in the formation of free coumarin.

Kosuge and Conn (1961) isolated the enzyme in crude form from extracts of an acetone powder of *M. alba* Desr. plants and purified this β-glucosidase approximately 40-fold by fractionation through precipitation techniques, dialysis, and liquid chromatography on carboxymethyl cellulose. This enzyme preparation was shown to be effective in catalyzing the hydrolysis of coumarinic acid β-glucoside but not *o*-coumaric acid β-glucoside as indicated by measuring the amount of glucose produced in the reaction mixture. Further enzymatic studies with the partially purified sweet clover β-glucosidase of Kosuge and Conn were reported by Kosuge (1961). By allowing the enzyme and the β-glucoside of coumarinic acid or *o*-coumaric acid, in concentrations approximating those obtained by extracting 1 gm of fresh seedling tissue of *M. alba* Desr. with 10 ml of water, to stand at 37°C, this investigator showed that 92% of the coumarinic acid β-glucoside was hydrolyzed in 30 minutes, whereas only a trace of the trans isomer was converted to the corresponding acid and glucose.

From results of metabolism studies with *Melilotus, Hierochloë,* and to a limited extent, *Trigonella* and *Lavandula* species, the sequence of biochemical reactions leading to the formation of coumarin in higher plants has been developed. Indirect evidence indicates that L-phenylalanine is synthesized in *M. alba, M. officinalis,* and *H. odorata* by way of the shikimic acid pathway. Once formed, this amino acid undergoes deamination catalyzed by L-phenylalanine deaminase to yield *trans*-cinnamic acid. Hydroxylation of the latter compound in the ortho position results in the formation of *o*-coumaric acid, which subsequently is conjugated as the β-glucoside. Evidence has also been given for an alternate route by which *trans*-cinnamic acid is converted directly to *o*-coumaric acid β-glucoside. Trans–cis isomerization of the β-glucoside of *o*-coumaric acid, mediated by photocatalysis, gives coumarinic acid β-glucoside. The cis compound then undergoes enzymatic hydrolysis to form the unstable coumarinic acid which spontaneously lactonizes to give coumarin. The course of this biosynthesis is summarized in Fig. 3.

B. Biosynthesis of Coumarin by *Phytophthora infestans*

Although coumarin has not been identified as a fungal metabolite in nature, this compound has been formed in the mold race 4 of *Phytophthora infestans* grown on a synthetic minimal medium containing glucose as the principal carbon source (Austin and Clarke, 1966). Umbelliferone (7-hydroxycoumarin) and herniarin (7-hydroxy-6-methoxycoumarin) also were formed as metabolites in this mold culture. These compounds were identified by paper, thin-layer, and gas-liquid chromatography, characteristic fluorescence and reactions with base, ultraviolet spectra, and comparison of properties of the isolated compounds with those of authentic compounds.

The pathway by which coumarin is synthesized in *P. infestans* also was investigated by Austin and Clarke (1966). Generally labeled L-phenylalanine-^{14}C, generally labeled L-tyrosine-^{14}C, and *trans-p*-coumaric acid-2-^{14}C were administered individually to 1-day-old surface cultures of *P. infestans* growing on synthetic minimal media. After the mold was allowed to grow for 8 days, the mycelium was filtered off and extracted with acetone. The aqueous culture filtrate also was extracted, but with

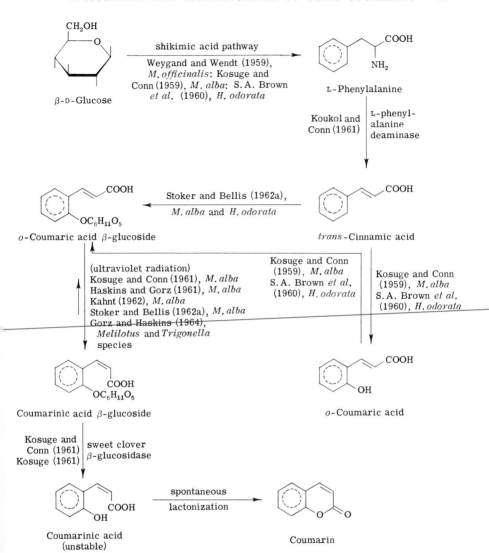

FIG. 3. Biosynthetic pathway to coumarin operative in higher plants as determined by experiments with *Melilotus*, *Hierochloë*, and *Trigonella* species.

the less water-soluble ethyl acetate. Coumarin, umbelliferone, and herniarin were isolated from extracts by thin-layer chromatography. The separated compounds were eluted from the adsorbent, fractionally sublimed, and crystallized from water. The incorporation of the radioactivity from each labeled compound into coumarin and umbelliferone was

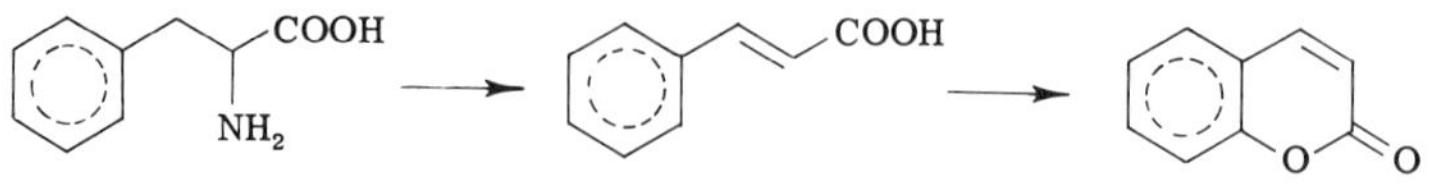

L-Phenylalanine   *trans*-Cinnamic acid   Coumarin

FIG. 4. Proposed reaction sequence for the biosynthesis of coumarin in *Phytophthora infestans*.

measured by the inverse dilution technique, but the incorporation of the ^{14}C into herniarin was not measured quantitatively. The results for coumarin showed that radioactive L-phenylalanine was incorporated to the extent of $9.60 \times 10^{-2}\%$, whereas ^{14}C-labeled L-tyrosine and *trans-p*-coumaric acid were poorly utilized. The umbelliferone acquired $12.47 \times 10^{-2}\%$ and $8.13 \times 10^{-2}\%$, respectively, of the ^{14}C from the administered radioactive L-phenylalanine and *trans-p*-coumaric acid. L-Tyrosine was also ineffective as a precursor of umbelliferone. From these results, it was reasoned that the fungal metabolic sequence leading to umbelliferone included the deamination of L-phenylalanine and para hydroxylation of the resulting *trans*-cinnamic acid. It was similarly proposed that coumarin was formed via L-phenylalanine and *trans*-cinnamic acid (Fig. 4). The biosynthetic route by which coumarin is formed in *P. infestans*, suggested from this limited study, is similar to that operative in higher plants.

C. TOXICITY OF COUMARIN

Coumarin is a slightly to moderately toxic compound. Ito *et al.* (1951) gave the acute oral LD_{50} of coumarin for mice as 196 mg/kg when calculated by the method of van der Waerden (1940) and as 227 mg/kg when determined by the method of Behrens and Kärber (Kärber, 1931). The acute oral LD_{50} for male albino Carworth strain rats has been reported by Hazleton *et al.* (1956) as 290 ± 30 mg/kg when administered in a propylene glycol medium and 520 ± 30 mg/kg when given in corn oil. More recently Jenner *et al.* (1964) established acute oral LD_{50} values for both young adult Osborne-Mendel rats and guinea pigs. The LD_{50} for the rats, evenly divided by sex, was given as 680 mg/kg with a range of 505-920 mg/kg when dosed in corn oil. The LD_{50} for guinea pigs consisting of both sexes was found to be 202 mg/kg with a range of 179-228 mg/kg. The coumarin was administered to the guinea pigs in a propylene glycol medium.

The effect of sublethal doses of coumarin on the animal body has been studied. Hazleton *et al.* (1956) conducting 90-day studies with male albino Carworth strain rats found that at a concentration of 1000 ppm of coumarin in the diet, growth was significantly retarded, but no histopathological effects were observed. In separate 90-day paired feeding experiments involving both male and female rats, a dietary level of 2500 ppm of coumarin resulted in pathological changes in the liver, primarily as fatty degeneration. In the females, a suppression of food efficiency also occurred. Studies by Hagan *et al.* (1967) have shown that Osborne–Mendel rats of both sexes receiving 1000 ppm of coumarin in the diet for 28 weeks suffered no adverse effects. When coumarin was consumed at a level of 2500 ppm for 29 weeks, however, toxic effects occurred resulting in metabolic changes and pathological alterations in the liver. This liver condition manifested itself as fatty infiltration and was more pronounced in the males. Hazleton *et al.* (1956) also studied the toxic effect of coumarin on mongrel dogs. Coumarin was administered orally to these animals by capsule, either in a powdered form or as a 4% solution in propylene glycol, in daily doses of 100 mg/kg for periods ranging from 8 to 19 days. At this dosage level, pathological changes were observed, notably liver centrilobular necrosis and acute kidney glomerular nephritis. This toxic effect of coumarin in dogs, however, was shown to be reversible. After becoming clinically toxic from coumarin, two dogs were observed for 40 and 61 days, respectively, following the administration of the last dose of coumarin. These animals were found to recover from toxic effects as evidenced by clinical findings, laboratory studies, and histopathological examination. No work was reported by these researchers to establish the threshold level of toxicity for dogs. Hagan *et al.* (1967) studied the toxic action of coumarin on dogs, both purebred beagles and mongrels, using lower doses. Administered orally by capsule, coumarin at a dose level of 10 mg/kg for a period ranging from 297 to 350 days gave no definite effects. When the dose was increased to 25 mg/kg for 133 to 330 days, however, pathological changes occurred, principally in the liver.

D. ANIMAL DETOXICATION OF COUMARIN

The chemical transformations involved when coumarin is detoxified by metabolic processes in the animal body were first investigated by Mead *et al.* (1958). Metabolites obtained from the urine of rabbits, to which measured doses of coumarin suspended in water were given by stomach tube, were identified as conjugates of 3-, 7-, and 8-hydroxycoumarins. 3-Hydroxycoumarin was excreted and isolated as the glucuronide and as the potassium sulfate. The sulfate upon acid hydrolysis gave the free 3-hydroxycoumarin. Positive chemical tests were obtained for 7-hydroxycoumarin glucuronide and potassium 7-hydroxycoumarin sulfate, but neither compound was isolated. Upon hydrolysis, however, both of these conjugates yielded 7-hydroxycoumarin. Conjugated 8-hydroxycoumarin and its hydrolyzed product were identified by paper chromatographic techniques but were not isolated. Conjugates of 3-, 7-, and 8-hydroxycoumarins also were found as metabolic products in the urine of ferrets, guinea pigs, mice, and rats to which coumarin had been administered. The urine of ferrets, guinea pigs, and mice, having been given coumarin, also yielded 5-hydroxycoumarin after acidic hydrolysis. This compound was not isolated but identified by paper chromatography. Thus, the only observed chemical changes resulting from the metabolism of coumarin were those of hydroxylation and conjugation. The metabolites measured, however, accounted for not more than 25% of the coumarin administered; the metabolic fate of the rest of the coumarin was not known. Reporting on the metabolism of coumarin by rabbits and rats, Booth *et al.* (1959) showed that in addition to hydroxylated derivatives, other organic products were obtained from the urine which indicated that the lactone ring of coumarin was opened. After feeding a diet containing 0.5% coumarin to rabbits, these researchers isolated *o*-hydroxyphenylacetic acid, 3-hydroxycoumarin, and 7-hydroxycoumarin from the urine of these animals. The compounds were identified by their paper chromatographic and crystallographic properties. Because of a hydrochloric acid treatment in the isolation procedure, 3-hydroxycoumarin and 7-hydroxycoumarin were obtained free of conjugation. Melilotic acid and *o*-hydroxyphenyllactic acid also were detected in the urine by paper chromatography. After acid hydrolysis of a fraction of chromatographed urine extract, *o*-coumaric acid and melilotic acid were separated from *o*-hydroxyphenyllactic acid by further chromatography. The precursors of these hydrolyzed products, and thus metabolites of coumarin, were suggested as *o*-hydroxyphenylacrylic acid and melilotoylglycine, respectively. When adult rats were given a 100 mg dose of coumarin dissolved in ethanol by intubation, no 3-hydroxycoumarin or 7-hydroxycoumarin was found in the urine. Urinary metabolites detected by chromatographic methods included *o*-hydroxyphenylacetic acid, melilotic acid, *o*-coumaric acid, and *o*-hydroxyphenyllactic acid. The identity of *o*-hydroxyphenylacetic acid again was confirmed by comparison of the crystallographic properties of the isolated compound with those of an authentic sample. An acid hydrolysis of a fraction of the chromatographed rat urine extract identical with that conducted on the rabbit urine gave *o*-coumaric acid and melilotic acid. Chromatographic methods revealed the identity of these compounds and also showed that coumaroylglycine and melilotoylglycine were their respective urinary precursors.

III. 3,3′-Methylenebis(4-hydroxycoumarin)

3,3′-Methylenebis(4-hydroxycoumarin) is produced by the growth of certain fungi on improperly cured hay or silage made from any of the common varieties of the sweet clovers *Melilotus alba* and *Melilotus officinalis* or sweet vernal grass *Anthoxanthum odoratum*. Cattle and sheep feeding on such moldy forage are afflicted with a hemorrhagic disease, known in agricultural practice as "sweet clover disease," which is characterized by a prolonged coagulation time of the blood and injury to the blood vessels. Many cattle having this disease undergo fatal hemorrhage after being subjected to surgical treatment or incurring accidental injuries. Other sickened animals that die show massive internal hemorrhage when submitted to postmortem examination. Although the cause of this disease is now clearly understood, research over two decades following the first observation of the disorder was conducted before the mycotoxin 3,3′-methylenebis(4-hydroxycoumarin) was isolated and identified as the causative agent.

A. ISOLATION AND IDENTIFICATION

Outbreaks of a new disease of livestock that fed on improperly cured sweet clover hay occurred in September and December of 1919 (Paulman, 1923). This disease having previously unobserved hemorrhagic symptoms was fatal to several cattle and sheep and a horse. It was suggested that mold growth found in the haystack was involved in the toxic condition of the hay.

Many cases of sweet clover disease of cattle were observed in the winter months of 1921–1922 both in Canada and in the United States (Schofield, 1922; Roderick, 1929). In a preliminary report, Schofield (1922) noted that in all cases of the malady, moldy sweet clover hay or silage had been consumed by the disease victim. In a later publication, Schofield (1924) described clinical observations, micropathological findings, and the experimental production of the toxicosis. This Canadian veterinarian proved a definite relationship between mold growth on sweet clover and the concurrent development of hemorrhagic properties of this forage crop. In three different samples of moldy hay, the clean stalks were carefully separated from the moldy stalks. When each of the six separate portions of hay was fed to six individual rabbits, the three rabbits receiving the moldy stalks died, while the three animals eating the clean stalks remained healthy. Several mold species were isolated from toxic sweet clover, and these molds were innoculated and grown individually in pure culture on healthy sweet clover stalks. When sweet clover artificially spoiled by certain of these molds was fed to rabbits, the disease symptoms of the stricken cattle were reproduced in the experimental animals. Feeding sweet clover on which was grown a variety of mold *Aspergillus* A to rabbits resulted in hypoprothrombinemia, hemorrhage, and death. Rabbits fed on sweet clover damaged by a *Penicillium*, strain D. B., suffered a 50% destruction of red blood cells and a slight increase in clotting time. Sweet clover spoiled by a *Mucor*, strain D. W., ingested by rabbits gave similar but less severe results. When the isolated molds apart from the sweet clover were fed to rabbits, they were found not to cause the hemorrhagic disease. Aqueous extracts of moldy sweet clover were prepared and fed to six rabbits. Three of these animals suffered a very acute anemia with a destruction of 50% of the red blood cells in less than a week. A delay in the clotting time of the blood also was produced. The other three rabbits remained normal. The results of these studies showed that certain molds have the ability to alter sweet clover from a harmless forage crop to a highly toxic substance.

Earlier, sweet clover had been listed as a poisonous plant by Pammel (1911) with the implication that the coumarin present in the plant is the toxic constituent. Once the causal relationship between mold growth on forage made from sweet clover and the toxicity of this animal food had been demonstrated, coumarin was not regarded further by Schofield as the causative principle of sweet clover disease. Roderick and Schalk (1931) made repeated attempts to extract the toxic substance from moldy sweet clover hay of proven toxicity. One experiment involved the extraction of 8 kg of chopped toxic hay with about 40 liters of water at room temperature for 16 to 18 hours. This extract was concentrated under reduced pressure at 48–50°C, and the concentrate then was fed to three rabbits over a period of about a month. The clotting power of the blood of these animals was not altered. Thereupon, the residual hay from the extraction process was fed to the rabbits and the usual symptoms of the hemorrhagic disease appeared. This experiment failed to confirm the findings of Schofield (1924) that the toxic agent is water soluble. Other attempts to extract the causative agent using 95% ethanol, ethyl ether, and 1% acetic acid as solvents also failed, and this phase of experimentation was suspended.

J. M. Brown *et al.* (1933) also prepared an aqueous extract of spoiled sweet clover hay, which had been shown to produce disease in cattle, by soaking the hay in distilled water for a period of 14 days. This aqueous extract was fed to two rabbits for a period of 25 days without producing any deleterious effects. Thus, the findings of these investigators were in agreement with those of Roderick and Schalk (1931) that the toxin contained in spoiled sweet clover hay is not soluble in water.

In an experiment to determine whether coumarin would induce the hemorrhagic disease, Roderick and Schalk (1931) administered this compound encapsulated in varying quantities to a group of five rabbits. Although three of the rabbits died, no hemorrhagic lesions were found. The two surviving rabbits were given up to 0.5 gm of coumarin daily over a period of 1 month without ill effect. It was thereby concluded that coumarin was not a likely cause of the hemorrhagic disease.

Although coumarin itself does not give rise to sweet clover disease, Smith and Brink (1938) reported that the addition of coumarin to partially cured alfalfa hay that then was allowed to spoil resulted in a slightly molded hay having an anticoagulative property. That moldy alfalfa hay alone induces the hemorrhagic disease had not been reported. This study was prompted by the observed relationship between the bitterness of sweet clover, attributed at least in part to coumarin, and the toxicity of spoiled sweet clover hay. In this work, freshly mown alfalfa first was dried to approximately 50% moisture, passed through a silage cutter, and divided into two equal portions. Ground coumarin was added through an 80 mesh screen to one lot of the alfalfa, with frequent turning, to the extent of 1.4% on a dry weight basis. The other part of the alfalfa was untreated. The quantities then were placed in separate piles and covered with paper, burlap and canvas. After 8 days the piles were uncovered and inspected. The untreated hay was heavily molded but was not examined for toxicity. Although only slightly moldy, the coumarin-treated hay when fed to two rabbits for 10 days induced an increase in the coagulation time of the blood. Another rabbit fed well-cured, mold-free alfalfa hay was found to be normal. From these results coumarin was implicated in the formation of the disease-producing agent — a conclusion not warranted, since the untreated moldy alfalfa hay had not been tested for toxicity. Thus, the significance of the mold on the hemorrhagic forage first observed by Schofield was lost, and two decades intervened before the relationship of fungi to the sweet clover disease was reestablished.

In 1934, Link and his collaborators at the University of Wisconsin Agricultural Experiment Station began an intensive program designed to isolate and identify the toxic agent in spoiled sweet clover. Five years later, on June 28, 1939, Campbell crystallized the hemorrhagic agent from cyclohexanone on a microscope slide (Campbell and Link, 1941; Link, 1943–1944). This isolation extended the previous work reported by Campbell et al. (1940) by which a solid concentrate of the hemorrhagic agent in artificially spoiled sweet clover hay (M. alba) had been prepared by very elaborate extraction and fractionation techniques.

This concentrate, which possessed an anticoagulative activity in rabbits approximately 200-fold over that of the original spoiled hay, contained as impurities water-insoluble, ethanol-, methanol-, and ether-soluble acidic substances including chlorophyll degradation products, fatty acids, phenolic compounds, and organic acids.

In the preparation of the hemorrhagic concentrate, the spoiled sweet clover hay first was milled to pass through a screen having openings approximately 0.5 mm in diameter. Extraction of this ground hay with Skellysolve A in a Soxhlet extractor for 24 hours removed neutral fats and waxes and certain other lipoidal materials which were physiologically inactive. The Skellysolve A-extracted hay then was steeped at 25–30°C for 48 hours in dilute hydrochloric acid, the concentration of which was maintained in the range of 0.01 to 0.1 M. The acid-steeped hay subsequently was filtered off. This treatment hydrolyzed the glycosides and separated inactive sugars, amines, alkaloids, peptides, dextrins, water-insoluble acids, and certain inorganic salts from the hay. The acid-steeped tissue fibers next were steeped in sodium hydroxide solution maintained at pH 11–12 at 25–30°C for 48 hours to digest the cellulose and to convert pectins, gums, hemicelluloses, and water-insoluble acids into soluble sodium salts. Concentrated hydrochloric acid subsequently was added to the leaching solution until pH 3 was attained to precipitate the pectins, gums, hemicelluloses, and water-insoluble acids on the fibers which then were collected on a filter. The inactive filtrate contained certain plant pigments. The tissue residue containing the precipitated material was leached with 90–95% ethanol at 25–30°C for 48 hours. The ethanolic extract, which contained the total activity, was separated by filtration. The leaching process was repeated three times, and the extracts were combined. The bulked extracts, which contained acids, chlorophyll degradation products, and some ethanol-soluble lipoidal material, then were concentrated to a syrup under reduced pressure at 25–30°C. This residue was dissolved in 0.5% sodium hydroxide, and the solution was acidified to pH 3 with hydrochloric acid, whereupon a brownish green, flocculent precipitate was formed. The precipitate was filtered off, and after this active material was dissolved in methanol, ether was added to the solution. A flocculent precipitate resulted which was separated by filtration and then redissolved in methanol. Addition of ether to this methanolic solution gave an inactive gummy precipitate and an active brownish green solution. Following removal of the volatile solvents from the active solution under reduced pressure at 25–30°C, the residue was dissolved in 0.5% sodium hydroxide solution. Upon acidification with hydrochloric acid, a fine precipitate was formed. Extraction with ether resulted in the transfer of some of the active material into the ether layer. The insoluble material remaining in the aqueous layer was dissolved by the addition of sodium hydroxide solution, again precipitated with hydrochloric acid, and extracted with ether. The insoluble material still remaining again was dissolved in methanol and subjected to the subsequent operations of the purification procedure to extricate the remaining active material. The ether extracts derived from this treatment were combined with the two ether extracts obtained previously. The solid hemorrhagic agent then was obtained by removing the ether under reduced pressure.

The isolation of the pure hemorrhagic agent by Campbell (Campbell and Link, 1941; Link, 1943-1944) involved the same initial procedure that gave the ether solution of the hemorrhagic concentrate and a further purification of this solution to remove contaminants including thermal degradation products of chlorophyll pigments, fatty acids, phenolic compounds, and organic acids.

The chlorophyll degradation products were removed by extracting the ether solution of the hemorrhagic agent concentrate four times with concentrated hydrochloric acid. After reextracting the combined acid layers with fresh ether and bulking the ether solutions, the resulting solution was concentrated by distillation at atmospheric pressure. The remaining ether was removed by distillation under reduced pressure at 25-28°C in an atmosphere of nitrogen or carbon dioxide. The active residue was dissolved in benzene, and the solution was diluted with Skellysolve A, whereupon a brown solid separated. After this inactive solid material was filtered off, the filtrate was concentrated to dryness under reduced pressure. The residue was subjected to this treatment a second time to remove more inactive solid material. The solvent again was removed by distillation under reduced pressure at 25-28°C, leaving a residue containing nonacidic substances as impurities. These substances were removed by extracting the residue, dissolved in 0.5% sodium hydroxide solution, several times with ether. The alkaline solution then was acidified with dilute hydrochloric acid, and the acidic organic materials were extracted into ether. The active ether solution was concentrated to a small volume. In attempts to crystallize the hemorrhagic agent, a small amount of the concentrated ether solution was placed on a microscope slide. With the addition of a drop of cyclohexanone, crystals were formed. This marked the first isolation of the hemorrhagic agent in crude form. Thereafter, the hemorrhagic agent was separated from the concentrated ether solution by fractional crystallization from ethanol. One recrystallization from a benzene-methanol mixed solvent and two recrystallizations from acetone gave the pure hemorrhagic agent which melted at 288-289°C. The molecular formula of this compound was determined to be $C_{19}H_{12}O_6$.

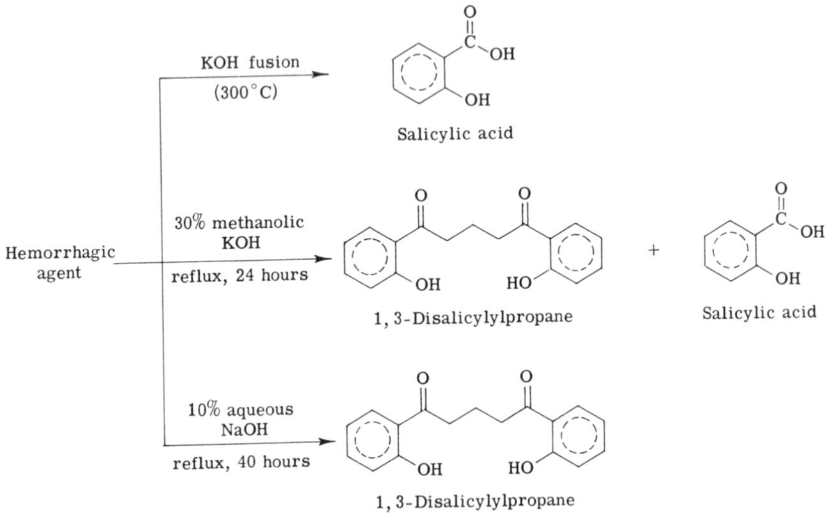

FIG. 5. Degradation reactions of the hemorrhagic agent.

Stahmann (Stahmann *et al.*, 1941; Link, 1943-1944) subsequently developed a substantially simplified isolation procedure that gave a higher yield of the hemorrhagic agent from spoiled sweet clover hay. Using this procedure, 1800 mg of the hemorrhagic agent was amassed by repeated isolations from spoiled hay. This quantity of the hemorrhagic agent was used for characterization studies.

Huebner (Stahmann *et al.*, 1941; Link, 1943-1944) subjected the hemorrhagic agent to three different alkaline degradation reactions to give some insight into the identity of the compound. Fusion of the hemorrhagic compound with potassium hydroxide at 300°C produced salicylic acid, $C_7H_6O_3$, quantitatively (102% of theoretical conversion) with a loss of five carbon atoms. Refluxing the hemorrhagic agent with 30% potassium hydroxide in 90% methanol for 24 hours gave a δ-diketone, 1,3-disalicylylpropane, along with some salicylic acid. By refluxing the hemorrhagic agent with 10% aqueous sodium hydroxide for 40 hours, 1,3-disalicylylpropane, $C_{17}H_{16}O_4$, was formed in 98% of theory with the loss of two carbon atoms and two oxygen atoms. These degradation reactions are shown in Fig. 5.

The isolated products of these reactions indicated that the structure of the hemorrhagic compound is characterized by two benzene nuclei linked by a five carbon δ-diketone. The existence of a hydroxyl group on each benzene nucleus of the degradation product, 1,3-disalicylylpropane, when the original anticoagulant exhibited no phenolic properties suggested the presence of a phenolic ester at each of these sites prior to hydrolysis. It was postulated that each of the ester carbonyl groups bridged the phenolic oxygen atom with the carbon atom adjacent to the respective keto group of the δ-diketone to form lactonic structures. Under mild alkaline degradation conditions, the ester carbonyl groups would be eliminated from a compound having such a structure. Thus, the loss of two carbon atoms and two oxygen atoms upon treatment of the hemorrhagic agent with 10% aqueous sodium hydroxide was in accordance with the expected behavior of the proposed lactonic linkages. It was reasoned further that the salicylic acid formed under the most severe alkaline degradation conditions resulted from the cleavage of the double bonds of the enol form of the intermediate δ-diketone. This supposition would account for the loss of five carbon atoms during the potassium hydroxide fusion. The molecular structure of the hemorrhagic agent thus deduced from the degradation studies, when written in the enolic form, was identical to that of α-methylen-bis-β-oxycumarin, also designated α-methylen-bis-benzotetronsäure, which had been prepared synthetically by Anschütz (1903, 1909) (Fig. 6). The melting point of the previously synthesized compound was reported as 260°C, however, nearly 30°C below that of the isolated mycotoxin.

To confirm the identity of the hemorrhagic agent as α-methylen-bis-benzotetronsäure, 4-hydroxycoumarin and from it α-methylen-bis-benzotetronsäure were prepared from salicylic acid, the primary degradation product of the hemorrhagic agent. Addition of metallic sodium to molten acetylsalicylic acid methyl ester, formed from salicylic acid, in accordance with the procedure of Pauly and Lockemann (1915) yielded 4-hydroxycoumarin that upon condensation with formaldehyde as described by Anschütz (1903) gave α-methylen-bis-benzotetronsäure. This reaction sequence is shown in Fig. 7. On April 1, 1940, it was reported by Huebner and Stahmann that the mixed melting point of the synthetic product and the hemorrhagic agent showed no depression (Link, 1943–1944). Other physical properties and the anticoagulative activity of the synthetic product were identical with those of the hemorrhagic agent isolated from spoiled sweet clover (Stahmann et al., 1941).

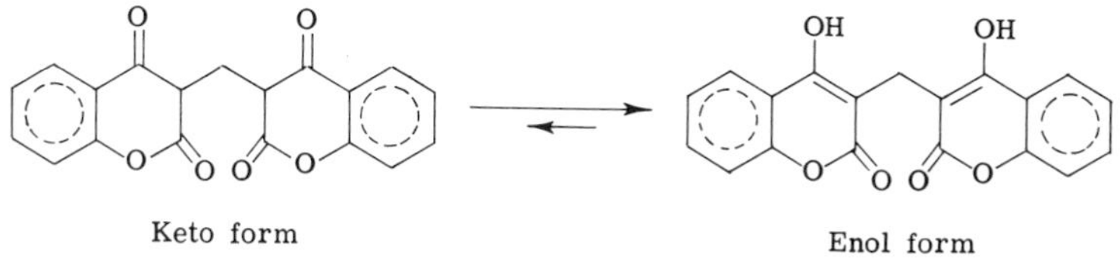

Keto form   Enol form

FIG. 6. α-Methylen-bis-benzotetronsäure.

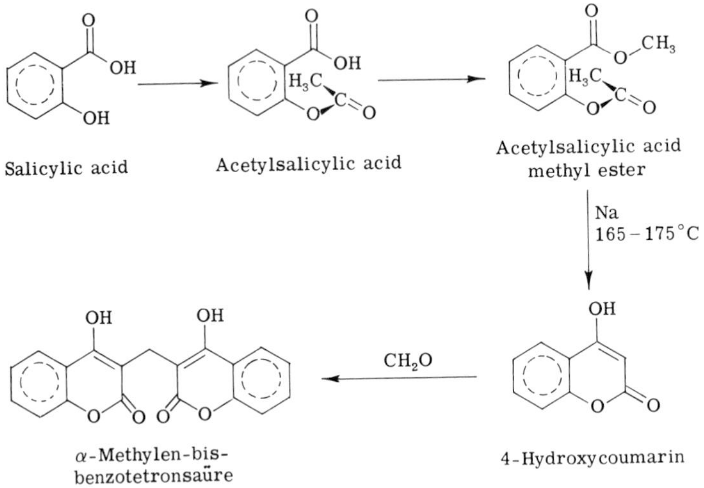

Salicylic acid   Acetylsalicylic acid   Acetylsalicylic acid methyl ester

Na
165–175 °C

α-Methylen-bis-benzotetronsaüre   CH₂O   4-Hydroxycoumarin

FIG. 7. Reaction sequence for the preparation of the hemorrhagic agent from salicylic acid.

Since the hemorrhagic agent was believed to be formed in spoiled sweet clover from coumarin, Stahmann and his co-workers named this compound as a coumarin derivative, 3,3′-methylenebis(4-hydroxycoumarin).

B. BIOSYNTHESIS OF 3,3′-METHYLENEBIS(4-HYDROXYCOUMARIN) IN FUNGI

3,3′-Methylenebis(4-hydroxycoumarin) is not a normal constituent of unspoiled sweet clovers (M. alba and M. officinalis) or of unspoiled sweet vernal grass (A. odoratum). It is only when these plants are infected with certain fungi that the toxin is produced. Research has revealed that 4-hydroxycoumarin is formed first in fungi and that this compound then is condensed with formaldehyde to give 3,3′-methylenebis(4-hydroxycoumarin). Biogenic studies have shown that the formation of the intermediate 4-hydroxycoumarin in fungi follows a route different from that leading to coumarin in the host plant.

After the toxic principle in spoiled sweet clover (M. alba) had been isolated and identified as 3,3′-methylenebis(4-hydroxycoumarin), it was suggested that the biosynthesis of this compound during spoilage proceeds by the conversion of coumarin to 4-hydroxycoumarin and the subsequent condensation of the latter compound with formaldehyde (Stahmann et al., 1941). It seemed, however, that only a very small amount of the coumarin present in the Melilotus species was transformed into 3,3′-methylenebis(4-hydroxycoumarin). The maximum concentration of the toxin found in the spoiled sweet clover hays was equivalent to only 0.0026% coumarin in these hays. The total coumarin content of the hays ranged from 0.75 to 1.58% as determined by the method of Roberts and Link (1937b).

In a preliminary report on the biological oxidation of coumarin and related compounds, Bellis (1958) presented evidence that two fungi, *Penicillium jenseni* and *Penicillium nigricans*, isolated from the soil, could produce 4-hydroxycoumarin along with traces of 3,3'-methylene-bis(4-hydroxycoumarin) from *o*-coumaric acid but not from coumarin in a modified Czapek medium. When the medium contained sucrose, the formation of a small amount of 3,3'-methylenebis(4-hydroxycoumarin) from 4-hydroxycoumarin by either *Penicillium* species was indicated. This study was the first to demonstrate experimentally that certain fungi possessed the capability of synthesizing 4-hydroxycoumarin and 3,3'-methylenebis(4-hydroxycoumarin) and thus marked the rediscovery of the relationship between fungi and the formation of the hemorrhagic principle in spoiled sweet clover.

More recently, Bellis and his colleagues (1967) reported that, in addition to the *Penicillium* species reported earlier, three thermophilic fungi, *Humicolor stellata*, *Humicolor languinosa*, and *Mucor pusillus*, grown on mineral salts media containing *o*-coumaric acid produced 4-hydroxycoumarin and 3,3'-methylenebis(4-hydroxycoumarin). These products were isolated from the culture media by chloroform extraction and identified by paper chromatographic mobilities, color reactions on paper chromatograms, and ultraviolet absorption spectra. None of the molds produced 4-hydroxycoumarin or 3,3'-methylenebis(4-hydroxycoumarin) when coumarin was used as the sole carbon source.

Spring and Stoker (1969) also identified 4-hydroxycoumarin and 3,3'-methylenebis(4-hydroxycoumarin) in a synthetic culture medium in which *o*-coumaric acid had been fed to *P. jenseni*. After the addition of *P. jenseni* mycelium to a liquid mineral salts medium containing *o*-coumaric acid, the culture was incubated at 24°C for 2 days. The medium subsequently was examined for metabolites. An aliquot was removed and extracted three times with equal volumes of an ether–chloroform mixed solvent (1:1 v/v). The solvent then was evaporated from the combined extracts, and the residue was chromatographed on paper and thin-layer plates. 4-Hydroxycoumarin and 3,3'-methylenebis(4-hydroxycoumarin) were identified by comparison of R_f values with those of authentic compounds in various solvent systems.

Davies and Ashton (1964) studied the formation of 3,3'-methylenebis-(4-hydroxycoumarin) in sweet clover hay (*M. alba*), perennial sweet vernal hay (*A. odoratum*), and a hay consisting of a mixture of oatgrass

(*Arrhenatherum elatius*) and cocksfoot (*Dactylis glomerata*). Spoilage of the sweet clover and sweet vernal hays produced 3,3'-methylenebis-(4-hydroxycoumarin) and traces of 4-hydroxycoumarin. Spraying these hays with formaldehyde resulted in an increase of 3,3'-methylenebis(4-hydroxycoumarin), with the 4-hydroxycoumarin becoming undetectable. Inoculation of the spoiled hays with the organism *P. jenseni* led to higher 3,3'-methylenebis(4-hydroxycoumarin) and 4-hydroxycoumarin contents than existed in the untreated spoiled hays. Addition of formaldehyde solution to these inoculated hays effected a still greater increase in the 3,3'-methylenebis(4-hydroxycoumarin) along with the disappearance of the 4-hydroxycoumarin. Neither 4-hydroxycoumarin nor 3,3'-methylenebis(4-hydroxycoumarin) was present in the hay prepared from oatgrass and cocksfoot and allowed to spoil. Treatment of this spoiled mixed hay with coumarin and formaldehyde and subsequent inoculation with *P. jenseni* failed to produce either 4-hydroxycoumarin or 3,3'-methylenebis(4-hydroxycoumarin). Inoculation of the spoiled mixed hay with *P. jenseni* following the addition of *o*-coumaric acid and formaldehyde, however, resulted in the formation of 3,3'-methylenebis(4-hydroxycoumarin) and a trace amount of 4-hydroxycoumarin. From these experiments it was concluded that *o*-coumaric acid and not coumarin is the precursor of 4-hydroxycoumarin and thence 3,3'-methylenebis(4-hydroxycoumarin).

Shieh and Blackwood (1967) showed that the fungus *Fusarium solani*, which was isolated from the soil, when grown on a simple salts medium containing *o*-coumaric acid at room temperature for 90 days, effected a greater than 50% conversion of the *o*-coumaric acid to 4-hydroxycoumarin. The latter compound, isolated from the culture by ether extraction, was identified by the melting point, absorption spectrum, paper chromatographic mobility, and color reactions. The 4-hydroxycoumarin formed in the *Fusarium* species, however, was found not to be converted further to 3,3'-methylenebis(4-hydroxycoumarin). Contrary to the findings of Booth *et al.* (1959) that *o*-hydroxyphenylhydracrylic acid is an intermediate in the metabolism of *o*-coumaric acid to 4-hydroxycoumarin by rats and rabbits given *o*-coumaric acid orally, Shieh and Blackwood found no evidence of this intermediate in the fungal metabolism of *o*-coumaric acid. In the following year, Bye *et al.* (1968) showed that 3-hydroxy-3-(*o*-hydroxyphenyl)propionic acid is an intermediate in the fungal metabolism of *o*-coumaric acid to 4-hydroxycoumarin in the organism *Aspergillus fumigatus* Fresenius. Growth of the fungus in a Czapek-Dox liquid medium containing 1% glucose and generally labeled *o*-coumaric acid-^{3}H resulted in the formation of the radioactive intermediate, which was isolated by extracting the acidified medium with ether and identified by its ultraviolet spectrum and thin-layer chromatographic techniques. Further incubation of the intermediary compound with the fungus in the presence of glucose gave 4-hydroxycoumarin. In contrast to these findings of Bye *et al.*, Spring and Stoker (1969) found that 3-hydroxy-3-(*o*-hydroxyphenyl)propionic acid was not converted to 4-hydroxycoumarin by *P. jenseni* culture. After incubation of *P. jenseni* mycelium with 3-hydroxy-3-(*o*-hydroxyphenyl)propionic acid in a liquid mineral salts medium at 24°C, the culture medium was examined for 4-hydroxycoumarin. An aliquot of the medium was extracted with an ether–chloroform solution, and the solvent was evaporated. Upon subjection of the residue to chromatographic techniques, no 4-hydroxycoumarin was detected. The inconsistency in these findings regarding 3-hydroxy-3-(*o*-hydroxyphenyl)propionic acid as an intermediate in the microbiological conversion of *o*-coumaric acid to 4-hydroxycoumarin could be the result of a fungal species difference.

Bocks (1967) showed that *o*-coumaric acid underwent an approximately 72% conversion to 4-hydroxycoumarin when incubated with pregrown mycelial mats of *Aspergillus niger* van Tiegh (mulder strain) at 25°C for 5 days. The 4-hydroxycoumarin was not further metabolized to 3,3′-methylenebis(4-hydroxycoumarin) during this 5-day incubation period. Under the same conditions, coumarin was metabolized by the fungus to melilotic acid along with smaller amounts of *o*-coumaric acid and traces of 4-hydroxycoumarin, catechol, and two unidentified products. The 4-hydroxycoumarin was explained as a product resulting from the further conversion of *o*-coumaric acid formed as a fungal metabolite of coumarin. The metabolites were extracted from the acidified culture media with chloroform and identified by paper and thin-layer chromatographic techniques. The metabolism of *o*-coumaric acid to 4-hydroxycoumarin in the fungus *A. niger* was compared with that occurring in the rabbit after receiving a sodium hydrogen carbonate solution of *o*-coumaric acid by stomach tube as described by Mead *et al.* (1958). These investigators showed that part of the *o*-coumaric acid was metabolized *in vivo* to 4-hydroxycoumarin and postulated that this conversion proceeds by an incomplete β-oxidation process. Such a route involves the initial α,β-unsaturation of melilotic acid to form *o*-coumaric acid. Specific hydrolysis of this unsaturated compound then affords 3-hydroxy-3-(*o*-hydroxyphenyl)propionic acid that upon subsequent dehydrogenation gives rise to 3-(*o*-hydroxyphenyl)-3-oxopropionic acid. Enolization of the β-keto acid yields β-hydroxycoumarinic acid. The spontaneous lactonization of this unstable compound to give 4-hydroxycoumarin terminates the β-oxidation sequence, and salicylic acid, the expected terminal β-oxidation product of *o*-coumaric acid, is not formed.

Bye *et al.* (1968) reported that melilotic acid in addition to *o*-coumaric acid was metabolized to 4-hydroxycoumarin by *A. fumigatus* in Czapek-Dox medium. These investigators also suggested this conversion undergoes the β-oxidation route similar to the one advanced previously by Mead *et al.* (1958). In the 1968 research, however, evidence was presented that the carboxyl group of melilotic acid or *o*-coumaric acid was initially transformed into the reactive coenzyme A ester, which was retained throughout the reaction sequence until lactonization of 3-(*o*-hydroxyphenyl)-3-oxopropionic acid coenzyme A ester occurred with the concurrent release of reduced coenzyme A.

The two suggested schemes for the metabolic transformations of melilotic acid or *o*-coumaric acid to 4-hydroxycoumarin are shown in Fig. 8. The only compound that has been isolated and identified as an intermediate in either of these reaction sequences, however, is 3-hydroxy-3-(*o*-hydroxyphenyl)propionic acid. This compound is common to both proposed schemes as is the succeeding intermediary compound 3-(*o*-hydroxyphenyl)-3-oxopropionic acid, formed with the loss of two hydrogen atoms. No experimental evidence has been presented which shows whether this keto acid enolizes to β-hydroxycoumarinic acid that in turn lactonizes to 4-hydroxycoumarin as postulated by Mead *et al.* (1958) or the keto acid cyclizes prior to undergoing enolization to form 4-hydroxycoumarin according to the route suggested by Bye *et al.* (1968).

Thus, contrary to the early belief of Stahmann *et al.* (1941), 4-hydroxycoumarin is derived not from coumarin directly but rather from melilotic acid or *o*-coumaric acid. Both the latter compounds have been shown by Roberts and Link (1937a) to be present in the common sweet clovers *M. alba* and *M. officinalis*, which upon spoilage cause the sweet clover disease.

Once formed by the metabolic action of fungi growing on forage prepared from sweet clover, the 4-hydroxycoumarin then reacts with formaldehyde to produce the toxin, 3,3′-methylenebis(4-hydroxycoumarin) as originally proposed by Stahmann *et al.* (1941). The origin of the formaldehyde that enters into this reaction only recently has been considered. Bellis (1958) and again later Bellis and his co-workers (1967) showed that 4-hydroxycoumarin and 3,3′-methylenebis(4-hydroxycoumarin) were formed in cultures of *Penicillium* species. Based on the observation that 4-hydroxycoumarin appeared in the medium long before any 3,3′-methylenebis(4-hydroxycoumarin) was detected, Bellis *et al.* (1967) suggested that exogenous formaldehyde reacted with the 4-hydroxycoumarin nonenzymatically. After studying the biosynthetic route by which 4-hydroxycoumarin is formed from melilotic acid or *o*-coumaric acid in *A. fumigatus*, Bye *et al.* (1968) also postulated a nonenzy-

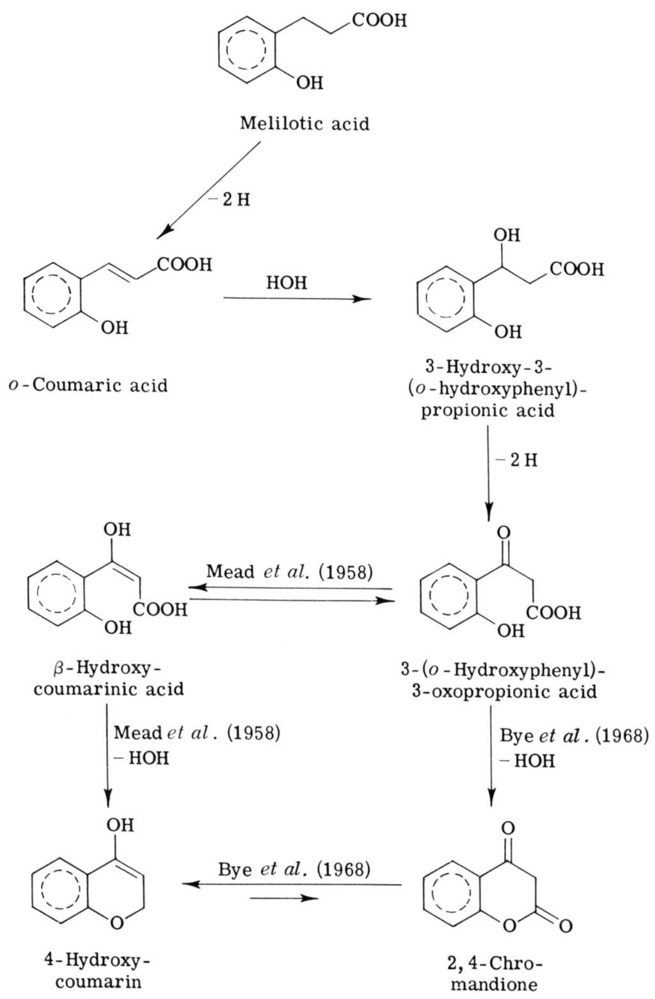

FIG. 8. Proposed reaction sequences for the conversion of melilotic acid and o-coumaric acid to 4-hydroxycoumarin.

matic condensation of 4-hydroxycoumarin with formaldehyde but gave no indication as to the origin of the formaldehyde.

In the same year, Spring and Stoker (1968) showed experimentally that 4-hydroxycoumarin derived from o-coumaric acid in *P. jenseni* culture is not metabolized further by this organism to 3,3'-methylenebis-(4-hydroxycoumarin) in a formaldehyde-free atmosphere. To ensure the absence of exogenous formaldehyde in the culture medium, these workers bubbled formaldehyde-free oxygen through this mineral salts solution containing o-coumaric acid and sucrose for 24 hours. *P. jenseni* mycelium then was added, and the culture was incubated at 24°C for 14 days while being supplied continuously with formaldehyde-free oxygen. A control experiment was conducted in which formaldehyde-free oxygen was bubbled through a sterile mineral salts solution containing 4-hydroxycoumarin and sucrose under the same conditions as those used in the metabolic experiment. In a second control experiment, a similar sterile solution of 4-hydroxycoumarin was placed in a vessel that then was sealed for 24 hours. At the end of the incubation period, the mycelium was filtered from the mold culture. The mycelium then was freeze-dried, and the metabolically formed 4-hydroxycoumarin retained in the mycelium was extracted into chloroform. After evaporation of the chloroform, the residue was refluxed with an excess quantity of formaldehyde-^{14}C to convert the 4-hydroxycoumarin to 3,3'-methylenebis(4-hydroxycoumarin)-methylene-^{14}C. The aqueous filtrate, which contained 4-hydroxycoumarin released from the mycelium, also was refluxed with excess ^{14}C-labeled formaldehyde. These reflux solutions were combined. Then the 4-hydroxycoumarin in both control solutions was transformed similarly into ^{14}C-labeled 3,3'-methylenebis(4-hydroxycoumarin) which was isolated from each solution by chloroform extraction and chromatographic techniques. Relative specific activity measurements of the radioactive 3,3'-methylenebis(4-hydroxycoumarin) obtained from the metabolic experiment and the control experiments indicated that no more than 0.9% of the 4-hydroxycoumarin—a value reported to be within experimental error—was converted into 3,3'-methylenebis(4-hydroxycoumarin) in *P. jenseni*. Consequently, it was concluded that 4-hydroxycoumarin, shown once again to be a metabolite of o-coumaric acid in *P. jenseni*, reacts with exogenous formaldehyde to form the toxin 3,3'-methylenebis(4-hydroxycoumarin).

The natural sources of such exogenous formaldehyde are varied. It has been reported that formaldehyde exists in the atmosphere, and it was inferred that this compound arises from the combination of carbon dioxide and water vapor in the presence of ultraviolet radiation (Dhar and Ram, 1932). On the mainland of Europe, a mean concentration of atmospheric formaldehyde occurring near the surface of the earth has been reported by Cauer (1951) as 0.5 μg/m^3 with a range of 0 to 16 μg/m^3. Formaldehyde has been shown to be present also in freshly collected rainwater at high altitudes in quantities ranging from 0.00015 to 0.001 gm/liter and in dew in concentrations approximating 0.0015 gm/liter (Dhar and Ram, 1933). More recently, formaldehyde has been identified as a product of undesirable fermentation processes in some silages.

Neumark *et al.* (1964) found that Minnesota green oat silage contained traces to 28 ppm (on a dry weight basis) of formaldehyde. A trace of formaldehyde also was found in one silage prepared from pearl millet (*Pennisetum glaucum*). In a later study, Neumark (1965) showed that a silage consisting of clover and 3% molasses contained 0.8 ppm formaldehyde. It remains to be reported, however, whether formaldehyde is formed by fermentation in sweet clover hay or silage.

Although suggestive evidence reported to date indicates that the formaldehyde which condenses with 4-hydroxycoumarin to form 3,3'-methylenebis(4-hydroxycoumarin) originates exogenously, the mechanism of this condensation reaction has yet to be studied *in vivo*.

C. TOXICITY OF 3,3'-METHYLENEBIS(4-HYDROXYCOUMARIN)

The toxicity of 3,3'-methylenebis(4-hydroxycoumarin) has been determined for mice, rats, and guinea pigs by Rose *et al.* (1942). The compound was administered orally or intravenously as an aqueous solution of the disodium salt, but all toxicity values were calculated on the basis of 3,3'-methylenebis(4-hydroxycoumarin). The acute median oral LD_{50} expressed with the standard error determined on 48 animals of each species was reported as 232.8±46.56 mg/kg for mice and 541.6±67.7 mg/kg for rats. The acute median intravenous LD_{50} based on 32 animals each was given as 64.30±6.11 mg/kg, 52.13±1.79 mg/kg, and 58.60± 2.29 mg/kg, respectively, for mice, rats, and guinea pigs. Sublethal doses administered intravenously resulted in slowing the respiratory process for prolonged periods. The route of entry of this fungal toxin to produce the sweet clover disease of cattle and sheep is oral, and since the toxic symptoms of all animal species afflicted with the disease are similar, the oral toxicity is the more meaningful.

In cumulative toxic dose studies, mice and rats were given daily doses of 3,3'-methylenebis(4-hydroxycoumarin) in the diet. 3,3'-Methylenebis-(4-hydroxycoumarin) in concentrations of 0.005, 0.01, 0.05, 0.1, and 0.5% in the food was fed to respective groups of five animals of both species. A sixth group of five rats was fed a diet containing 1% 3,3'-methylenebis(4-hydroxycoumarin). Of the twenty-five mice constituting the study, only two survived for 30 days, and these were from the group which received the lowest dose. One rat of the group ingesting a diet containing 0.01% 3,3'-methylenebis(4-hydroxycoumarin) and three rats of the group consuming the toxin at the 0.005% level survived for 30 days; the remaining twenty-six rats involved in the study died. Rabbits were subjected to daily intravenous injections of 3,3'-methylenebis(4-hydroxycoumarin). Five groups of five rabbits each were given respective daily doses of 0.1, 0.25, 0.5, 1, and 2 mg/kg. Twelve of the twenty-five rabbits survived the 42-day experiment — three of the group which received 0.5 mg/kg, four of the group injected with 0.25 mg/kg, and all five given 0.1 mg/kg. Pathology of the animals that died was characterized by hemorrhage, pulmonary edema, and in some cases, central necrosis of the liver. All animals surviving the cumulative toxic dose studies were sacrificed, and postmortem examinations showed no pathological changes.

IV. Psoralens

The psoralens are a group of furocoumarins characterized by a linearly annulated structure formed by the fusion of a furan ring in the 2 and 3 positions with carbon atoms 7 and 6, respectively, of the coumarin structure. The parent compound in this family of coumarin derivatives is psoralen. Several chemical nomenclatures for this furocoumarin are in use. Perhaps the most explicative of these, adopted by Späth and Pesta (1934), is based on the conventional numbering of both the coumarin structure and the furan ring from the respective hetero atom with primed number designations used in the furan ring. This numbering system for psoralen is shown in Fig. 9. Accordingly, psoralen becomes furo-2′,3′, 7,6-coumarin.

In 1963, *Chemical Abstracts* began to index psoralen as δ-lactone of 6-hydroxy-5-benzofuranacrylic acid. Since 1967, this compound has been listed as 7*H*-furo [3,2-*g*] [1] benzopyran-7-one. These latter names, however, have not as yet come into common usage.

The fundamental investigations of Kuske (1938, 1940) demonstrated that many psoralens possess potent phototoxic properties. In a review, Pathak *et al.* (1962) showed that photosensitizing psoralens occur as natural constituents of many plants belonging principally to the botanical families *Umbelliferae, Rutaceae, Leguminosae*, and *Moraceae*. In addition, the growth of the mold *Sclerotinia sclerotiorum* on the celery plant has been reported by Scheel *et al.* (1963) to result in the biosynthesis of two phototoxic psoralen derivatives, 8-methoxypsoralen and 4,5′,8-trimethylpsoralen, in the affected area of this plant. These compounds were

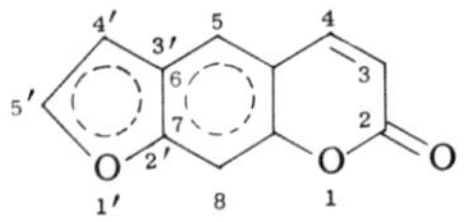

FIG. 9. Psoralen.

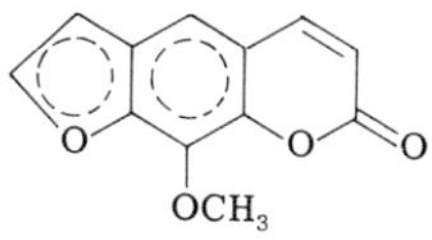

FIG. 10. 8-Methoxypsoralen.

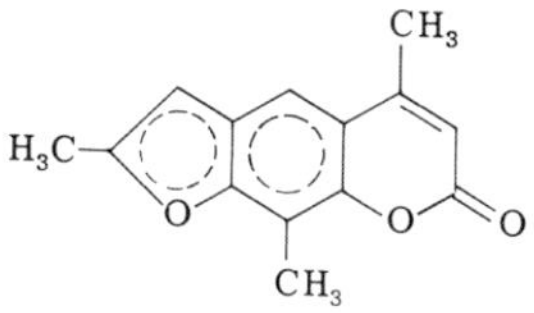

FIG. 11. 4,5′,8-Trimethylpsoralen.

shown to be the causal agents of a bullous dermatitis suffered by numerous workers engaged in handling celery parasitized by the fungus. Earlier, it had been proved clinically by Birmingham *et al.* (1961) that the outbreak of the cutaneous lesions occurred only when exposure to sunlight or ultraviolet irradiation having wavelengths in the 320 to 370 nm region followed contact with the diseased plant.

A. PHOTOTOXINS PRODUCED IN CELERY BY *Sclerotinia sclerotiorum*

Two photoreactive furocoumarins, 8-methoxypsoralen (Fig. 10) and 4,5′,8-trimethylpsoralen (Fig. 11) have been shown to be products of mold action. These phototoxins were isolated from celery infected by the fungus *S. sclerotiorum,* but not from normal celery (Scheel *et al.,* 1963). Attack of *S. sclerotiorum* on celery results in a disease known variously as pink rot, *Sclerotinia* rot, foot rot, or watery soft rot. This disease originally was attributed to the mold *Sclerotinia libertiana* (Lib.) Fckl., but later the correct name of the causal fungus was given as *Sclerotinia sclerotiorum* (Lib.) de Bary (Brooks, 1928). Contact with pink-rot celery and subsequent exposure to sunlight results in a blistering cutaneous disorder.

Celery dermatitis had long affected workers engaged in handling this plant. Early investigators attributed this condition to the skin-sensitizing capacity of celery oil. The first incident of dermatitis from celery was reported in France by Legrain and Barthe (1926). The disorder, which affected the fingers, hands, and forearms of a market gardener engaged in gathering celery, was characterized by erythema, edema, and vesicles with serous discharge. Following a complete cure, the gardener resumed work handling celery and suffered a recurrence of the dermatitis, which was manifested by simple erythema on the hands and intense erythematous, erysipeloid, hard edematous eruptions on the forearms. This acute dermatitis was attributed to the aromatic oil in celery. In England, Henry (1933, 1938) described a dermatitis response in workers preparing celery for commercial canning. This disease, originally confined to the upper limbs, began as an erythematous, vesicular, or papular rash. The more severe lesions described by Legrain and Barthe also were observed in some cases. Limonene, a constituent of celery oil, was suggested as the etiologic factor. In the United States, the first case of celery dermatitis appeared as a rash on both hands of a fruit company employee in New York (Gelfand, 1936). From skin tests using the patch method, it was concluded that the volatile oil of celery was the responsible irritant. Wiswell *et al.* (1948) studied celery dermatitis among farm workers and employees of a food packing plant in Massachusetts. Results of patch tests using the oily residues obtained from individual ether extracts of ground celery stalks and leaves led to the conclusion that the dermatitis was caused by the celery oil. Patch tests with pure *d*-limonene indicated that this terpene might be an active principle within the oil. Based on positive reactions to patch tests with volatile celery oil applied to the upper arms of six Massachusetts celery workers, Palumbo and Lynn (1953) also credited the oil to be the causative agent responsible for the dermatitis.

Klaber (1942), in England, was the first to suggest that celery dermatitis might be dependent on the action of ultraviolet radiation. This theory was based, in part, on the pattern of incidence of celery dermatitis among workers in a commercial cannery. Only employees engaged in the inspection of celery hearts and working under the brightest natural illumination available developed bullous eruptions. Such working conditions were favorable for the development of photosensitization. Musajo *et al.* (1954) isolated 5-methoxypsoralen from celery (*Apium graveolens*) and after showing this furocoumarin to possess phototoxic properties concluded that it was the causative agent in celery dermatitis. The isolation was accomplished by first extracting air-dried and finely ground celery three times with boiling alcohol for 5 hours. The alcohol extracts then were combined, concentrated, and treated with an equal volume of 10% aqueous potassium hydroxide. The resulting basic solution was extracted with ether, acidified with hydrochloric acid, and allowed to stand for 12 hours, whereupon the acid solution was extracted with ether. Dried with sodium sulfate, the ether extract was distilled under reduced pressure to yield 5-methoxypsoralen.

In a comprehensive study by Birmingham *et al.* (1961), experimental evidence was first obtained which showed that green Pascal celery infected by *S. sclerotiorum* contained a phototoxic material capable of producing vesicular and bullous lesions on human skin. Although not isolated, the causative agent was identified as a furocoumarin by spectrophotometric analysis of an ether extract of the diseased celery. Application of the same ether extract to human skin resulted in the formation of cutaneous lesions after exposure to natural sunlight or ultraviolet radiation having wavelengths ranging from 320 to 370 nm. It was left for speculation whether this furocoumarin was 5-methoxypsoralen, as shown previously by Musajo and co-workers (1954) to be present in normal celery. Earlier, Wiswell *et al.* (1948) had related the belief of many celery workers that the dermatitis resulted from handling only that celery having soft rot disease. While admitting an apparent relationship, in some instances, between the dermatitis and diseased green Pascal celery, these investigators dismissed the importance of the rot with the explanation that disintegration of the celery by the soft rot organism released the celery oil which was thought to be the causative principle. This belief was founded on results of patch tests on human subjects in which oily residues from individual ether extracts of celery stalks and celery leaves were used.

It was not until the isolation of 8-methoxypsoralen and 4,5′,8-trimethylpsoralen from celery infected with pink rot disease by Scheel *et al.* (1963) that the causative agents in celery dermatitis were identified conclusively. In this work, diseased celery tissue was first homogenized in a blendor, and the water was removed by lyophilization. The resulting dry product was extracted twice with petroleum ether at room temperature, and after the combined extracts were concentrated by evaporation at room temperature, crystals of 8-methoxypsoralen and 4,5′,8-trimethylpsoralen separated. These crystals were filtered off and dissolved in hot absolute methanol. On cooling, crystals of 4,5′, 8-trimethylpsoralen formed while the 8-methoxypsoralen remained in solution. After collecting these crystals on a filter, water was added to the solution which then was allowed to stand in a refrigerator. Crystals of 8-methoxypsoralen slowly separated from the mixed solvent and were collected on a filter.

The 8-methoxypsoralen isolated from the diseased celery, after recrystallization from water using decolorizing carbon, melted at 145–148°C. Both the melting point of authentic 8-methoxypsoralen and the mixed melting point of the isolated material with the genuine compound were in the range of 145–148°C. Further, the melting point of the isolated compound was in close agreement with the values of 144°C and 145–

146°C reported, respectively, by Priess (1911) and Thoms (1911) who, working together, first isolated 8-methoxypsoralen from the fruits of *Fagara xanthoxyloides* Lam. and the value of 146°C given by Späth and Pailer (1936) who first synthesized the compound. The identity of the isolated compound was confirmed by carbon and hydrogen analyses and by spectrophotometric examination. The ultraviolet spectrum of the material isolated from celery, dissolved in absolute ethanol, was identical with that of the pure 8-methoxypsoralen. The molar absorptivity at 250 nm was 3.03×10^4. Comparison of the infrared spectrum of the isolated compound with that of the authentic 8-methoxypsoralen indicated that the two compounds were identical.

Following recrystallization from absolute methanol, the 4,5′,8-trimethylpsoralen obtained from pink-rot celery melted at 226-228°C, a value in close agreement with that obtained for an authentic sample of 4,5′,8-trimethylpsoralen, but which was somewhat lower than the 234.5-235.0°C reported for the compound synthesized by Kaufman (1961). Carbon and hydrogen analyses and results of spectrophotometric examination supported the identity of the compound isolated from diseased celery with 4,5′,8-trimethylpsoralen. The ultraviolet spectra of ethanolic solutions of the compound obtained from celery and of authentic 4,5′,8-trimethylpsoralen were identical; the molar absorptivity at 250 nm was 2.81×10^4. Infrared analysis showed a close similarity of the isolated compound and genuine 4,5′,8-trimethylpsoralen.

Scheel and his co-workers (1963) did not find 5-methoxypsoralen in either normal or diseased celery. An explanation only can be suggested as to the reason for the failure to obtain this compound found earlier by Musajo *et al.* (1954) to be present in the celery plant. In a previous discussion on the biosynthesis of coumarin, it was pointed out that much of the coumarin isolated from various plants existed in those plants as coumarinic acid β-glucoside, and only through the conditions of the isolation procedure was the coumarin itself obtained. Insofar as it is known, no work has been published concerning the form 5-methoxypsoralen takes in the celery plant, but if it too exists conjugated as the glucoside, it would not be removed by the room-temperature extraction with petroleum ether used by Scheel and his associates (1963). The classic method for plant extraction utilizing boiling ethanol used by Musajo *et al.* (1954), however, would remove the glucoside from plant tissue. Treatment of the extract with 10% potassium hydroxide would effect hydrolysis of this glucoside, and subsequent acidification with hydrochloric acid would facilitate lactonization to form 5-methoxypsoralen.

B. Phototoxic Action of Mycotoxins Produced by *Sclerotinia sclerotiorum*

Quantitative bioassay methods have been developed to evaluate the phototoxic cutaneous response of psoralens. Since conditions to produce toxic photosensitization reactions in skin require a chemical sensitizer and exposure to ultraviolet radiation, the bioassay methods have been based on either the minimum chemical agent dose or minimum radiation dose necessary to evoke erythema in human or animal skin.

Musajo *et al.* (1954), after epicutaneous application of an ethanolic solution containing 5 μg of the furocoumarin to 1 square inch areas of human skin, used the time of ultraviolet irradiation to produce erythema under controlled experimental conditions as a measure of the photodynamic activity of the compound. The reactivities of compounds evaluated were expressed relative to that of psoralen, the most photoreactive of the furocoumarins. The test sites, which were located on the arms and backs of the subjects, were irradiated with a Philips HPW 125 lamp, emitting ultraviolet radiation concentrated at 365.5 nm, at a distance of 15 cm. Separate skin sites to which 8-methoxypsoralen and psoralen were applied produced erythema after 16 and 6 minutes, respectively. Thus, by this bioassay procedure, 8-methoxypsoralen has an activity of 37.5% relative to psoralen.

Pathak and Fitzpatrick (1959a) developed a quantitative bioassay method to determine the phototoxicity of psoralens on guinea pig skin and human skin, based on the minimum quantity of furocoumarin necessary to induce an erythema after ultraviolet irradiation for a constant time. Various concentrations of each psoralen, dissolved in 95% ethanol, were applied topically to separate 1 square inch sections of smooth skin on the clean-shaven backs of albino guinea pigs. After a lag period of 30 to 45 minutes, the skin sites were subjected to Wood's light radiation, having a primary emission of 365 nm, at a distance of 14 cm for 45 minutes. The quantity, called minimum effective concentration, of the furocoumarin required to produce a clearly visible erythema 18 and 36 hours after irradiation of the zone of treatment was taken as a measure of biological activity. The two phototoxins 8-methoxypsoralen and 4,5′,8-trimethylpsoralen, which were later shown to be products of *S. sclerotiorum* growth on celery, were among the furocoumarins evaluated. The minimum effective concentration of 8-methoxypsoralen on guinea pig skin was 15 μg/square inch of skin surface, whereas that of 4,5′,8-trimethylpsoralen was only 2.5 μg/square inch. When the cutaneous response to topical application of these compounds on human skin was determined by this bioassay procedure, the irradiation time was reduced to 15 minutes. The minimum effective concentration of 8-methoxypsoralen to induce a positive response in the test areas, located on the shaved inner side of the forearms, was 25 μg/square inch. The minimum effective concentration of 4,5′,8-trimethylpsoralen on human skin was 10 μg/square inch.

In another publication, Pathak and Fitzpatrick (1959b) described more fully a bioassay method by which the minimum effective concentrations of psoralens required to evoke a photosensitive response on guinea pig and human skin were determined. In this procedure, however, slightly different experimental conditions were used. A 250-W, model 70 Glo-Craft ultraviolet lamp, filtered to emit radiation having wavelengths ranging from 320 to 400 nm, was positioned 12 cm from the psoralen-treated skin areas. Each test site was a 2.5 cm square area of skin surface (approximately 1 square inch). By this method, 10 μg (4.6 × 10^{-5} mmole) of 8-methoxypsoralen/2.5 cm square gave a clearly visible erythematous response on guinea pig skin 36 hours after ultraviolet exposure. Under identical conditions, 5 μg (2.2 × 10^{-5} mmole) of 4,5′,8-trimethylpsoralen/2.5 cm square resulted in a similar response. In human subjects, the minimum effective concentration of 8-methoxypsoralen was 25 μg (11.5 × 10^{-5} mmole)/2.5 cm square of skin surface and that of 4,5′,8-trimethylpsoralen was 5 μg/2.5 cm square.

Using a similar bioassay procedure, Perone *et al.* (1964) studied the toxic action of 8-methoxypsoralen and 4,5′,8-trimethylpsoralen on rabbit skin. In this procedure, varying concentrations of each compound dissolved in acetone were applied topically to 1 square inch areas on the shaved backs of albino rabbits. Following a 30-minute lag period, the test sites were irradiated with Wood's light at a distance of 6.5 inches for 1 hour. Seventy-two hours after ultraviolet exposure, skin sites treated with 1.5 μg of 8-methoxypsoralen/square inch reacted to give an erythematous response; no cutaneous effect was produced by 1.0 μg/square inch. At a concentration of 0.1 μg/square inch, 4,5′,8-trimethylpsoralen produced a positive erythematous reaction, but 0.05 μg/square inch failed to elicit a response.

V. Summary

Certain coumarin derivatives originating as metabolites of fungal growth on food materials produce toxicoses in both animals and man. The pharmacological and physiological actions of such mold-produced toxins vary with the molecular structure. The isolation of the toxic coumarins from mold-infested food crops and the subsequent identification of these compounds have been achieved by the systematic research endeavors of many investigators.

Coumarin itself has been identified as a metabolite of the fungus *Phytophthora infestans* but only in laboratory culture. Results of limited research indicate that the biosynthetic pathway to coumarin in this organism parallels the well-established route operative in higher plants. No implication of a relationship of the slightly to moderately toxic coumarin with a naturally occurring mycotoxicosis has been made.

The fatal hemorrhagic disease of cattle and sheep fed on moldy hay or silage prepared from the common sweet clovers *Melilotus alba* and *Melilotus officinalis* or sweet vernal grass is caused by the presence of 3,3'-methylenebis(4-hydroxycoumarin) in the forage. This toxicant is formed in the plant tissue by the reaction of formaldehyde with mold-produced 4-hydroxycoumarin. A variety of fungal species metabolize melilotic and *o*-coumaric acids (normal constituents of the unspoiled forage crops) in laboratory culture to 4-hydroxycoumarin by an incomplete β-oxidation process. Only single species of *Aspergillus, Penicillium,* and *Mucor*, however, have been isolated from naturally occurring toxic sweet clover. The source of the formaldehyde that reacts with 4-hydroxy-coumarin to form 3,3'-methylenebis(4-hydroxycoumarin) is most likely exogenous. It remains, however, for the mechanism of this condensation reaction to be studied *in vivo*.

A photodermatitis prevalent among celery harvesters results from sensitization of the skin to solar radiation by contact with celery tissue infected with the mold *Sclerotinia sclerotiorum*. Growth of *S. sclerotiorum* on celery tissue produces two furocoumarins, 8-methoxypsoralen and 4,5',8-trimethylpsoralen, which possess phototoxic activity. The cutaneous disorder arises when the skin of the celery worker comes in contact with the diseased celery containing the photodynamically active psoralens and subsequently is exposed to sunlight. The activating wavelengths of the solar radiation lie in the near ultraviolet region and extend from 320 to 370 nm. The acute and persistent dermatitis is characterized by lesions ranging from simple erythema in mild cases to intense bullous eruptions in the most severe cases.

REFERENCES

Anschütz, R. (1903). *Ber. Deut. Chem. Ges.* **36**, 463.
Anschütz, R. (1909). *Ann. Chem.* **367**, 169.
Austin, D. J., and Clarke, D. D. (1966). *Nature* **210**, 1165.
Bellis, D. M. (1958). *Nature* **182**, 806.
Bellis, D. M., Spring, M. S., and Stoker, J. R. (1967). *Biochem. J.* **103**, 202.

Birmingham, D. J., Key, M. M., Tubich, G. E., and Perone, V. B. (1961). *Arch. Dermatol.* 83, 73.

Bocks, S. M. (1967). *Phytochemistry* 6, 127.

Booth, A. N., Masri, M. S., Robbins, D. J., Emerson, O. H., Jones, F. T., and DeEds, F. (1959). *J. Biol. Chem.* 234, 946.

Bourquelot, E., and Hérissey, H. (1920). *Compt. Rend. Acad. Sci.* 170, 1545.

Brooks, F. T. (1928). "Plant Diseases," p. 129. Oxford Univ. Press, London and New York.

Brown, J. M., Savage, A., and Robinson, A. D. (1933). *Sci. Agr.* 13, 561.

Brown, S. A. (1960). *Z. Naturforsch.* 15b, 768.

Brown, S. A. (1962). *Science* 137, 977.

Brown, S. A., and Neish, A. C. (1955). *Can. J. Biochem. Physiol.* 33, 948.

Brown, S. A., Towers, G. H. N., and Wright, D. (1960). *Can. J. Biochem. Physiol.* 38, 143.

Bye, A., Ashton, W. M., and King, H. K. (1968). *Biochem. Biophys. Res. Commun.* 32, 94.

Campbell, H. A., and Link, K. P. (1941). *J. Biol. Chem.* 138, 21.

Campbell, H. A., Roberts, W. L., Smith, W. K., and Link, K. P. (1940). *J. Biol. Chem.* 136, 47.

Cauer, H. (1951). *In* "Compendium of Meteorology" (T. F. Malone, ed.), p. 1126. Waverly Press, Baltimore, Maryland.

Charaux, C. (1925). *Bull. Soc. Chim. Biol.* 7, 1056.

Davies, E. G., and Ashton, W. M. (1964). *J. Sci. Food Agr.* 15, 733.

Dhar, N. R., and Ram, A. (1932). *Nature* 130, 313.

Dhar, N. R., and Ram, A. (1933). *J. Indian Chem. Soc.* 10, 287.

Edwards, K. G., and Stoker, J. R. (1967). *Phytochemistry* 6, 655.

Gelfand, H. H. (1936). *J. Allergy* 7, 590.

Gorz, H. J., and Haskins, F. A. (1964). *Crop Sci.* 4, 193.

Guérin, P., and Goris, A. (1920). *Compt. Rend. Acad. Sci.* 170, 1067.

Hagan, E. C., Hansen, W. H., Fitzhugh, O. G., Jenner, P. M., Jones, W. I., Taylor, J. M., Long, E. L., Nelson, A. A., and Brouwer, J. B. (1967). *Food Cosmet. Toxicol.* 5, 141.

Haskins, F. A., and Gorz, H. J. (1961). *Biochem. Biophys. Res. Commun.* 6, 298.

Haskins, F. A., Williams, L. G., and Gorz, H. J. (1964). *Plant Physiol.* 39, 777.

Hazleton, L. W., Tusing, T. W., Zeitlin, B. R., Thiessen, R., Jr., and Murer, H. K. (1956). *J. Pharmacol. Exptl. Therap.* 118, 348.

Henry, S. A. (1933). *Brit. J. Dermatol. Syphilis* 45, 301.

Henry, S. A. (1938). *Brit. J. Dermatol. Syphilis* 50, 342.

Ito, Y., Kitagawa, H., Suzuki, Y., and Yamagata, M. (1951). *J. Pharm. Soc. Japan* 71, 596.

Jenner, P. M., Hagan, E. C., Taylor, J. M., Cook, E. L., and Fitzhugh, O. G. (1964). *Food Cosmet. Toxicol.* 2, 327.

Kahnt, G. (1962). *Naturwissenschaften* 49, 207.

Kärber, G. (1931). *Arch. Exptl. Pathol. Pharmakol.* 162, 480.

Kaufman, K. D. (1961). *J. Org. Chem.* 26, 117.

Klaber, R. (1942). *Brit. J. Dermatol. Syphilis* 54, 193.

Kosuge, T. (1961). *Arch. Biochem. Biophys.* 95, 211.

Kosuge, T., and Conn, E. E. (1959). *J. Biol. Chem.* 234, 2133.

Kosuge, T., and Conn, E. E. (1961). *J. Biol. Chem.* 236, 1617.

Koukol, J., and Conn, E. E. (1961). *J. Biol. Chem.* 236, 2962.

Kuske, H. (1938). *Arch. Dermatol. Syphilis* 178, 112.

Kuske, H. (1940). *Dermatologica* 82, 273.

Legrain, and Barthe, R. (1926). *Bull. Soc. Franc. Dermatol. Syphilig.* 33, 662.

Link, K. P. (1943-1944). *Harvey Lectures* 39, 162.

Lutzmann, H. (1940). *Ber. Deut. Chem. Ges.* 73B, 632.

Mead, J. A. R., Smith, J. N., and Williams, R. T. (1958). *Biochem. J.* **68**, 67.
Musajo, L., Caporale, G., and Rodighiero, G. (1954). *Gazz. Chim. Ital.* **84**, 870.
Neumark, H. (1965). *J. Sci. Food Agr.* **16**, 668.
Neumark, H., Bondi, A., and Volcani, R. (1964). *J. Sci. Food Agr.* **15**, 487.
Palumbo, J. F., and Lynn, E. V. (1953). *J. Am. Pharm. Assoc., Sci. Ed.* **42**, 57.
Pammel, L. H. (1911). "A Manual of Poisonous Plants Chiefly of Eastern North America,
    with Brief Notes on Economic and Medicinal Plants, and Numerous Illustratious,"
    Part II, p. 552. Torch Press, Cedar Rapids, Iowa.
Pathak, M. A., and Fitzpatrick, T. B. (1959a). *J. Invest. Dermatol.* **32**, 255.
Pathak, M. A., and Fitzpatrick, T. B. (1959b). *J. Invest. Dermatol.* **32**, 509.
Pathak, M. A., Daniels, F., Jr., and Fitzpatrick, T. B. (1962). *J. Invest. Dermatol.* **39**, 225.
Paulman, V. C. (1923). *Vet. Med.* **18**, 734.
Pauly, H., and Lockemann, K. (1915). *Ber. Deut. Chem. Ges.* **48**, 28.
Perone, V. B., Scheel, L. D., and Meitus, R. J. (1964). *J. Invest. Dermatol.* **42**, 267.
Priess, H. (1911). *Ber. Deut. Pharm. Ges.* **21**, 267.
Roberts, W. L., and Link, K. P. (1937a). *J. Biol. Chem.* **119**, 269.
Roberts, W. L., and Link, K. P. (1937b). *Ind. Eng. Chem., Anal. Ed.* **9**, 438.
Roderick, L. M. (1929). *J. Am. Vet. Med. Assoc.* **74**, 314.
Roderick, L. M., and Schalk, A. F. (1931). *N. Dakota Agr. Expt. Sta., Tech. Bull.* **250**.
Rose, C. L., Harris, P. N., and Chen, K. K. (1942). *Proc. Soc. Exptl. Biol. Med.* **50**, 228.
Rudorf, W., and Schwarze, P. (1958). *Z. Pflanzenzuecht.* **39**, 245.
Sargeant, K., Sheridan, A., O'Kelly, J., and Carnaghan, R. B. A. (1961). *Nature* **192**, 1096.
Schaeffer, G. W., Haskins, F. A., and Gorz, H. J. (1960). *Biochem. Biophys. Res. Commun.*
    3, 268.
Scheel, L. D., Perone, V. B., Larkin, R. L., and Kupel, R. E. (1963). *Biochemistry* **2**, 1127.
Schofield, F. W. (1922). *Can. Vet. Record* **3**, 14.
Schofield, F. W. (1924). *J. Am. Vet. Med. Assoc.* **64**, 553.
Shieh, H. S., and Blackwood, A. C. (1967). *Can. J. Biochem.* **45**, 2045.
Smith, W. K., and Brink, R. A. (1938). *J. Agr. Res.* **57**, 145.
Späth, E., and Pailer, M. (1936). *Ber. Deut. Chem. Ges.* **69B**, 767.
Späth, E., and Pesta, O. (1934). *Ber. Deut. Chem. Ges.* **67B**, 853.
Spring, M. S., and Stoker, J. R. (1968). *Can. J. Biochem.* **46**, 1247.
Spring, M. S., and Stoker, J. R. (1969). *Can. J. Biochem.* **47**, 301.
Srinivasan, P. R., Shigeura, H. T., Sprecher, M., Sprinson, D. B., and Davis, B. D. (1956).
    J. Biol. Chem. **220**, 477.
Stahmann, M. A., Huebner, C. F., and Link, K. P. (1941). *J. Biol. Chem.* **138**, 513.
Stoermer, R. (1909). *Ber. Deut. Chem. Ges.* **42**, 4865.
Stoker, J. R., and Bellis, D. M. (1962a). *J. Biol. Chem.* **237**, 2303.
Stoker, J. R., and Bellis, D. M. (1962b). *Can. J. Biochem. Physiol.* **40**, 1763.
Thoms, H. (1911). *Ber. Deut. Chem. Ges.* **44**, 3325.
van der Waerden, B. L. (1940). *Arch. Exptl. Pathol. Pharmakol.* **195**, 389.
Weygand, F., and Wendt, H. (1959). *Z. Naturforsch.* **14b**, 421.
Wiswell, J. G., Irwin, J. W., Guba, E. F., Rackemann, F. M., and Neri, L. L. (1948).
    J. Allergy **19**, 396.

CHAPTER 2

The Biological Action of the Coumarins

LESTER D. SCHEEL

I. Introduction

Knowledge of the biological activity of coumarins dates back to ancient times (Bloomfield, 1897) as evidenced by writings in the Indian sacred book Atharva Veda (1400 B.C.) that refer to a herb treatment for leukoderma. Hoernle (1893-1912) in a translation of the Bower manuscript, an old Buddhist writing dating approximately A.D. 200, found mention of a cure for leukoderma using a poultice application of material from a plant that is now classified as *Psoralea corylifolia*.

Isolation, characterization, and identification of the coumarin compounds occurred in the 1800's, while interest in the chemical action of this family of compounds has continued to the present time. Their use has flourished as a result of their unique properties. Because of their pleasant and persistent odor, they have been used in the perfume and cosmetic trade. Their flavor has made them popular as condiments for centuries. The use of these compounds as a fish poison also stems from ancient practices. Current interest in coumarins as a special family of compounds associated with mold metabolism is an outgrowth of numerous observations of the toxic effects associated with food spoilage. Evidence of this interest is scattered throughout the literature but new developments arising from this interest have been sporadic until recently. The metabolism of various mold species, under certain conditions, changes in response to the environment, and, in so doing, produces coumarin compounds. At the present time it is not possible to relate this adaptive change to a physiological or metabolic function, but it is known that these coumarin compounds are linked to both carbohydrate and amino acid metabolism. Recent evidence indicates that the primary active form of these compounds in some plants is coumarinic acid glycoside (Kosuge and Conn, 1961). Our purpose in writing this chapter is to summarize the biological activity of coumarins originating from mold metabolism and to relate this biological activity to the effects of these compounds on living cells.

As presented in Chapter 1, the metabolic pathways leading to the formation of coumarin compounds from either amino acids or sugars have been explored with great ingenuity. As a result of this work the three hypotheses for the biological synthesis of the coumarin compound that remains identifiable and is functional in some living cells are (1) the acetate condensation method; (2) the shikimic acid pathway; and (3) the amino acid pathway. The first two involve fundamental systems in plant metabolism and constitute photosynthesis-linked systems. The third system is not limited to plants but is also active in nonphotosynthetic cells such as bacteria and molds. The organization of the material in this chapter starts with the identification of coumarin as a lactone and follows the development of the knowledge of its biological activity from 1820 to the present, in a somewhat chronological order, as the identity of the causative agents became known.

II. Coumarin

In a careful study with frogs, rabbits, and dogs, Ellinger (1908) showed, without doubt, that coumarin could produce deep narcosis in both cold- and warm-blooded animals, demonstrating that this effect was on the central nervous system and not on the heart or lungs as had been previously suggested by Kohler (1875). He also showed that the cause of death in the frog was due to respiratory inhibition mediated through the central nervous system.

In warm-blooded animals, subcutaneous injections of coumarin in rabbits (150 to 200 mg/kg in oil solution) produced narcosis characterized by loss of voluntary muscular control, loss of the corneal reflex, slower and deeper respiration, a slower pulse, and a 4°C decrease in temperature. The narcosis lasted 3 to 4 hours, and was followed by a slow recovery, with recovery usually complete by the next day.

The lethal dose for rabbits was found to be 300 to 400 mg/kg of body weight. In the case of lethal doses in the rabbit, the kymograph record showed that death was due to respiratory arrest. Just before death, there is a sharp increase in heart rate and an increase in blood pressure.

A direct application of coumarin solution to the heart muscle did not change the heart rate or blood pressure.

In dogs, 80 mg of coumarin/kg produced a narcosis similar to that observed in rabbits. When the dose given was lethal, three out of five dogs studied showed small hemorrhagic ulcers in the stomach, duodenum, and jejunum. The urine contained blood and sugar. Death occurred only after a prolonged comatose state and, therefore, it was possible that the hemorrhagic effects might have been due to metabolites rather than to coumarin. This work was extended and confirmed by de Moura Campos (1935) and Marolda (1936).

Bose (1958), in reviewing the biochemical action of coumarins, points out that coumarin, which he characterized as a common constituent found in 80 different plants, has some very specific effects. It can suppress seed germination at low concentrations and will inhibit root growth and sprout formation from potato eyes.

The blastocholine (inhibitory) effect of coumarin was also observed in tadpoles by Grimm (1951) when they were exposed to concentrations of 1 to 4×10^{-6} M/liter.

In developing an analytic technique for chromosomes, Sharma and Bal (1953) showed that pretreatment in coumarin solution for short periods results in clarification of plant chromosomes, but prolonged treatment caused chromosome fragmentation and shrinkage. Quercioli (1955), studying the cytohistological effects of coumarin and its derivatives, observed pre-prophasic inhibition, stickiness, and mutagenic and mitotic effects.

Mead et al. (1958a,b) studied the metabolism of coumarin in animals (ferret, guinea pig, mouse, rabbit, rat) and was able to recover about 25% of the administered dose as 3-, 5-, 7-, and 8-hydroxycoumarins. Salicylic acid and the 4- or 6-hydroxycoumarins were not found, indicating that oral administration of coumarin does not result in lactone ring opening or that beta oxidation of cinnamic acid is not one of the metabolic sequences in the metabolism of coumarin by these animals. The work of Booth et al. (1959) demonstrates lactone ring opening as well as oxidation of the side chain, by identification of metabolites in rabbit urine.

Although such metabolites as glucuronides, glycine conjugates, and ethereal sulfates were all found in small quantities, the hydroxylation reaction is probably the main detoxication step for coumarin in mammals as demonstrated by the research of Mead et al. (1958a,b) and Booth et al. (1959).

In reviewing the toxicity of coumarin, Jacobs (1953) points out that the effect of coumarin on man and animals has been studied for over 100 years and that he had not been able to find in the literature a single instance of severe physiological change resulting from coumarin ingestion in human beings even though coumarin had been used as a common flavoring ingredient during this time. However, Dominici (1948) cites three cases of coumarin intoxication attributable to the use of coumarin as a therapeutic agent in man.

The review of the physiological activity of the naturally occurring coumarins by Soine (1964) lists 31 "pharmacological and physiological properties of coumarins." This list includes the activities of the coumarin molecule itself, the hydroxycoumarins, the psoralens, and the glycosides of some of these compounds. Soine mentions the work of Bose (1958), which emphasized that the wide variety of chemical structures involved in such a listing pointed up the specific pharmacological effects observed for certain chemical structures. These effects include the photodynamic effect of the linear furanocoumarins, the anticoagulant activity of the 4-hydroxycoumarins, and the narcotic effects of coumarin and furocoumarins (especially in cold-blooded animals). The hemostatic property of herniarin (7-methoxycoumarin) and ayapin (6,7-methyl diether of coumarin) illustrates the fact that physiological action cannot always be predicted by chemical structure in this group of compounds.

Austin and Clarke (1966) investigated the ability of the potato blight organism *Phytophthora infestans* to produce coumarin during growth on a synthetic medium. Using ^{14}C-labeled L-phenylalanine, L-tyrosine, and *p*-coumaric acid, they showed that L-phenylalanine was the precursor of choice used by this organism in the synthesis of coumarin. Insofar as we have been able to determine, the synthesis of coumarin by *P. infestans* reported here is the first demonstration that a fungus can synthesize coumarin from known carbon compound substrates.

In addition to coumarin the synthesis of 7-hydroxycoumarin (umbelliferone) and 7-methoxycoumarin (herniarin) were found and identified by Austin and Clarke. However, in potato tuber lesions, the accumulation of scopoletin (6-methoxy-7-hydroxycoumarin) and esculetin (6,7-dihydroxycoumarin) has been demonstrated along with that of umbelliferone. These authors suggest that the infecting organism contributes to the accumulation of coumarins and that the intermediate compounds synthesized by the infecting organism may be further metabolized by the host tissue.

In studying the effect of coumarin on mold spores, Knypl (1963) showed that concentrations of coumarin greater than 100 ppm inhibited the growth of *Aspergillus niger* and *Penicillium glaucum* but stimulated sporulation of *Rhizopus nigricans*. Similar effects on the mold *Sclerotinia rolfsii* were reported by Lavee (1959); here, 50 to 100 ppm produced only slight effects, but 500 ppm prevented growth.

The metabolism of coumarin by *Aerobacter sporogenes* reported by Levy and Weinstein (1964) indicates that this organism may be able to utilize coumarin to form L-tyrosine.

Bellis (1958), in studying the soil molds *Penicillium jenseni* and *Penicillium nigricans,* showed that when coumarin is supplied as the sole carbon source in modified Czapek medium, the molds grow, but only slowly. However, when sucrose and coumarin are present, the growth is vigorous and umbelliferone (7-hydroxycoumarin) is detectable in the medium. Bocks (1967), using *Aspergillus niger,* has reported that the medium yielded melilotic acid as the main metabolite with small amounts of *o*-coumaric acid, 4-hydroxycoumarin, and catechol also detected.

III. Hydroxycoumarins

Work on the antifungal activity of the *Alternaria* species of molds grown on a variety of liquid media led Brian *et al.* (1949, 1951) to isolate a compound which they called alternaric acid. They showed that concentrations of 1 μg/ml prevented the germination of *Absidia glauca* spores. This compound was also phytotoxic. Raistrick *et al.* (1953; Thomas, 1961) isolated and characterized Alternariol and its monomethyl-ether from the mycelium of *Alternaria tenuis.* The structure of this compound was found to be dihydroxybenzo umbelliferone, with the added benzene ring fused at the 3,4 position of the umbelliferone.

Bose (1958), in a brief review of the biochemical properties of some of the natural coumarins, points out that these compounds affect the living cells of plants and animals in a variety of ways. Best (1944), who isolated scopoletin from the roots of a tobacco plant infected with a tomato spotted wilt virus, found that the scopoletin formed in the leaf tissue in the area of the lesions was transported by the plant to the root tissue for storage. Later, the use of ^{14}C-labeled compounds by Reid (1958) showed that scopoletin occurs in the tobacco plant both in a free form and as the glycoside. This work demonstrates that the precursor to this compound is probably phenylalanine.

In scorbutic guinea pigs capillary fragility is decreased by the injection of aesculin. Similar observations have been made in humans following a 20-mg dose (Bose, 1958).

While the 4-hydroxycoumarin compounds have anticoagulant activity, herniarin and ayapin, which are esters of 7- and 6,7-hydroxycoumarins, have been found to possess a hemostatic activity *in vivo* as well as *in vitro.* The corresponding phenolic compounds are inactive in this regard (Bose, 1958).

Soine (1964), in a later review of the literature on the physiological properties of this family of compounds, points out that, except for coumarin and dicumarol, all other naturally occurring coumarins bear an oxygen substituent at the 7 position, so that, in reality, one could say that umbelliferone (7-hydroxycoumarin), rather than coumarin, is the parent compound. Thus, of his list of 31 physiological properties, it is possible to identify 26 of these properties with particular hydroxycoumarins. Although there is some structural overlap in some of the activity, it is frequently very limited, and so the biological activity of this family of compounds shows a remarkable variety of specific effects without marked untoward reactions as, for example, the anticoagulant coumarins, which do not produce side reactions in man. Experience has shown that the dose of the anticoagulant for each individual is relatively constant for a given effect over a long period of time. This constancy of effect demonstrates that stimulation to produce a metabolic tolerance does not occur at the effective dose in man.

In 1955 Smith *et al.* (1956) announced the discovery of a new antibiotic, Streptonivicin, which was found to be a 3 amino-4,7,hydroxycoumarin derivative. This compound is produced by *Streptomyces nivens,* a soil organism isolated from a soil sample obtained in Queen's Village, New York. It is active against both gram-positive and gram-negative organisms *in vitro.* When grown on a medium containing sucrose only a trace of the antibiotic is produced but when grown on media containing the monosaccharides glucose or fructose it yields 100 to 400 μg of antibiotic/ml of broth.

Studies concerning the behavior of Streptonivicin in animals and man showed that the compound was absorbed into the bloodstream and distributed to the tissues in 2 to 4 hours at levels of 40 μg/gm of wet weight of tissue (Taylor *et al.,* 1956). Brain tissue showed the lowest value, 20 μg/gm. These studies showed that 30% of the dose is excreted in the feces and only 1% in the urine. Larson *et al.* (1956), in their studies on Streptonivicin toxicology, showed that the acute toxicity of this compound in the mouse was tenfold less than in the guinea pig (LD_{50}, 262 mg/kg as against 11.5). In cumulative dosage studies in dogs it was shown that 300 mg/kg administered intravenously produced severe vascular irritation, hemorrhage, and edema as well as degenerative changes in the liver and kidneys. This relatively low toxicity is evidence that this compound might be a useful antibiotic for use in man.

In the investigation of the relationship between the phenolase activity of potato tissue and flavonoid compounds, Baruah and Swain (1959) reported the presence of aesculin, scopolin, and scopoletin in potato tissue. They observed that the free phenols were oxidized but the glycosides were not.

Austin and Clarke (1966), in studies with *Phytophthora infestans* grown on a synthetic media, found that this organism produced umbelliferone, herniarin, and coumarin. They showed that scopoletin and scopolin, and aesculetin and umbelliferone, accumulate in the tuber tissue surrounding the blight lesions of potato tubers infected by *P. infestans*. They suggest that the infecting organism contributes to the accumulation of coumarins in the tissue and that the intermediates synthesized by the infecting organism are further oxidized by the surrounding potato tuber tissue. Thus, the symbiotic existence of the parasite and the host indicate in this case a natural similarity in metabolic products. However, as Austin and Clarke (1966) point out, definitive data to prove that the coumarin compounds are involved as primary control factors in the host-parasite relationship are not yet available.

The work of Uritani (1967) in identifying and investigating the growth of *Ceratocistis fimbriata*, *Thielaviopsis basicola*, *Fusarium oxysporum*, or *Helicobasidium mompa* on sweet potato tubers showed that the bitter substances generated in the diseased tissue area are a complex mixture of organic compounds which include ten kinds of terpene compounds previously identified by Akazawa (1960). It was postulated that these compounds were produced by the potato tissue itself because it was observed that they accumulated in the tuber tissue after exposure to toxic chemicals or following attack of the plant by insects. The ipomeanine, a sesqui terpene, inhibits mold growth and thereby participates in the defense action of the plant.

When sweet potato tubers are attacked by pathogens such as *C. fimbriata*, umbelliferone, scopoletin, aesculetin, scopolin, and skimmin accumulate in the same region as the terpenes (Uritani, 1967). In view of the work of Austin and Clarke, it is possible that these compounds may represent the products of mold metabolism in the infected sweet potato tissue.

Coumestrol, a normal metabolite found in clover and alfalfa leaves, was found to be increased in amount by infection with *Pseudopegia medicaginis* or *Leptosphaerulina briosiana*. The estrogenic activity of coumestrol present in cured forage for cattle was suggested as a cause of complications in breeding programs (Uritani, 1967).

IV. 4-Hydroxycoumarin

Vogel (1820) isolated a substance from *Melilotus officinalis* and compared its properties with those of coumarin obtained from the tonka bean and showed them to be identical. Reinsch (1867) in similar experiments obtained coumarin from *Melilotus vulgaris* and thus confirmed the observations of Vogel.

It happened that in the following year (1868) Perkin synthesized coumarin from the condensation of sodium salicylate and acetic anhydride.

This synthesis proved the structure of coumarin and led to his suggestion that it belonged to the family of cinnamic acids.

In 1868 Fittig reported that the correct structure of coumarin should be that of a lactone and proved this by showing that coumaric acid obtained from coumarin by alkaline hydrolysis can be hydrogenated to form melilotic acid, which was then known as orthohydroxy hydrocinnamic acid.

In 1901 Ludwig Wolff showed that the condensation of a tetronic acid lactone in the presence of an aldehyde resulted in the formation of a bis type structure on the aldehyde carbon.

Anschütz reported in 1909 that he had synthesized a bis condensation product from reacting phenyltetronic acid lactone (β-oxycoumarin) in the presence of formaldehyde to obtain methylenebis-β-oxycoumarin. This chemical synthesis was not found to be biologically significant until 1941 when Link and his co-workers (Stahmann et al., 1941) reported the isolation and characterization of this compound as the hemorrhagic agent present in spoiled sweet clover hay.

The reports of Schofield (1922, 1924) describe a hemorrhagic disease in cattle fed moldy sweet clover hay. The pathological change observed included massive subcutaneous swellings caused by blood extravasation into the tissue, kidney nephrosis, and cloudy swelling of the liver cells. Due to the large amount of blood lost from the circulation the animals also had a severe anemia.

In his investigation Schofield also proved the etiology of the disease by feeding the spoiled hay to rabbits and recreating symptoms of the disease in the rabbit. The observations were confirmed by feeding calves on moldy sweet clover hay and other calves on properly cured sweet clover hay. In these studies the moldy hay caused the death of the calves through hemorrhagic disease, but those calves fed well-cured hay remained well. He showed that the toxic effect was not related to the age of the plant. Since the rabbits fed either fresh or well-cured sweet clover hay did not become sick, the sweet clover plant could not be considered a poisonous plant.

Following a careful study of several outbreaks of the disease in different herds of cattle, Schofield observed that in every case of the disease the cattle had been fed moldy hay. This caused Schofield to remark that "moldy hay is a useful danger signal."

In work conducted to evaluate the microflora of "spoiled hay," a greater constancy of species of molds was found than had been demonstrated in bacterial flora. Isolation experiments identified an *Aspergillus* mold and a *Penicillium* mold, both of which, when grown on sweet clover and fed to rabbits, produced the typical hemorrhagic disease. This production of the disease by specific molding of sweet clover tissue was considered proof that the disease involved an interaction of the mold with the plant tissue.

In 1929 and 1931 Roderick published his studies concerning the pathology of the sweet clover disease in which he described a detailed examination of the vascular bed in the area of hemorrhage that failed to demonstrate vascular damage or rupture. Careful and extensive blood chemistry done before and during the development of the disease did not reveal abnormal findings other than delayed coagulation. Careful examination of organs and tissues revealed only the peculiar fact that the hemorrhage occurred subcutaneously in areas where normal activity produced mild trauma; these included the flanks, joints, and areas of the side where the animal's body exerts pressure when the animal is lying down. The peritoneal covering of the stomach also was frequently found to be hemorrhagic. Examination of the liver, kidneys, pancreas, and lungs revealed only minor hemorrhage but no observable toxic lesions. In consideration of the extensive hemorrhage in the animals, the examination of the liver, and of the spleen, which showed no hemosiderin deposition, indicated to Roderick that little or no destruction of red cells occurred before the hemorrhage. The only abnormal observation found in this extensive examination of the diseased animal was the fact that the blood coagulation of the animals fed moldy sweet clover was progressively delayed before the terminal hemorrhagic event.

A detailed study of the anticoagulant effect on spoiled hay fed to cattle was reported by Roderick in 1931. This study reports changes in fibrinogen, antithrombin, blood platelets, calcium, and prothrombins. The development of quantitative techniques for these studies and the arduous task of gathering these data on cattle fed moldy sweet clover hay finally yielded positive evidence that the failure of blood coagulation was due to the loss of prothrombin in the blood of the animal. This was proved by blood transfusion experiments using blood from normal and diseased animals.

By using the Howell acetone precipitation of prothrombin (Howell and Cekada, 1926), Roderick (1931), demonstrated that when prothrombin was added to the blood of diseased cattle the coagulability of the diseased animal's blood was restored. Thus, loss of coagulability of the blood coupled with mild traumatic injury were the identifiable causes of hemorrhage in this disease.

During the 20 years covered by the above research, etiological and experimental evidence that the disease associated with moldy sweet clover hay was not of an infective nature had accumulated. Therefore, in the papers of both Roderick and Schofield the opinion is offered that a toxic substance is created in the moldy sweet clover hay.

In 1932 Dr. Link, at the University of Wisconsin Agricultural Bio-chemistry Department, was invited to assist in the investigation of an epi-demic of cattle poisoning due to sweet clover hay in northern Wisconsin. By 1934 he had organized a group of graduate student chemists to work on the isolation of the toxic material from sweet clover hay. The initial effort depended on the development of a reliable semiquantitative biolog-ical assay technique for evaluating the toxic property of the spoiled hay and any extractable fractions of the plant.

In order to achieve a uniformity of response in the test animal, the rab-bit, it was found necessary to breed a susceptible strain of rabbits for use in the bioassay tests. An arbitrary hay sample was chosen as a standard of comparison, and the isolation of fractions from the spoiled hay were as-sayed for anticoagulant activity routinely.

The quantity of blood available for tests from a rabbit was limited, and results from the Lee-White clotting time test proved to be too variable for a reliable assay of clotting deficiency. The publication of the one-stage prothrombin assay by Quick (1937) increased the sensitivity and preci-sion of the bioassay. However, since the normal plasma clotting time does not increase when plasma is diluted to double its volume, the method of Quick was modified in Link's laboratory by the use of diluted plasma (12.5%) in the bioassay of toxic hay samples and isolated fractions of ex-tractables from hay. This modification increased the sensitivity and the precision of the bioassay technique so that even small changes in pro-thrombin activity could be detected.

By 1939 Dr. Link and his team of chemists, Campbell, Schoeffel, Smith, Roberts, Huebner, Overman, Sullivan, and Stahmann, success-fully isolated 1.8 gm of crystalline hemorrhagic agent from 200 pounds of spoiled sweet clover hay. Although the spoiled hay used in these studies contained 0.75 to 1.5% coumarin, only 0.26% of the total coumarin was found to act as the hemorrhagic agent. Under the careful guidance of Dr. Link, Huebner, Stahmann, and Sullivan (1943-44) successfully synthe-sized 4-hydroxycoumarin from salicylic acid, acetic anhydride and methyl alcohol, via the Pauly condensation.

Following the Anschutz formaldehyde addition reaction, a compound identical to the material isolated from spoiled sweet clover was obtained by these workers. This compound, for which the structure was elucidated by Huebner (Stahmann et al., 1941) as 3,3'-methylenebis-4-hydroxycou-marin, was shown to have identical anticoagulant and physical properties with the isolated hemorrhagic agent of spoiled sweet clover hay. In addi-tion to the isolation and identification work, Link and his students pre-pared over 300 organic structure modifications of the anticoagulant and showed that maximum activity was obtained only when an intact 3 substituted 4-hydroxycoumarin residue was present in the molecule. Substituents added to the benzene ring or esters of the hydroxyl group induced a loss of anticoagulant activity (Overman *et al.*, 1944).

When a Michael condensation of benzalacetone and 4-hydroxycoumarin was performed by Sullivan *et al.* (1943) and Ikawa *et al.* (1944), the new monocoumarin derivative obtained had an anticoagulant activity that exceeded the potency of the methylenebis type compounds in all animals except the rabbit. The resistance of the rabbit to the activity of this compound was rationalized on the basis that the vitamin K content of the diet of this animal inhibited the anticoagulant activity of this family of compounds (Overman *et al.*, 1942). In subsequent work with this family of compounds the optical isomers of the Michael condensation products have been resolved, and the levorotatory isomer has been shown to be seven times more potent as an anticoagulant than the racemic mixture (West *et al.*, 1961).

In studies concerning the metabolism of the coumarin anticoagulants in man Burns *et al.* (1953a,b) have shown that 3,3'-carboxymethylenebis-(4-hydroxycoumarin) ethyl ester is hydroxylated in the 7 position of only one of the 4-hydroxycoumarin units in the bis structure before being excreted in the urine. Pulver and von Kaulla (1948) studied the metabolism of this compound in the rabbit and identified the de-esterified free acid and a free acid derivative in which one of the lactone rings had opened. Roseman *et al.* (1954), in studying the metabolism of 4-hydroxycoumarin in the dog, showed that 50% of the dose was excreted unchanged, and 25% was excreted as the 4-hydroxycoumarin β-D-glucopyranosiduronic acid. The remaining 25% could not be identified. This observation was confirmed by Mead *et al.* (1958a,b) in their extensive studies on the metabolism of hydroxycoumarins. During their extensive studies on coumarin in five species of animals they were able to detect and identify the 3-, 7-, and 8-hydroxycoumarins in the urine of one or more of the animal species studied. In no case did they detect coumaric acid, melilotic acid, salicylic acid, or 4-, or 6-hydroxycoumarin.

In studying the metabolism of the monocoumarin in the rat, Barker *et al.* (1969) have been able to identify the 6-, 7-, and 8-hydroxycoumarins and the glucuronide of 4-hydroxy as well as the 7-hydroxycoumarin. These compounds accounted for 93% of the dose administered to rats. Thus, it appears that adequate evidence indicates that not only is coumarin detoxified by hydroxylation in the animal body but so also is 4-hydroxycoumarin and its derivatives.

Overman *et al.* (1942) and Link (1948) in early reports concerning the biological interaction of vitamin K, vitamin C, and dicumarol provided the first evidence that the anticoagulant properties of these compounds are grossly modified by the presence of these vitamins in the animal. In an extension of this line of reasoning, Chmielewska and Cieslak (1958) have suggested that tautomerism involving enolization of the coumarin carbonyl group might provide a chemical explanation of the vitamin K type of antagonism. They have been able to identify, isolate, and characterize the keto-enol tautomers in support of their hypothesis. This hypothesis, developed within the framework of the known properties of the numerous anticoagulant compounds that have been tested (Overman et al., 1942, 1944), has been extended by Arora and Mathur (1963) who confirmed its validity. However, much additional data must be accumulated in this area before a clear mechanism is demonstrated.

The acute toxicity of dicumarol was studied by Rose et al. (1942) in mice, rats, and guinea pigs. The results show that the intravenous LD_{50} for the three species is very similar (rats, 52; mice, 64; guinea pigs, 59 mg/kg). However, when the compound was given by stomach tube the dose was much larger (mice, 233; rats, 542 mg/kg) and much more variable (15-20% variation). When dicumarol was administered intravenously to rabbits at a level of 0.1 mg/kg daily, all animals survived 42 days, but if given at 0.25 mg/kg two out of five animals died after 29 doses. In rats given 1.25 mg/day with their food, three out of five animals survived 30 days of continuous dosage. The major pathological change found was hemorrhage into various tissues, especially subcutaneous tissues, accompanied by pulmonary edema. Wakim et al. (1943) using large doses of dicumarol (40-60 mg/kg) observed that the signs of the toxic syndrome were extreme vasodilation with dyspnea and shock. There was no change in the blood coagulation or prothrombin times during these experiments.

In very careful studies by McCarter et al. (1944) in dogs, the pathological change caused by toxic doses of dicumarol was described. The morphological changes found were hemorrhages due to the development of toxic lesions in the small blood vessels and a toxic lymphoid reaction. No liver necrosis was found.

The findings in dogs were compared to findings in man, and they were found to be nearly similar. The vascular damage described accounts for the hemorrhagic effect of these anticoagulants, but does not prove that the anticoagulant and the sclerotic effects of the compound are related.

V. Psoralens

As a result of observations on sunbathers coming to his office for treatment, Freund (1916) noted that certain of the cases examined displayed lesions which resembled a contact dermatitis. He suggested that exposure to sunlight activated and made the presumed sites of contact with some photoactive agent more reactive. Thus, a photoactivated contact dermatitis was recognized and discussed in medical circles for about 20 years before Kuske (1940) reported the first bioassay for psoralens. During this discussion the persistant pigmentation following the healing of the acute lesions suggested that these materials might be the same as those contained in plant poultices used in the treatment of vitiligo (Pathak and Fitzpatrick, 1959a).

The bioassay developed by Kuske (1940) using 5-methoxypsoralen (bergapten) proved that the lesions were produced by contact with the chemical and that exposure to ultraviolet radiation shortly after this contact was necessary to elicit a response. He also showed that the extent and intensity of the lesion depended upon the amount of chemical administered, using the forearm of human volunteers for his tests. In his report he also mentioned that tests were conducted to determine if sunlight passing through window glass would produce the lesions.

These tests proved the true nature of the photosensitization because ultraviolet radiation having wavelengths shorter than 3200Å does not pass through window glass. Under the test conditions of 30 minutes' exposure to long wavelength ultraviolet light, only the sites where 5-methoxypsoralen had been placed reacted, thereby demonstrating the cause and effect relationship of both the chemical and the radiation exposure to produce the lesions.

During the next 15 years additional sources of photosensitizing chemicals were discovered, and, in 1954, Musajo et al. reported a biological assay using exposure to the radiant energy of a mercury vapor lamp filtered to give a radiation exposure of 3655Å. Musajo, in his studies, showed that the response was proportional to the radiation dosage. When 5 μg/cm^2 of the test compound was placed on human skin and irradiated under the fixed exposure conditions of light, intensity, and distance for various lengths of time, the minimum light exposure required to produce erythema in 24 to 48 hours was recorded as the response index. In testing psoralen at a concentration of 5 μg/cm^2 on human skin, 6 minutes of light exposure were required to obtain an erythema response. Under the same test conditions 8-methoxypsoralen (xanthotoxin) required 16 minutes of exposure and 5-methoxypsoralen (bergapten) required 22 minutes.

The idea of a bioassay for this type of action was improved by Pathak and Fitzpatrick (1959b) when they used a Woods lamp which filtered out the short wavelength ultraviolet light (less than 3200Å). Using this technique on albino guinea pigs, Pathak et al. (1967) evaluated a large series of furanocoumarins for their photosensitizing properties. As a result of this work it was possible to establish the fact that substitution of the 3 position of the furanocoumarin or substitution of the furanocoumarins with any group which interferes with the resonance of the ring structure will reduce the photosensitizing capability of the furanocoumarin.

In an extension of this work, Perone *et al.* (1964) showed that the growth of the mold *Sclerotinia sclerotiorum* on celery produced a photosensitization of the skin of rabbits. Their work provided evidence that the toxic compounds were present in the plant tissue only in the region infected by the mold and that this "phototoxic" effect could not be demonstrated when other molds were grown in the celery tissue.

The isolation, characterization, and phototoxic effect of 8-methoxypsoralen and 4,5′,8-trimethylpsoralen from the *Sclerotinia sclerotiorum*-infected celery tissue was reported by Scheel *et al.* (1963). Based upon the bioassay results above, it was shown that the 4,5′,8-trimethylpsoralen was 15 times more active as a photosensitizer than the 8-methoxypsoralen. When compared with unsubstituted psoralen the 4,5′,8-trimethylpsoralen showed equal activity. The bioassay method for the photodynamic action devised by Pathak and Fitzpatrick (1959b) has given reproducible results regardless of the animal species used. In an extension of the bioassay work, Pathak *et al.* (1961) exposed solutions of lactic or succinic dehydrogenase and cytochrome oxidase to ultraviolet radiation, to show that the furanocoumarins form a complex with the enzymes which inhibits the enzymatic activity. This work suggests that furanocoumarin can act as an adjunct in the radiation denaturation of proteins by stabilizing radiation-induced alteration of protein structure following interaction of the psoralen with the protein.

El Mofty *et al.* (1959a,b) studied the effect of 8-methoxypsoralen on copper and glutathione levels in the blood and liver of rats. In both adrenalectomized animals and normal animals, the oral administration of 8-methoxypsoralen produced an increase in copper in the blood with a coincident decrease in copper in the liver. When the pituitary gland was removed from these rats the changes in copper distribution did not take place following 8-methoxypsoralen treatment. These observations indicate that 8-methoxypsoralen affects copper metabolism, which appears to be under the control of the pituitary gland.

Studies by Khadzhai and Kuznetsova (1962) with imperitorin (8-isoamylene oxypsoralen) showed this compound to have a spasmolytic effect about equal, on a dose basis, to that of papaverine. In further studies, using 8-methoxypsoralen and 5-methoxypsoralen, these authors showed that a solution of 10 ppm applied to the isolated rabbit heart preparation lowered arterial pressure and smooth muscle tonus and stimulated the rhythm and decreased the amplitude of the heart contractions (Khadzhai and Kuznetsova, 1963).

In clinical trials with 8-methoxypsoralen Hoekenga (1959) found that the only adverse effects were epigastric distress, vertigo, and nausea.

Working on 8-methoxypsoralen photoactivation, Judis (1960) has shown that dopa oxidation is stimulated in the presence of this compound by long wavelength ultraviolet light (3660Å), but that this compound protected dopa against oxidation caused by short wavelength ultraviolet light (2540Å).

Rodighiero (1954) has shown that 8-methoxypsoralen, psoralen, and angelicin inhibit seed germination and root and seedling growth. When psoralen or 5-methoxypsoralen solutions ($5 \times 10^{-3}M$) was applied to onion root tips, a 40% inhibition of mitosis was induced during a 4-hour incubation at 20°C. Higher concentrations of these compounds caused total inhibition of mitosis in onion root tips (Musajo, 1955). Chakraborty *et al.* (1957) showed that psoralen and some of its substituted derivatives are actually inhibitors of mold growth.

In studies with mice Hakim *et al.* (1960) have shown that mice with pigmented skin are more resistant to the photosensitization of 8-methoxypsoralen and respond by an increase in pigmentation. In the case of albino mice severe erythema and tumor formation on the ears of the mice treated with 8-methoxypsoralen was observed.

Coumarins have been used as fish poisons for a number of years. In the work of Späth and Kuffner (1936) the activities of ten coumarins were tested, and it was shown that while coumarin itself is active at a dilution of 1:6800, 5 methoxypsoralen is active at a dilution of 1:93,000 using *Lebistes verticulatus* as the test organism. The nonlinear psoralen angelicin did not kill the fish but did induce narcosis.

VI. Discussion

Our purpose here is to identify the limits of our knowledge concerning the coumarin compounds discussed in Chapter 1 and this chapter. In addition, it is the hope of the author that this summary will lead to further research concerning the nature and conditions under which the coumarin compounds become the products of mold metabolism. Although coumarin was isolated more than 150 years ago, the physiological function of this compound in plant or mold metabolism has not been elucidated. It was not until recently (Kosuge and Conn, 1961) that the data showed that the precursor of coumarin the coumarinic acid glycoside was the primary form of the coumarin in plant tissue. Since these glycosides are readily hydrolyzed during autolysis of plant tissue, as reported by Roberts and

Link (1937), it would appear that the plant tissue transport system may involve coumarin as an intermediate metabolic product that is rapidly utilized by the plant.

The lack of knowledge concerning the function of coumarin in mold metabolism is pointed up by the fact that the only identification of coumarin from molds was found by Austin and Clarke (1966) in their studies on *Phytophthora infestans*. However, the many references in the literature citing the occurrence of hydroxycoumarins resulting from mold growth indicates that these compounds are intimately involved in mold metabolism. That the mechanism is not always clearly functional is indicated by the observations of Bellis (1958) showing that stress was a primary factor in the ability of the mold to synthesize coumarin. The elucidation of the stress response mechanism in mold growth remains, for the most part, an unexplored area for future research.

The metabolic sequential use of the coumarin compounds by the parasite and the host was suggested by Austin and Clarke (1966) when they identified coumarin, umbelliferone, and herniarin as products of mold metabolism on synthetic media but found umbelliferone, scopoletin, and aesculetin in the infected potato tuber. This symbiotic susceptibility of plant tissue is frequently suggested by various studies with mold organisms, but definitive quantitative studies in this area are lacking.

Although 7-hydroxycoumarin is the most common hydroxycoumarin found in nature, it is still not known if this coumarin is a metabolic intermediate or a detoxification product of plant or mold metabolism. In addition, the 7-hydroxyglycoside is known in plants but references to the occurrence of 7-hydroxycoumarinic acid glycoside in plants have not been found in our literature search. In view of the recent observations relating to the coumarinic acid glycosides in plant tissue, this area should be investigated with regard to the hydroxycoumarins. In general, the data indicate that all the hydroxycoumarins (except 4-hydroxycoumarin) are less toxic to animals than is coumarin. Therefore, in the area of condiment or flavoring substances, additional research may yield useful information. In addition, the use of the glycosides as condiments has not received the attention that would allow even a cursory evaluation.

In the case of esters of the hydroxycoumarins the data concerning the pharmacological action of herniarin and ayapin (Bose, 1958) indicate that the esterification of the hydroxy group changes the pharmacological properties of the compound. This area of evaluation has only begun to yield data.

The relationship of chemical structure to anticoagulant activity of 4-hydroxycoumarin has received a great deal of attention, and, to date, two pharmacological activities have been identified: (1) the anticoagulant activity and (2) the capillary sclerotic activity. Little or no work concerning the sclerotic activity of the bis type compounds has been reported. The increased therapeutic safety index and the absence of bleeding complications in patients given monocoumarin anticoagulants is due to the much lower capillary sclerotic activity of this type of compound. However, the data concerning the interrelationship of blood coagulation, capillary integrity, and anticoagulants remain sketchy and disorganized and therefore leave the treatment of the thrombotic diseases in a trial and error state.

As a result of metabolic studies associated with the coumarin precursors in molds, it appears that hydroxylation of coumarin in the 4 position does not take place and that the utilization of coumarin, as such, by molds does not occur readily. However, utilization of the coumarinic acid glycoside was not tested, and therefore the data concerning the mechanism of 4-hydroxylation are incomplete. In addition, the synthesis of the bis type structure to form dicumarol by the formal condensation by *Pencillium* and *Aspergillus* molds is "assumed" to be spontaneous. Further studies concerning the hydroxylation steps in mold metabolism might be very revealing, since it is known that these hydroxycoumarins act in the natural environment as inhibitors of mold growth.

The limited knowledge concerning the coumarinic acid glycosides and the position these compounds occupy in the metabolism of molds becomes more complex when the metabolic function of the psoralens is considered. In the generation of the furanocoumarin compounds the substituted and esterified structures appear to be the rule rather than the exception. Again mold metabolism appears to be host specific, and symbiosis may actually depend on a host-derived precursor. Proof for these suggestions is entirely lacking at the present time, but the validity is supported by the observations that *Sclerotinia* growth on carrots or lettuce does not produce the same degree of dermatitis that occurs in celery harvesters.

The occurrence of the furanocoumarinic acid glycosides has been described, but their pharmacological properties and metabolic function are not known. However, the range of organisms in which these compounds have inhibitory effects on proliferation or metabolism has led to the suggestion that they may be growth regulators in plants. The method by which metabolic inhibition takes place is inadequately described at the present time but may involve the action of these compounds as metabolic sulfhydryl-blocking agents. Additional work in this area and quantitative bioassay work on the phototoxic action of various materials derived from molds would greatly enhance our understanding of the species specificity of the mold-host symbiotic relationship.

In evaluating the uses of the furanocoumarins and an understanding of their mechanisms of action, we again find a lack of data. For instance, the mechanism of action of these compounds as fish poisons, the effect of 8-methoxypsoralen on the copper and glutathione levels in the blood of rats, and the spasmolytic effect of 8-isoamylene oxypsoralen on smooth muscle tissue are still only isolated observations. If they could be explored and understood these compounds would most likely be found to have a practical use.

The increase in number and types of pharmacological responses reported in the literature covering the furanocoumarins indicates that a very large potential for use exists in this family of compounds provided sufficient knowledge concerning their mechanisms of action can be developed.

VII. Summary

The biochemical function of the coumarins, hydroxycoumarins, and furanocoumarins in mold or plant metabolism remains obscure. However, it has been demonstrated that mold grown on synthetic media produces coumarin or hydroxycoumarins, some of which are highly toxic. In the case of *Sclerotinia sclerotiorum* production of furanocoumarins, a definite symbiotic relationship between the host and the infecting organism has been observed.

The wide variety of specific pharmacological activities observed with these substances offers a unique opportunity for exploitation of useful effects in the management of man's environment. Some examples have been cited but it is obvious from the above discussion that further study of the metabolism and occurrence of these compounds and their function in mold metabolism will provide valuable knowledge for the development of new uses for coumarin compounds.

General knowledge concerning the toxic action in food spoilage has been quantitatively increased by the isolation and identification of toxic coumarin compounds from moldy food products. The extension of the present knowledge to a full understanding of the origin of these compounds and the symbiotic mold-host relationship will lead to an understanding of the mechanism by which these compounds are produced by the molds.

REFERENCES

Akazawa, T. (1960). *Arch. Biochem. Biophys.* **90**, 82.
Anschütz, R. (1909). *Ann. Chem.* **367**, 169.
Arora, R. B., and Mathur, C. N. (1963). *Brit. J. Pharmacol.* **20**, 29.
Austin, D. J., and Clarke, D. D. (1966). *Nature* **210**, 1165.
Barker, W. M., Hermodson, M. A., and Link, K. P. (1969). *Federation Proc.* **28**, 290.

Baruah, P., and Swain, T. (1959). *J. Sci. Food. Agr.* **10**, 125.

Bellis, D. M. (1958). *Nature* **182**, 806.

Best, R. J. (1944). *Australian J. Exptl. Biol. Med. Sci.* **22**, 251.

Bloomfield, M., transl. (1897). "The Sacred Books of the East" (F. M. Muller, ed.), Vol. XLII, pp. 36 and 302. Oxford Univ. Press (Clarendon), London and New York.

Bocks, S. M. (1967). *Phytochemistry* **6**, 127.

Booth, A. N., Masri, M. S., Robbins, D. J., Emerson, O. H., Jones, F. T., and DeEds, F. (1959). *J. Biol. Chem.* **234**, 946.

Bose, P. K. (1958). *J. Indian Chem. Soc.* **35**, 367.

Brian, P. W., Curtis, P. J., Hemming, H. G., Unwin, C. H., and Wright, J. M. (1949). *Nature* **164**, 534.

Brian, P. W., Curtis, P. J., Hemming, H. G., Jefferys, E. G., Unwin, C. H., and Wright, J. M. (1951). *J. Gen. Microbiol.* **5**, 619.

Burns, J. J., Wexler, S., and Brodie, B. B. (1953a). *J. Am. Chem. Soc.* **75**, 2345.

Burns, J. J., Weiner, M., Simson, G., and Brodie, B. B. (1953b). *J. Pharmacol. Exptl. Therap.* **108**, 33.

Chakraborty, D. P., DasGupta, A., and Bose, P. K. (1957). *Ann. Biochem. Exptl. Med. (Calcutta)* **17**, 59.

Chmielewska, I., and Cieslak, J. (1958). *Tetrahedron* **4**, 135.

de Moura Campos, F. A. (1935). *Ann. Fac. Med. Univ. S. Paulo* **11**, 165.

Dominici, G. (1948). *Rass. Clin.-Sci.* **24**, 233.

Ellinger, A. (1908). *Arch. Exptl. Pathol. Pharmakol. Suppl.*, p. 150.

El-Mofty, A., El-Mofty, A. M., Abdelal, H., and El-Hawary, M. F. S. (1959a). *J. Invest. Dermatol.* **32**, 645.

El-Mofty, A., El-Mofty, A. M., Abdelal, H., and El-Hawary, M. F. S. (1959b). *J. Invest. Dermatol.* **32**, 651.

Fittig, R. (1868). *Z. Chem.* **4**, 595.

Freund, E. (1916). *Dermatol. Wochschr.* **63**, 931.

Grimm, H. (1951). *Biol. Zentr.* **70**, 367.

Hakim, R. E., Griffin, A. C., and Knox, J. M. (1960). *Arch. Dermatol.* **82**, 572.

Hoekenga, M. T. (1959). *J. Invest. Dermatol.* **32**, 351.

Hoernle, A. F. R. (1893-1912). "Bower Manuscript." Govt. Printing, Calcutta, India.

Howell, W. H., and Cekada, E. B. (1926). *Am. J. Physiol.* **78**, 500.

Ikawa, M., Stahmann, M. A., and Link, K. P. (1944). *J. Am. Chem. Soc.* **66**, 902.

Jacobs, M. B. (1953). *Am. Perfumer Essent. Oil Rev.* **62**, 53.

Judis, J. (1960). *J. Am. Pharm. Assoc., Sci. Ed.* **49**, 447.

Khadzhai, Ya. I., and Kuznetsova, V. F. (1962). *Farmatsevt. Zh.* **17**, 57.

Khadzhai, Ya. I., and Kuznetsova, V. F. (1963). *Farmakol. i Toksikol.* **26**, 219.

Knypl, J. S. (1963). *Nature* **200**, 800.

Kohler, H. (1875). *Centr. Med. Wiss.* **13**, 867 and 881.

Kosuge, T., and Conn, E. E. (1961). *J. Biol. Chem.* **236**, 1617.

Kuske, H. (1940). *Dermatologica* **82**, 273.

Larson, E. J., Connor, N. D., Swoap, O. F., Runnells, R. A., Prestrud, M. C., Elbe, T. E., Freyburger, W. A., Veldkamp, W., and Taylor, R. M. (1956). *Antibiot. Chemotherapy* **6**, 226.

Lavee, S. (1959). *J. Exptl. Botany* **10**, 359.

Levy, C. C., and Weinstein, G. D. (1964). *Nature* **202**, 596.

Link, K. P. (1943-1944). *Harvey Lectures* **39**, 162.

Link, K. P. (1948). *Chicago Med. Soc. Bull.* **51**, 57.

McCarter, J. C., Bingham, J. B., and Meyer, O. O. (1944). *Am. J. Pathol.* **20**, 651.

Marolda, C. I. (1936). *Rev. Asoc. Med. Arg.* **49**, 1567.

Mead, J. A. R., Smith, J. N., and Williams, R. T. (1958a). *Biochem. J.* **68**, 61.

Mead, J. A. R., Smith, J. N., and Williams, R. T. (1958b). *Biochem. J.* **68**, 67.

Musajo, L. (1955). *Farmaco (Pavia), Ed. Sci.* **10**, 2.

Musajo, L., Caporale, G., and Rodighiero, G. (1954). *Gazz. Chim. Ital.* **84**, 870.

Overman, R. S., Stahmann, M. A., and Link, K. P. (1942). *J. Biol. Chem.* **145**, 155.

Overman, R. S., Stahmann, M. A., Huebner, C. F., Sullivan, W. R., Spero, L., Doherty, D. G., Ikawa, M., Graf, L., Roseman, S., and Link, K. P. (1944). *J. Biol. Chem.* **153**, 5.

Pathak, M. A., and Fellman, J. H. (1961). *Proc. 3rd Intern. Congr. Photobiol., Copenhagen, 1960*, p. 552. Elsevier, Amsterdam.

Pathak, M. A., and Fitzpatrick, T. B. (1959a). *J. Invest. Dermatol.* **32**, 255.

Pathak, M. A., and Fitzpatrick, T. B. (1959b). *J. Invest. Dermatol.* **32**, 509.

Pathak, M. A., Fellman, J. H., and Kaufman, K. D. (1960). *J. Invest. Dermatol.* **35**, 165.

Pathak, M. A., Daniels, F., Jr., and Fitzpatrick, T. B. (1962). *J. Invest. Dermatol.* **39**, 225.

Pathak, M. A., Worden, L. R., and Kaufman, K. D. (1967). *J. Invest. Dermatol.* **48**, 103.

Perkin, W. H. (1868). *J. Chem. Soc.* **21**, 53.

Perone, V. B., Scheel, L. D., and Meitus, R. J. (1964). *J. Invest. Dermatol.* **42**, 267.

Pulver, R., and von Kaulla, K. N. (1948). *Schweiz. Med. Wochschr.* **78**, 956.

Quercioli, E. (1955). *Atti Accad. Nazl. Lincei, Rend., Classe Sci. Fis., Mat. Nat.* **18**, 313.

Quick, A. J. (1937). *Am. J. Physiol.* **118**, 260.

Raistrick, H., Stickings, C. E., and Thomas, R. (1953). *Biochem. J.* **55**, 421.

Reid, W. W. (1958). *Chem. & Ind. (London)* p. 1439.

Reinsch, H. (1867). *Neues Jahrb. Pharm.* **28**, 65.

Roberts, W. L., and Link, K. P. (1937). *Ind. Eng. Chem., Anal. Ed.* **9**, 438.

Roderick, L. M. (1929). *J. Am. Vet. Med. Assoc.* **74**, 314.

Roderick, L. M. (1931). *Am. J. Physiol.* **96**, 413.

Rodighiero, G. (1954). *Giorn. Biochim.* **3**, 138.

Rose, C. L., Harris, P. N., and Chen, K. K. (1942). *Proc. Soc. Exptl. Biol. Med.* **50**, 228.

Roseman, S., Huebner, C. F., Pankratz, R., and Link, K. P. (1954). *J. Am. Chem. Soc.* **76**, 1650.

Scheel, L. D., Perone, V. B., Larkin, R. L., and Kupel, R. E. (1963). *Biochemistry* **2**, 1127.

Schofield, F. W. (1922). *Can. Vet. Record* **3**, 74.

Schofield, F. W. (1924). *J. Am. Vet. Med. Assoc.* **64**, 553.

Sharma, A. K., and Bal, A. K. (1953). *Stain Technol.* **28**, 255.

Smith, C. G., Dietz, A., Sokolski, W. T., and Savage, G. M. (1956). *Antibiot. Chemotherapy* **6**, 135.

Soine, T. O. (1964). *J. Pharm. Sci.* **53**, 231.

Späth, E., and Kuffner, F. (1936). *Monatsh. Chem.* **69**, 75.

Stahmann, M. A., Huebner, C. F., and Link, K. P. (1941). *J. Biol. Chem.* **138**, 513.

Sullivan, W. R., Huebner, C. F., Stahmann, M. A., and Link, K. P. (1943). *J. Am. Chem. Soc.* **65**, 2288.

Taylor, R. M., Miller, W. L., and Vander Brook, M. J. (1956). *Antibiot. Chemotherapy* **6**, 162.

Thomas, R. (1961). *Biochem. J.* **80**, 234.

Uritani, I. (1967). *J. Assoc. Offic. Agr. Chemists* **50**, 105.

Vogel, A. (1820). *Ann. Physik* [1] **64**, 161.

Wakim, K. G., Chen, K. K., and Gatch, W. D. (1943). *Surg., Gynecol. Obstet.* **76**, 323.

West, B. D., Preis, S., Schroeder, C. H., and Link, K. P. (1961). *J. Am. Chem. Soc.* **83**, 2676.

Wolf, L. (1901). *Ann. Chem.* **315**, 145.

Chapter 3

The Natural Occurrence and Uses of the Toxic Coumarins

Vernon B. Perone

I. Introduction

There are various categories into which coumarins may be classified, depending upon their derivation, chemical configuration, mode of action, toxicity, etc. For the purpose of this chapter, they have been grouped into three principal categories: (1) coumarins, (2) furocoumarins, and (3) hydroxycoumarins. The aflatoxins, a mixed group of highly toxic metabolites produced by the fungus *Aspergillus flavus,* are discussed in detail in Chapter 1 of Volume VI of this treatise. It is among these that the most extensive research has been done, and therefore it is these about which the most is known. This chapter will discuss the more important compounds which exist in nature, the uses that have been made of them, and the toxic properties revealed by their usage. It would be impractical to cite every investigator who has contributed to the vast coumarin literature that has accumulated during the past 150 years. Therefore, the appended references represent those whose studies are believed to be pertinent to the topics discussed; any omission has been unintentional.

II. Coumarins

History of Occurrence, Toxicity, and Uses

Coumarins are widely distributed in the plant families of legumes, orchids, grasses, and citrus fruits and have been isolated from their roots, trunks (bark), leaves, and fruits (Spath, 1937; Reppel, 1954). They occur in plants predominately in the free state, but may also exist as the glyco- sides or glucosides of the corresponding coumaric acid (Dean, 1952). Natural coumarins are either neutral or phenolic, and, with the exception of coumarin itself, 4-hydroxycoumarin, and dicoumarol, all have an ox- ygen atom at position 7. They are derivatives of umbelliferone, one of the most widely distributed compounds of this class. They may be extracted from plants by selective organic solvents or by steam distillation.

Some complex coumarins have been isolated as metabolic products of bacteria and fungi, e.g., *Sclerotinia sclerotiorum* infects celery to produce pink rot (Birmingham *et al.*, 1961). The isolation and identification of two phototoxic coumarins, 8-methoxypsoralen and 4,5',8-trimethylpsoralen, has been made from diseased celery (Scheel *et al.*, 1963; Perone *et al.*, 1964). Also Spensley (1963) has studied aflatoxins resulting from *Asper- gillus flavus* infection of oilseed cake derived from peanuts. Several cou- marins have been found outside plant life, the most notable being the ben- zocoumarins isolated from the scent glands of the beaver (Lederer, 1949).

The simplest member of the natural coumarins is coumarin, the name of which is derived from coumarona, the common name of *Coumarouna odorata* (Soine, 1964). It was first isolated by Vogel (1820) from the tonka bean (*Dipteryx odorata*). In 1833, coumarin was isolated from sweet clover (*Melilotus alba*). Later it was discovered in vanilla leaf (*Trilisa odoratissima*) and in sweet woodruff (*Asperula odorata*) (Jacobs, 1953), and it has been isolated from lavender (*Lavendula officinalis*) (Brown, 1963) as herniarin, 7-methoxycoumarin. Since its discovery in 1820, coumarin has been isolated from more than 80 plants. It has a dis- tinctive odor, but, as the chemical configuration of coumarin becomes more complex, the odor becomes less characteristic (Bose, 1958). Simi- larly, as complexity increases, the plant family occurrence decreases (Soine, 1964).

When sweet potato roots are attacked by the fungi *Ceratocistis fim- briata, Thielaviopsis basicola, Fusarium oxysporum,* or *Helicobasidium mompa* certain coumarin derivatives, such as umbelliferone, scopoletin, aesculetin, scopolin, and skimmin, are produced (Uritani, 1967). They may take part in the defense action against fungus infection and are some- times harmful to animals and humans. The organism causing potato blight *Phytophthora infestans* has long been recognized. The affected tissues surrounding the lesions fluoresce strongly under ultraviolet light, and this fluorescence is due in part to an accumulation of oxygenated coumarins and their glucosides including scopoletin (Hughes and Swain, 1960), aes- culetin, umbelliferone (Austin and Clarke, 1966), and coumarin itself. Although the potato tuber alone is capable of producing these compounds

(Barauh and Swain, 1959), the possibility that they could also be produced by *P. infestans* was investigated by Austin and Clarke in their 1966 study. Using a French bean extract inoculated with *P. infestans*, they were able to isolate coumarin, herniarin, and umbelliferone. Their studies suggested that *P. infestans* may contribute to the accumulation of coumarins that occurs in infected potato tubers.

Van Sumere *et al.* (1957) demonstrated an inhibitory effect of coumarin on the germination of wheat stem rust uredospores. Bellis (1958), using two soil molds, *Penicillium jenseni* and *Penicillium nigricans*, showed that these molds were able to utilize coumarin as the whole carbon source, and, during growth, the content decreased with no other coumarin derivatives being isolated. Knypl (1963) conducted studies using *Aspergillus niger*, *Penicillium glaucum*, and *Rhizopus nigricans*, each containing coumarin in various concentrations, to determine its inhibitory effect on these molds. In general, he found that coumarin at concentrations of 100–1000 ppm inhibited or markedly retarded the germination of spores. The most sensitive were spores of *A. niger*, which did not germinate in yeast water containing 1000 ppm coumarin; at 100 ppm concentration, germination was 20–30 hours later than the controls. However, in a synthetic medium containing coumarin at a concentration of 1–10 ppm, a slight stimulating effect on *A. niger* was noted. The most resistant were spores of *P. glaucum*. Coumarin at 100 ppm concentration only slightly retarded their germination, and, at 1000 ppm, germination was 20–30 hours later than the controls. Germination of *R. nigricans* spores was considerably stimulated by 5–10 ppm of coumarin dissolved in yeast water, and slight retardation occurred only at a concentration of 500–1000 ppm. On the basis of his results, Knypl concluded that coumarin is a fungistatic agent but not a fungitoxic one. A physiological adaptation to coumarin occurs, although the nature of the adaptation is not known. Molds are much more sensitive to coumarin when cultured on yeast water, and Knypl assumed that this substance blocks the synthesis of unknown metabolites fundamental to later stages of spore germination.

In studies on the metabolic fate of coumarin in rabbits, Mead *et al.* (1958) showed that *o*-coumaric acid was partly converted to 4-hydroxycoumarin, but coumarin itself remained unchanged. When Bellis (1958) substituted *o*-coumaric acid for coumarin as the sole carbon source for *P. jenseni* and *P. nigricans* molds, he observed that as the concentration of *o*-coumaric acid fell, relatively large amounts of 4-hydroxycoumarin were formed. These studies also suggested that dicoumarol (3,3'-methylene-bis-4-hydroxycoumarin) occurred in the medium extract. *o*-Coumaric acid and not coumarin may be converted to dicoumarol in cultures of *Penicillium* isolated from the soil. This is significant when considering the investigations conducted by Stahman *et al.* (1941) that led to the identification of dicoumarol as the hemorrhagic agent in spoiled sweet clover hay. Conversion of *o*-coumaric acid to dicoumarol by contamination of the hay with soil fungi may lead to concentrations of dicoumarol of the order of those found.

Following its isolation in 1820, coumarin, because of its agreeable odor, was widely used in the perfume industry. It has also been used as a flavoring in chocolate, imitation vanilla extract, ice cream, and tobacco. Jacobs (1953) stated that coumarin as a compound suitable for use in food has been withdrawn by manufacturers of aromatic chemicals. This action was based on results of toxicity tests demonstrating the toxicity of coumarin in animals. In Great Britain, Hagan *et al.* (1967) conducted toxicity studies of coumarin in dogs and concluded it was not acceptable for use as food because of toxic effects observed in chronic feeding studies. Goldberg (1964) emphasizes the toxic potentialities of coumarin because of its wide distribution in the plant kingdom and subsequent ingestion by animals and man. The acute toxicity arises from its narcotic, sedative, and hypnotic action and its anticoagulant activity. He believes that it may be used safely as a food additive if a tolerance could be proposed, but cautions that one would have to specify which coumarin, because the data for one member of this group could not categorically be applied to all members.

Rupprecht (1969) used a combination of thin-layer and column chromatographic techniques to isolate 17 coumarin derivatives used in nickel plating solutions containing coumarin additives. Those isolated were 4-hydroxy-, 5-hydroxy-, 6-hydroxy-, and 7-hydroxycoumarin, and 4,6,7-trihydroxycoumarin. Cadmium deposition from solutions of varying ionic strengths and coumarin concentrations were studied polarographically by Partridge *et al.* (1969). Under certain conditions, two polarographic waves were obtained, both corresponding to cadmium deposition. This effect was explained by the variation in potential of the orientation of absorbed coumarin molecules.

Spath and Kuffner (1936) studied the action of coumarins on fish (guppies) and found that those whose average weight was 130 mg died in 8 hours when exposed to a concentration of 1 mole of coumarin per 1000 liters of water. These "fish poisons" are contained in plants as coumarones and are particularly effective as insecticides (Brodersen and Kjaer, 1946).

Zweidler *et al.* (1964) have listed coumarin derivatives used as optical brighteners or bleaching agents. Flax treated with aesculin, a glucoside of aesculetin, showed a brightening effect which disappeared upon washing and exposure to light. Some 7-aminocoumarin derivatives are incorporated into detergents and soap powders for brightening wool and nylon fabrics. Coumarin derivatives of 3-phenyl-7-amino type are used to ensure lightfastness of synthetic fibers and plastics, and amino- and phenyl-coumarins are effective brighteners for polyvinyl chloride type plastics.

In Great Britain a dentifrice containing coumarin for the prevention of dental caries has been developed (Mamalis, 1969). In a study using hamsters, 19% fewer cavities were observed with the coumarin-containing dentifrice than with a control.

III. Furocoumarins

A. HISTORY OF OCCURRENCE, TOXICITY, AND USES

The furocoumarins, more commonly known as psoralens, occur naturally, as do the coumarins, in free form; as glucosides, they occur in many tropical and subtropical plants, seeds, and fruits. Little is known concerning the function of these compounds in plant metabolism. The parent compound of this family is psoralen (ficusin), and the more important members in addition to psoralen are 5-methoxypsoralen (bergapten), 8-methoxypsoralen (xanthotoxin), 8-isopentyloxypsoralen (imperatin), 5,8-dimethylpsoralen (isopimpinellin), isopsoralen (angelican), and 4,5',8-trimethylpsoralen. In all, approximately 40 psoralen compounds have been identified. The use of these compounds in the treatment of skin depigmentation dates to antiquity. Scheel (1967) has traced the use of furocoumarins in the treatment of leucoderma to 1400 B.C. Pathak and Fitzpatrick (1959) refer to their cutaneous usage in ancient Egypt, India, and other countries of the orient.

In 1834 the first furocoumarin was isolated (from bergamot oil) and identified as 5-methoxypsoralen by Kalbrunner. Thoms (1911) was the first to isolate 8-methoxypsoralen, and the chemical structure of this compound was established by Spath and Holzen (1933) who synthesized it. The chemical structure of 4,5',8-trimethylpsoralen was established by Kaufman (1961) who synthesized it from 4,8-dimethyl-7-hydroxycoumarin and allyl bromide. The natural occurrence of 4,5',8-trimethylpsoralen was first reported by Scheel et al. (1963) who isolated it from pink rot celery which had been infected by the fungus Sclerotinia sclerotiorum.

Dermatologists around the world have long observed and been concerned with a variety of plants and plant extracts containing psoralens that produce contact dermatoses in the presence of or subsequent to exposure to sunlight. Stowers (1897) in England and White (1897) also in England both reported dermatitis from contact with wild parsnips. Straton (1912) in England observed several cases of dermatitis from handling cow parsnips and another case due to contact with the sap from a fig tree. Several publications during the next few years dealing with dermatitis from plants and from eau de cologne appeared in the foreign and American dermatological journals. Freund (1916) reproduced pigmentation on human skin by first applying eau de cologne and then having the subjects expose themselves to sunlight and sea-water. He obtained the same results when oil of bergamot was sub-stituted for the cologne. Legge (1921) observed skin lesions among workers who picked and packed figs. Rosenthal (1924), in describing cases of dermatitis resulting from plant contact, originated the word "berloque." In 1926, Oppenheim, in Vienna, presented cases of bul-lous linear lesions which had occurred 24-48 hours after contact with grass while the patients were sunbathing. The linear configuration of the eruption led to the disease being titled "dermatitis bullosa pratensis striata." Later, other workers changed it to the simpler "Oppenheim's disease." Oppenheim first believed these lesions were caused by a para-site similar to that causing scabies. In a later report, he attributed the eruptions to contact with meadow grass in sunlight (Oppenheim and Fes-sler, 1928).

Legrain and Barthe (1926) were the first to observe a skin eruption in a French farm worker who had handled celery. They believed the eruptions to be due to celery oil. Del Vivo (1930) studied the action of five essential oils of eau de cologne and demonstrated that oil of bergamot was the prin-cipal cause of pigmentation following exposure of the skin to sunlight. Berlin (1930) described cases of dermatitis in children who had handled green figs in whom a persistent pigmentation of the affected areas re-mained following the onset of the eruptions and their subsequent disap-pearance. Gross and Robinson (1930) and Goodman (1931) observed dermatitis in women following the use of perfume containing oil of ber-gamot as the odorant in which the active ingredient was 5-methoxypsor-alen. Phyladelphy (1931) was the first investigator to recognize the im-portance of sunlight in this form of reaction, while Kitchevatz (1934) was the first to recognize that this disease was a form of photosensitization. In later studies, Kitchevatz (1936) reproduced the disease on human skin by the application of fig extracts and subsequent exposure of the sites to sun-light. He believed that the offending agent was chlorophyll because of the variety of plants that caused the reaction. Hirschberger and Fuchs (1936) proved that bullous lesions could be produced on human skin when wild parsnips were rubbed on the area that was later exposed to sunlight, and Meischer and Burckhardt (1937) reported dermatitis in gardeners who worked with cow parsnips. Kuske (1938) defined further the nature of the reaction between certain plants, the skin, and sunlight. By using the juice of hogweed, wild parsnips, and figs, supplemented by exposure to sunlight, he was able to produce varying degrees of lesions on the skin. Bechet (1939) reported that the incidence of dermatitis among fig workers was 10%.

All these observations have made it evident that numerous plants possess phytophotodermatitis potential, and that many of these plants belong to the Umbelliferae family. Included in this family are wild parsnip, cow parsnip, parsley, celery, angelica, wild carrot, fennel, and dill. In Denmark, Jensen and Hansen (1939) observed several patients with bullous lesions which occurred following contact with wild parsnips. Using the juice of this plant, they rubbed the skin and exposed these areas to sunlight. The resulting lesions were identical clinically to those which the patients had developed originally. They then conducted studies to determine the wavelengths of light involved by using a window glass to filter out those rays below 3200 Å emanating from a mercury vapor lamp. Jensen and Hansen concluded that the greatest activity from the light spectrum necessary to activate a phototoxin of this nature was between 3200 and 3600 Å. Sams (1941), in the United States, studied lime oil dermatitis using both Florida sunlight and a mercury vapor lamp to reproduce the lesions. His experiments demonstrated that the action spectra required was between 3100 and 3700 Å. It was nearly identical to the action spectra associated with photodermatitis due to parsnips.

Klaber (1942), in London, published a comprehensive paper on phytophotodermatitis. He observed many cases of dermatitis due to rue, wild parsnip, figs, and celery. His studies with rue and the juice of figs on skin subsequently exposed to sunlight and ultraviolet light confirmed the observations of Jensen and Hensen and Sams. Klaber also demonstrated that the long rays of light between 3200 and 3800 Å were responsible for the reaction. The short rays below 3200 Å were also capable of producing skin lesions of varying degrees of intensity, and therefore it was necessary to filter them out. He originated the term "phytophotodermatitis" to describe the variety of lesions resulting from exposure to plants and sunlight.

Chakraborty et al. (1957) studied several natural coumarins including psoralens for their antifungal activity and concluded that the psoralens were the most effective antifungal agents of those studied. Fowlks et al. (1958) studied bacteria exposed to furocoumarins in the presence of ultraviolet light and demonstrated a photosensitizing effect on these bacteria.

Pathak and Fitzpatrick (1959) developed a bioassay technique whereby the relative photodynamic activity of psoralens could be studied on laboratory animals. Using guinea pigs, alcoholic solutions of various concentrations of furocoumarins were applied to shaved skin areas and the test sites irradiated under ultraviolet light at wavelengths of 3200 Å and above. Of 18 compounds evaluated by this technique, 11 showed photo-dynamic action by producing reactions on the animal skin varying from mild erythema to vesiculation. The five most active compounds were psoralen, 4,5',8-trimethylpsoralen, 5',8-dimethylpsoralen, 5'-methyl-psoralen, and 8-methoxypsoralen. Methyl derivatives of psoralen were nearly as active as psoralen itself.

Buck *et al.* (1960) and Pathak (1961) have studied the action spectrum of 8-methoxypsoralen, and their observations indicated that the wave-lengths which most effectively evoke a photosensitization response are between 3400 and 3800 Å, with a maximum effectiveness at 3600 Å.

Scheel (1967) in his review of the furocoumarins has selected the word "phototoxic" to describe the effects of these compounds on mammalian skin and on paramecia, bacteria, and bacteriophage. The use of the word phototoxic in this context denotes a combination of a concentration of the compound and exposure to radiant energy such that cellular damage or cellular death will occur. In the absence of either the compound or the radiation, no toxic effect is observed.

B. CELERY DERMATITIS

The first published observation of skin lesions attributed to handling celery was by Legrain and Barthe (1926). They believed the eruptions they observed on a farm worker were due to celery oil. Henry (1933) examined workers in an English vegetable cannery who washed, chopped, and packed celery. Four percent of the workers presented skin eruptions which Henry believed were due to celery oil. Gelfand (1936), in the United States, observed a patient with dermatitis caused by handling celery in a canning factory. From a series of patch tests, he concluded that the offending material was celery oil. Henry (1938) revisited vegetable canneries and observed that the number of celery workers with dermatitis had increased to 30%. Again he concluded that the eruptions were due to celery oil. A curious observation which he did not pursue was that derma-titis was three times more prevalent among workers who handled pink celery, which was described as more pungent than the white or gold varie-ties also handled.

Wiswell *et al.* (1948) studied celery workers in Massachusetts who were growing a green Pascal variety introduced in 1939. They stated that celery dermatitis was not a problem until the introduction of this variety. Although most workers stated that their eruptions occurred only when they had been in contact with rotted celery, these investigators believed this to be a coincidence associated with the introduction of the new variety of celery. They also believed that the evidence was against the rot. Patch test results of the workers to normal and rotted celery in many instances demonstrated the most severe reactions were to rotted celery only; however, they concluded that the dermatitis was due to celery oil and that this was limited to the green Pascal variety. Palumbo and Lynn (1953) also conducted a study to determine the cause of celery dermatitis among Massachusetts farmers growing green Pascal celery. They, too, concluded that the cause was due to celery oil.

In the summer of 1959, Birmingham *et al.* (1961) undertook a study of celery dermatitis among workers in the celery fields of Michigan and Florida. The only factual knowledge at this time was that the dermatitis affected large numbers of workers who developed bullous lesions on the skin when they had been in contact with the plant. Conflicting theories existed regarding the cause, the most widely held theory being that it was due to an allergic hypersensitivity to celery oil. Other views included an allergic hypersensitivity to certain varieties of celery, phytophotodermatitis, photosensitization, or, as the Michigan farmers believed, contact with pink rot, the fungus disease of the celery plant caused by *Sclerotinia sclerotiorum*. It was Tucker (1959a) who suggested that the cause was due to photosensitivity reaction of furocoumarins or psoralens contained in the celery. Of 302 celery workers examined by Birmingham and his group, 163 displayed various stages of vesicular and bullous dermatitis along with areas of depigmentation and hyperpigmentation remaining from previous lesions. In a series of patch tests on 25 workers, Birmingham's group applied the juice and crushed leaves of both normal and pink rot celery to the skin and exposed the sites to sunlight. All normal celery test sites were negative; all the pink rot sites exhibited bullous lesions within 48 hours.

In the laboratory, these investigators reproduced the disease on the skin of noncelery workers using the same technique. Again, only the pink rot sites produced lesions; the normal celery sites remained negative. Similar results were obtained using rabbit skin. In a series of further experiments, irradiation of the test sites was made by a carbon arc lamp with filters to eliminate wavelengths below 3200 Å. These studies yielded identical results. The ultraviolet spectrum of an ether extract of pink rot celery gave absorption peaks at 2930, 2600, and 2430 Å, which closely resembled the absorption spectrum of irradiated 8-methoxypsoralen reported by Lerner *et al.* (1953). Examination of the ether extract of pink rot celery by infrared spectrophotometry provided further evidence that the active material was a furanocoumarin. Whether it was 5-methoxypsoralen (bergapten), which Musajo *et al.* (1954) showed existed in normal celery, was not conclusively established.

Birmingham and his co-workers demonstrated that only celery infected with pink rot contains a photoreactive furanocoumarin which produces lesions on the skin after the sites of application are irradiated by natural sunlight or ultraviolet light emitting wavelengths between 3200 and 3700 Å. These results gave impetus to a further study by Scheel *et al.* (1963) which resulted in the isolation and characterization of two phototoxic coumarins from pink rot celery. With lyophilized material from the pink areas only of rotted celery, extraction and crystallization techniques yielded two crystalline forms with differing melting points. Bioassay of the crystals on both human and animal skin indicated a very high activity. Following purification, the two compounds were identified through their melting points, carbon–hydrogen determinations, and infrared and ultraviolet spectrographic analyses. Compared with known synthetic compounds, they proved to be 4,5′,8-trimethylpsoralen and 8-methoxypsoralen. These investigators, using identical extraction procedures on lyophilized normal celery, were unable to isolate any similar compounds.

The early reports in the literature describing celery dermatitis made no distinction between rotted and normal celery, and the only conclusions that could be drawn from the results obtained from the investigators were that they used normal celery in their experiments and assumed that cases of dermatitis they observed were due to normal celery. The first suggestion that it might be other than normal celery that was responsible was Henry's observation (1938) that a pink variety in England was more pungent and caused an increased incidence of dermatitis than other varieties. Later, Wiswell *et al.* (1948) obtained more positive patch test results with rotted celery than normal celery, and even though the workers he studied attributed their eruptions to rotted celery, he discounted these findings. Birmingham *et al.* (1961) studies presented the first evidence that it was pink rot and not normal celery that contained a photoreactive furanocoumarin. However, they did not discount the possibility that normal celery, if concentrated sufficiently, could produce dermatitis, especially in considering Musajo's isolation of 5-methoxypsoralen from celery in Italy. Again, it was not clear whether Musajo worked with normal or diseased celery or a combination of the two. Unless rigid supervision of packing celery is maintained, rotted or partially rotted plants may be mixed with the healthy ones, and in several days much of it would be infected. This is a continual problem for all celery farmers and packers. A request to Musajo by this author regarding the possibility of such an occurrence with the celery he used did not result in information that would help clarify the problem. Scheel's isolation and characterization of two phototoxic coumarins from pink rot but not normal celery provided further evidence that normal celery was not an important cause of dermatitis among celery handlers.

To provide further evidence that normal and pink rot celery react differently on subsequently irradiated skin, Perone *et al.* (1964) conducted a study using an adaptation of Pathak's bioassay method to determine first, the location of the toxin in the celery plant; second, the quantitation and comparison of cutaneous reactions produced by the isolated furanocoumarins from pink rot celery, and, finally, the specificity of *S. sclerotiorum* in making celery reactive. Fifteen kilograms each of normal and pink rot celery were treated separately but identically. Two-inch sections from leaves to base were separated and extracted with petroleum ether. Each extract was concentrated to 5 ml and tested repeatedly on human and rabbit skin exposed to July sunlight and light from a mercury vapor lamp, at a wavelength of 3650 Å. In every test site, all normal celery extracts produced negative results. The extracts of pink rot and the adjacent 2-in. areas produced bullae on human skin and necrosis on rabbit skin. All other normal appearing areas of diseased celery produced negative results. It was concluded that only pink rot areas and areas adjacent to pink rot contained photoreactive furanocoumarins.

Acetone solutions of Scheel's purified crystals were assayed on rabbit skin along with synthetic 4,5′,8-trimethylpsoralen and 8-methoxypsoralen. The isolates and the synthetics gave identical reactions, i.e., isolated and synthetic 8-methoxypsoralen produced reactions in concentrations as low as 1.5 μg/sq in. of skin surface, and isolated and synthetic 4,5′,8-trimethylpsoralen produced reactions in concentrations as low as 0.1 μg/sq in. These compounds possessed potent biological activity.

It was well known that the fungus *S. sclerotiorum* caused pink rot disease of celery. Guba (1950) reported that other fungi can also cause celery to rot. Among these are *Botrytis cinera, Bacillus carotovorus, Rhizopus nigricans,* and *Penicillium expansum.* Using cultures of each of these organisms supplied by Guba, stalks of normal celery were separately inoculated and allowed to rot. The expressed juice from each was assayed on human and rabbit skin with negative results. It was concluded that only *S. sclerotiorum* caused the disease in celery that produces dermatitis. The mechanism by which the organism alters the chemical components of celery from an inactive to a photoreactive state has yet to be explained.

The solution to the problem of celery dermatitis began when the Michigan farmers blamed pink rot disease of the plants, and Tucker suggested that the cause might be due to psoralens in celery. The studies of Birmingham *et al.* (1961); Scheel *et al.* (1963); and Perone *et al.* (1964), clearly demonstrated that it is the fungus *S. sclerotiorum* which attacks a healthy celery plant and causes pink rot disease. It is only pink rot infected tissue which contains phototoxic furanocoumarins, and contact of the furanocoumarins with the skin in the presence of light containing wavelengths close to 3650 Å is required for a reaction to occur. The severity of the reaction depends upon the concentration of furanocoumarins and the length and intensity of the light exposure. It is concluded that those isolated cases of dermatitis caused by normal celery may be due to individual allergic responses and are not phototoxic responses.

C. PSORALENS AND VITILIGO

Vitiligo is an acquired progressive achromia of the skin due to a functional abnormality of the melanocytes (Pinkus, 1959); familial incidence is common (Lerner, 1955). It is manifested by scattered pigmentless areas which may appear on both the covered and exposed parts of the body. The incidence of vitiligo in several countries is estimated by Lerner (1959) to be 1% in the United States, 0.39% in Switzerland, 1.64% in Japan, and 3% in India. Ancient Hindus, Turks, Egyptians, and others from oriental countries were using psoralens extracted from native plants for the treatment of leukoderma or vitiligo more than 3000 years ago. Among the plants used for this purpose were the bavachee, *Psoralea corylifolia*, and *Ammi majus* (Fitzpatrick and Pathak, 1959). Considering the crudeness of the preparations and the uncontrolled dosages and exposures to sunlight, surprisingly good results were obtained, although extensive blistering and scarring did occur.

The first controlled use of psoralens in the treatment of vitiligo was by Uhlmann (1927), who used topical applications of oil of bergamot (bergapten) and exposed the areas to natural or artificial light. He treated approximately 50 patients by this method with what he considered good results. In Egypt, Fahmy and Abu-Shady (1947, 1948) performed extensive studies with the *Ammi majus* L. plant which grows abundantly along the Nile delta. From it they isolated three active compounds which they named (after the plant) ammoidin, ammidin, and majudin. These compounds were shown to correspond exactly to the known compounds xanthotoxin (8-methoxypsoralen), imperatorin (8-isoamyleneoxypsoralen) and bergapten (5-methoxypsoralen). Under Fahmy's direction, El Mofty (1948) conducted clinical studies using these compounds on vitiligenous patients with relatively good results. His studies showed ammoidin and ammidin to be the most active compounds in repigmentation, and these were made available commercially in combination as a tablet and liquid by a Cairo firm. El-Hefnawi (1950) used the commercial compounds both orally and topically with exposures to both artificial ultraviolet light and sunlight and reported some cures of vitiligo. In France, Sidi and Bourgeois-Gavardin (1952) also reported good results with these compounds.

Interest was stimulated in the United States from these results; although, only 8-methoxypsoralen (methoxsalen) was marketed for clinical trials in this country. Lerner *et al.* (1953) studied patients receiving methoxsalen orally with exposure to artificial ultraviolet light. They reported a decrease in the erythema response of these patients. Fitzpatrick *et al.* (1955) reported that volunteers with normal skin receiving 20-30 mg methoxsalen orally, but with uncontrolled exposures to sunlight showed increased tanning, and all subjects reported an increased tolerance to sunlight exposure. However, in a study of healthy adult males receiving 50 mg orally and exposure to sunlight controlled for 85 minutes, they observed erythema, edema, vesiculation, and scaling, which progressed to sloughing of the skin, and, in 22 of the 47 subjects, depigmentation occurred. In another group of volunteers receiving 10-75 mg methoxsalen orally, they found that after a dose of 10-30 mg and exposure to solar radiation limited to 60 minutes, erythema was produced which was followed by increased tanning without tenderness or exfoliation. In higher doses, the erythema was followed progressively to exfoliation. At the 75 mg level, blistering of the skin occurred. They concluded from their studies that although methoxsalen had a potential use in skin tolerance to sunlight, it did not have a primary protective action but rather augmented all cutaneous reactions.

In a histological study of changes following psoralen therapy, Becker (1959) concluded that the stratum corneum was thickened, pigment production correlated directly with the amount of inflammation developed, and psoralen tanning was produced by the large number of melanin granules retained in the stratum corneum and basal cell layer. It was his opinion that psoralens plus ultraviolet light did not produce any skin change that was not produced by ultraviolet light alone, but that the combination did markedly accentuate the reactions. Zimmerman (1959) also conducted a histological study of the skin of patients following oral treatment with methoxsalen and ultraviolet light exposures. His conclusions were in agreement with Becker's.

In a study to determine the toxicity of orally administered methoxsalen, Tucker (1959b) observed 120 male subjects following the ingestion of 10 mg daily for 21 days. Liver function tests were within normal limits, and only minor subjective side effects were reported by three subjects. He concluded that methoxsalen was nontoxic. London (1959) reported a study in which he compared the effectiveness of oral versus topical application of psoralens in the treatment of vitiligo. The oral administration was composed of a mixture of 10 mg of 8-methoxypsoralen and 5 mg of 8-isoamyleneoxypsoralen, and the topical cream contained 0.75% of the former and 0.25% of the latter. It was his conclusion that topical application was preferred to oral treatment, provided caution was used in sunlight exposure, because repigmentation occurred sooner and was longer lasting than with oral treatment. While occasional blistering occurred with topical applications, London found oral treatment resulted in toxic symptoms of severe nausea and vomiting, nervousness, blistering, and dermatitis which accounted for some of his patients discontinuing treatments.

Elliott (1959) treated 27 vitiligenous patients for 1½ to 32 months with orally ingested methoxsalen. No patient was cured of vitiligo; however, seven had complete repigmentation of selected areas, and eighteen believed the results justified the length and expense of the treatment. He concluded that the amount of repigmentation was more dependent on the total time of exposure to sunlight and the location of the lesion than on the duration of the disease or extent of involvement. The dorsum of the hands and feet were most resistant to repigmentation. Elliott believed that 100% repigmentation could be achieved if treatment were continued long enough with sufficient intensity.

Glatt (1961) treated nine vitiligenous patients using a combination of oral and topical applications of psoralens with irradiation by artificial ultraviolet light. He found that initial oral therapy prevented excessive local reaction, but when prolonged, it prevented mild erythemas that enhance repigmentation. He believed the topical phase of the treatment was the more important because it supplied a high concentration of psoralens to the skin, and increased the erythema response of the skin to ultraviolet light. Glatt emphasized that tests of liver function, blood counts, and urinalyses should be performed regularly, and hepatic and renal impairment are contraindications to oral therapy. Lerner (1968) stated that the treatment of vitiligo with methoxsalen yielded good results in only 15% of patients. A more recent medication is trimethylpsoralen (trioxsalen, trisoralen), which is made synthetically. Trioxsalen has fewer side effects than methoxsalen, but Lerner believes its repigmentation characteristics are about the same.

D. PSORALENS AND SUNTANNING

Because of the enthusiastic response generated for the use of psoralens in the treatment of vitiligo, and because of the preoccupation of a significant number of individuals with obtaining a healthy-appearing suntan, the use of psoralens was channeled in this latter direction. Several studies were undertaken for the purpose of establishing the practicability of people with normal skins using psoralens to obtain or enhance a suntan. The studies of Fitzpatrick in 1955 on normal skin treated with psoralens has been reviewed above in the section on the treatment of vitiligo.

Daniels et al. (1959) designed a study to test clinical claims that with methoxsalen and sunlight exposure subjects received a better tan. The oral dosage ranged from placebo to 40 mg with controlled exposure to natural sunlight. In contrast to testimonial evidence, their studies revealed no significant tanning or protective effects with the 20 mg dosage. Conspicuous effects were observed, however, with 30–40 mg levels. Imbrie et al. (1959) reported that a 30-minute exposure to natural sunlight 2 hours after ingesting 30 mg of methoxsalen increased the tolerance of exposed skin to subsequent exposures of ultraviolet light. It required twice the amount of ultraviolet light to produce erythema of the skin in individuals treated 5 weeks previously with a single oral dose of methoxsalen than in individuals treated orally with a placebo.

Kanof (1959) pointed out the difficulty of standardizing sun exposure following administration of methoxsalen due to the wide range of individual tolerance. Because the compound augments all of the skin's responses to sunlight, his exposure rule was to limit initially such sun exposure as the patient knew he could tolerate without burning. The exposure time was increased as hyperpigmentation and thickening occurred. A point emphasized was that methoxsalen does not allow limitless sun exposure without burning. Kanof treated more than 200 persons with oral methoxsalen, many of whom were examined periodically for liver function, blood counts, and urinalyses. He reported no toxic effects following doses of 40 mg for many months. Stegmaier (1959) reported that of 25 persons receiving oral methoxsalen prior to sunlight exposure, 23 were provided protection against sunburn, and increased pigmentation was observed in 18. Liver function tests and blood counts remained normal.

IV. Hydroxycoumarins

A. ANTICOAGULANT ACTIVITY

The 4-hydroxycoumarins are important because of their anticoagulant activity, and they are widely used therapeutically to control intravascular thrombosis. The coagulation of blood is a complex biological phenomenon, involving approximately twelve factors, four of which are affected by oral anticoagulant drugs. Of the oral anticoagulant drugs, the most important are the synthetic coumarins. Among the major drugs in use are acencoumarol, cyclocoumarol, dicoumarol, ethylbiscoumacetate, phenprocoumon, and warfarin sodium, all known by a variety of commercial or trade names. Estimation of the effect of oral anticoagulants is made by determining the prothrombin time of citrated plasma of withdrawn venous blood by the Quick et al. (1935) method or the many variations of this method. The indications for therapeutic use of anticoagulants may be considered short term, which include patients with varicose veins or those who have had a previous episode of venous thrombosis in the lower limb, those with an injury requiring immobility of a lower limb, those over age 45 who have had a major gynecological operation, those in congestive heart failure, and those who will be confined to bed for an extended period. Riggs (1950) disagrees with this last use of anticoagulants. He believes that routine anticoagulant therapy for long-term bed patients imposes an impossible burden on the hospital and laboratory staff and subjects many patients to an unnecessary risk of bleeding.

The use of anticoagulants may also be considered long term, which includes patients with coronary artery disease (Bjerkelund, 1957; Harvald et al., 1962; and Medical Research Council, 1959), those under age 50 who have survived an acute myocardial infarction, those who had an attack of acute coronary insufficiency (Wood, 1961), those with rheumatic heart disease with recurrent embolic episodes (Owren, 1961), and those with venous thrombosis and pulmonary embolism (Sevitt, 1962).

The chief contraindication to the use of anticoagulants is a history of bleeding. These drugs exert their toxic effect mainly through hemorrhage. Minor hemorrhages occur in 40% of patients, while major hemorrhages are rare, although they can occur. Continual vigilance by the physician managing anticoagulant therapy and accurate prothrombin time determinations are necessary. Coumarins taken orally have caused skin rashes and alopecia, but have not been responsible for any major toxic effects (Anticoagulants, 1963). The antidote for oral anticoagulants is vitamin K.

Douglas and McNicol (1964) consider the toxicity of anticoagulant drugs (hemorrhage) as resulting from the disturbance of the hemostatic mechanism, which is the primary action of the drugs; other side effects include disturbances other than hemorrhage. Minor hemorrhages such as microscopic hematuria, epistaxis, rectal bleeding, subconjunctival hemorrhage, bleeding from shaving cuts, and gum oozing are three times more common than major hemorrhages such as macroscopic hematuria, hemoptysis, hematemesis, or intracranial bleeding (Douglas and McNicol, 1964). These investigators have not observed any side effects from coumarin drugs.

Haan and Tilsner (1965) studied 125 patients during a 2-year period, 65 of whom were treated with an anticoagulant (phenprocoumon), while 60 were treated without anticoagulants. All patients had suffered myocardial infarction. In the anticoagulant-treated group, 34% continued to have symptoms of angina, while 66% remained asymptomatic. In the non-anticoagulant-treated group, 55% continued to have pain, while 45% remained free of symptoms.

B. DICOUMAROL AND SWEET CLOVER DISEASE

The first published report describing a new disease affecting cattle feeding on sweet clover (*Melilotus alba*) was by Schofield (1924); the observed mortality rate was nearly 80%. His studies clearly demonstrated that only spoiled sweet clover, i.e., the ensilage attacked by molds, actually produced the disease. He isolated several molds from the spoiled clover including *Aspergillus* and *Penicillium*, neither of which produced symptoms when cultured on media other than sweet clover and subsequently fed to rabbits. When the molds were grown on sweet clover, *Aspergillus* produced hemorrhage, while *Penicillium* produced a reduction of red blood cells with a slight alteration of clotting time. Schofield concluded that a toxic substance was produced in moldy sweet clover that when fed to cattle and experimental rabbits, produced fatal hemorrhage.

Working independently, Roderick (1929) confirmed the findings of Schofield that the disease was caused by spoiled clover only. In pathological studies of the affected animals, Roderick concluded that the characteristic lesion of the disease was a massive hemorrhage which resulted in cattle losses numerically greater than the increasing use of sweet clover as a forage. Death was preceded by a continual development of a delay in the coagulation of the blood. In a later study, Roderick (1931) performed detailed studies on the blood of hemorrhagic cattle and concluded that the delay in blood coagulation involved a reduction in the prothrombin, and the reduction parallelled the delay in the coagulation. Further evidence for this conclusion was presented by Roderick using the Howell prothrombin time assay.

These findings, together with requests from Wisconsin farmers concerned with sweet clover disease, stimulated Link and his associates to conduct a series of detailed studies in an attempt to isolate and identify the causative agent of sweet clover disease. Their first step was to develop a sensitive and reproducible prothrombin assay adaptable to laboratory animals (Campbell *et al.*, 1940). Having succeeded in perfecting their assay, Campbell and Link (1941) then isolated the anticoagulant in a pure crystalline form with the molecular formula $C_{19}H_{12}O_6$.

In later studies Stahman *et al.* (1941) and Huebner and Link (1941) established the chemical identity of the anticoagulant as 3,3'-methylenebis(4-hydroxycoumarin) and prepared it synthetically from a degradation product of salicylic acid. Finally, Overman *et al.* (1942) conducted a series of assays, administering the synthetic and the natural products to various species of laboratory animals, including rabbits, rats, guinea pigs, and dogs; they were confirmed as being identical. The common name given to this anticoagulant was dicoumarol. The use of dicoumarol as a prophylaxis against thrombosis attracted the attention of clinicians, although, in the beginning, there was controversy over its safety. Link maintained that vitamin K could effectively counteract the anticoagulant action of dicoumarol. Shapiro *et al.* (1943) were the first clinicians in the United States to confirm Link's claim. Others in this country, Canada, and Sweden soon followed.

From 1946 to 1948, one of Link's graduate students, Scheel, assayed the anticoagulant effect of two dozen synthetic coumarins in an attempt to find a suitable rat poison. Dicoumarol proved to be an ineffective rodenticide because of its low activity and because it competed with vitamin K in the rat's natural diet. One of the coumarin compounds assayed, identified as No. 42, proved to have superior rodenticidal properties and was named warfarin by Link through a combination of the first letters contained in the Wisconsin Alumni Research Foundation and the "arin" from coumarin. In 1950, Link proposed the use of warfarin as an anticoagulant for man, but the use of "rat poison" did not meet with a positive response. Strangely, in 1951, an impromptu human assay occurred which proved to be significant in demonstrating the applicability of this drug. An army inductee had attempted suicide by consuming an amount of warfarin lethal to rats. However, it was not fatal to him and with blood transfusions and vitamin K therapy he recovered (Kellum, 1952). As a result of this incident and the controlled studies with warfarin sodium by Shapiro (1953) and Clatanoff *et al.* (1954), these investigators concluded that it possessed properties not inherent in dicoumarol or other anticoagulants they had tested. After the manufacture of warfarin was perfected, it became available commercially under the trade name of Coumadin Sodium, and has been widely used by clinicians. Warfarin, or Coumadin, is claimed to be more potent than dicoumarol and has the advantage that it can be given orally, rectally, intravenously, or intramuscularly with the same absorption rate by all four routes (Shapiro and Ciferri, 1957).

In less than a quarter of a century after Schofield and Roderick first observed and described the death of cattle feeding on spoiled sweet clover hay, Link and his co-workers had derived from this problem of fatal hemorrhage a new drug, which, along with heparin, can substantially prevent and alleviate intravascular thrombosis.

C. SCOPOLETIN AND AESCULETIN

Scopoletin is a natural constituent of the bark of *Scopolia japonica*, the roots of *Gelsemium sempervirens*, and parts of other plants. The first synthetic product was made by Head and Robertson (1931). Chemically, scopoletin in 6-methoxy-7-hydroxycoumarin and is also known as chrysotropic acid, gelsemic acid, β-methylaesculetin, and coumarin lactone. It often occurs in plants in combination with its glucoside, scopolin.

Best (1936) first isolated a fluorescent compound from virus-infected tobacco plants which he later identified as scopoletin. He also found smaller quantities in healthy tobacco plants. Uritani (1953) isolated scopoletin in the uninjured portion of injured sweet potatoes. Reid (1958), working with fresh tobacco plants, isolated scopoletin and, using ^{14}C-labeled precursors, determined that scopoletin is directly derived from phenylalanine. Wender et al. (1958) have isolated scopoletin from tobacco smoke. Baruah and Swain (1959) using paper chromatographic techniques isolated approximately 20 fluorescent compounds from potato tubers (*Solanum tuberosum*). Among the isolates were scopoletin and aesculetin (6,7-dihydroxycoumarin). Aesculetin has been used in filters which absorb ultraviolet light.

V. Food Spoilage

Coumarins are known to be associated with food spoilage and the subsequent toxic action of spoiled food. After careful isolation and characterization of the toxic compounds from spoiled food, it has been shown that the toxicity of the food materials would be increased by factors of 1000 to 1,000,000 over the toxicity of the parent products present in unspoiled food materials (Scheel, 1967). Because of the large increase in toxicity caused by spoilage, and the hazard involved in the use of spoiled food for either animal or human consumption, it is important to emphasize the frequency with which various mold species are implicated in the production of toxic products.

An early observation which eventually yielded evidence of coumarin production as a result of mold infection was the investigations of Schofield and Roderick in the 1920's of cattle fed spoiled sweet clover hay. The spoiled hay was highly toxic to the cattle, many of which died of massive internal hemorrhages. Two varieties of *Aspergillus* were isolated from the moldy hay and were shown to produce toxic products when inoculated into and grown on freshly cut nontoxic sweet clover hay.

In the 1930's, Link and his associates at the University of Wisconsin performed extensive studies on spoiled sweet clover hay and isolated the hemorrhagic agent dicoumarol. Burnside (1957) observed a disease condition in swine, resulting from the ingestion of moldy corn, which produced a 22% mortality in the affected animals. Of several molds isolated from the toxic corn, two of these, *Aspergillus flavus* and *Penicillium rubrum*, produced toxic reactions in swine when cultured on sterile corn.

Toxicity of the moldy corn was manifested in the animals by acute hepatitis, degenerated kidney tubules, icteric discoloration of the carcass, and hemorrhages in the gastrointestinal tract and heart.

Ambrecht et al. (1963) state that toxic reactions to Aspergillus spoilage of food have been observed in many animal species. These investigators were able to isolate the toxic products from mold mycelia grown on culture media without the presence of coumarin or its precursors, demonstrating that coumarins are one of the metabolic products produced by mold growth. The toxicity is dependent on the amount of mold mycelia present in the spoiled food material. As a result of extensive culture and isolation studies by Nesbitt et al. (1962), De Iongh et al. (1962), Sargeant et al. (1963), and Asao et al. (1963), it has been shown that the toxic compounds formed by Aspergillus are two difuranocoumarins, aflatoxin B and aflatoxin G. These two compounds were highly toxic to day-old ducklings weighing 50 gm. Uritani (1953) isolated scopoletin, umbelliferone, and aesculetin from infected sweet potatoes. It was his view that these coumarins alter the defense mechanism of the plants against infection, which makes them harmful to animals and humans.

The psoralens produced in diseased celery as a result of infection with the fungus S. sclerotiorum have been shown to possess a high degree of skin toxicity through their photosensitizing capacity. These observations in some cases indicate a metabolic effect of the disease on surrounding tissues, and in others show that the metabolism of the infecting organism is involved in the development of the coumarin compounds.

Presently available data emphasize the need for a better understanding of the changes in coumarin synthesis, since the significant increase in toxic properties of spoiled food may be sufficient to produce death in animals and perhaps in humans. Although some research has been done regarding the generation of the coumarin molecule in some plant species, symbiotic relationship between the host tissue and the infecting organism should be elucidated. Because of the high degree of toxicity of some of the coumarin compounds and the problems connected with food spoilage in the tropical and semitropical regions of the world, more data in this area of research are needed.

VI. Summary

Coumarins are constituents of plants and have been isolated as metabolic products of bacteria and fungi. Some compounds have been isolated outside plant life. Excellent reviews on their natural occurrence have been published by Spath and Holzen (1933) and Dean (1952). Coumarin compounds have been used medicinally for 3000 years. Modern usage has extended into such diverse areas as flavoring agents for foods and tobaccos, perfume odorants, anticoagulants, antibiotics, rodenticides, insecticides, fungicides, sun screen agents, optical brighteners, and preventives of dental caries; in electroplating, printing inks, optical filters; and in the treatment of vitiligo, their oldest recorded usage. Of most benefit to man has been the use of coumarins in medicine; this utilizes the anticoagulant properties of the 4-hydroxycoumarins in therapeutically controlling intravascular thrombi. The furocoumarins, or psoralens, have had extensive application in the treatment of vitiligo and as protectives against cutaneous burning resulting from solar radiation.

Several investigators have cautioned against incorporating coumarins in foods following the demonstration of coumarin toxicity in animals in feeding experiments (Jacobs, 1953; Hagan et al., 1967). The toxicity of coumarins associated with food spoilage should not be minimized, as mortalities among several species of animals have been recorded when they were fed spoiled food. Human ingestion of spoiled plant food in an affluent society is not a major problem; however, in underdeveloped areas it may be a cause for concern. An approach to the problem of plant spoilage may be in the development of specific fungicides for soil treatment. Goldberg (1964) recognized the toxicity of coumarins manifested by their narcotic, hypnotic, sedative, and hemorrhagic affects; however, he believed that a human tolerance level could be developed with further research concerned with quantitative ingestion of coumarins.

The aflatoxins are considered to be the most toxic of the mycotoxins and have no known practical use (for a full discussion see Chapter 1, Volume VI, this treatise). Many other coumarins, however, have demonstrated a utility based upon their toxicity. Certain coumarin derivatives, which are also highly toxic to fish, have been used successfully as insecticides. The cutaneous toxicity of the furocoumarins or psoralens has been well documented by many investigators who have observed dermatoses from contact with a variety of plants containing these compounds. When the concentration of psoralens and exposure to sunlight are in sufficient combination, blistering of the skin resulting in hyperpigmentation occurs readily. It is this toxic property of psoralens, which has been investigated in controlled clinical studies, that led to a relatively nontoxic treatment for vitiligo and to the development of protective agents against sunburn. The majority of clinicians (and patients) using psoralens both orally and topically for these purposes have reported only minimal side reactions.

The most toxic property of 4-hydroxycoumarin is its hemorrhagic activity, the cause of deaths of cattle feeding on sweet clover hay. However, Campbell and Link (1941), by their discovery of dicoumarol, successfully used this activity to treat patients with intravascular disorders. Scheel

(1970) has suggested that anticoagulants may be useful in those over 40 years of age to prevent cardiovascular problems. The coumarins must, as must any chemical compound that possesses both beneficial and toxic properties, be evaluated by weighing benefits to be derived against toxicity produced. Those who have investigated and conducted clinical studies of the coumarins have proved repeatedly that the toxicity of the majority of these compounds have been minimized, making this family of compounds a valuable asset to man's well-being.

REFERENCES

Ambrecht, B. H., Hodges, F. A., Smith, H. R., and Nelson, A. A. (1963). *J. Assoc. Offic. Agr. Chemists* **46**, 805.
Anticoagulants. (1963). *Brit. Med. J.* **I**, 801.
Asao, T., Buchi, G., Abdel-Kader, W. M., Chang, S. B., Wick, E. L., and Wogan, G. N. (1963). *J. Am. Chem. Soc.* **85**, 1706.
Austin, D. J., and Clarke, D. D. (1966). *Nature* **210**, 1165.
Barauh, S., and Swain, T. (1959). *J. Sci. Food Agr.* **10**, 125.
Bechet, H. (1939). *Ann. Dermatol. Syphilig.* **10**, 125.
Becker, S. W. (1959). *J. Invest. Dermatol.* **32**, 263.
Bellis, D. M. (1958). *Nature* **182**, 806.
Berlin, C. (1930). *Dermatol. Wochschr.* **90**, 773.
Best, R. J. (1936). *Australian J. Exptl. Biol.* **14**, 199.
Birmingham, D. J., Key, M. M., Tubich, G. E., and Perone, V. B. (1961). *Arch. Dermatol.* **83**, 73.
Bjerkelund, C. J. (1957). *Acta Med. Scand.* **158**, 330.
Bose, P. K. (1958). *J. Indian Chem. Soc.* **35**, 367.
Brodersen, R., and Kjaer, A. (1946). *Acta Pharmacol.* **2**, 109.
Brown, S. A. (1963). *Phytochemistry* **2**, 137.
Buck, H. W., Magnus, I. A., and Porter, A. D. (1960). *Brit. J. Dermatol.* **72**, 249.
Burnside, J. E. (1957). *Am. J. Vet. Res.* **18**, 817.
Campbell, H. A., and Link, K. P. (1941). *J. Biol. Chem.* **138**, 21.
Campbell, H. A., Roberts, W. L., Smith, W. K., and Link, K. P. (1940). *J. Biol. Chem.* **136**, 47.
Chakraborty, D. P., DasGupta, A., and Bose, P. K. (1957). *Ann. Biochem. Exptl. Med.* (*Calcutta*) **17**, 57.
Clatonoff, D. V., Triggs, P. O., and Meyer, O. O. (1954). *Arch. Intl. Med.* **94**, 213.
Daniels, F., Jr., Hopkins, C. E., Imbrie, D. J., Bergerson, L., Miller, O., Crowe, F., and Fitzpatrick, T. B. (1959). *J. Invest. Dermatol,* **32**, 321.
Dean, F. M. (1952). *Fortschr. Chem. Org. Naturstoffe* **9**, 225.
De Iongh, H., Beerthius, R. K., Vees, R. O., Barrett, C. B., and Ord, W. O. (1962). *Biochim. Biophys. Acta* **65**, 548.
Del Vivo, G., (1930). *Giorn. Ital. Dermatol. Sif.* **71**, 467.
Douglas, A. S., and McNicol, G. P. (1964). *Practitioner* **194**, 62.
El-Hefnawi, H. (1950). *Trudeau Rev. Med. Res.* **3**, 1.
Elliott, J. A., Jr. (1959). *J. Invest. Dermatol.* **32**, 311.
El Mofty, A. M. (1948). *J. Egypt. Med. Assoc.* **31**, 651.

Fahmy, I. R., and Abu-Shady, H. (1947). *Quart. J. Pharm.* **20**, 281.

Fahmy, I. R., and Abu-Shady, H. (1948). *Quart. J. Pharm.* **21**, 499.

Fitzpatrick, T. B., and Pathak, M. A. (1959). *J. Invest. Dermatol.* **32**, 229.

Fitzpatrick, T. B., Hopkins, C. E., Delbert, D., Blickenstaff, M. S., and Swift, M. B. (1955). *J. Invest. Dermatol.* **25**, 187.

Fowlks, W. L., Griffith, D. G., and Oginsky, E. L. (1958). *Nature* **181**, 571.

Freund, E. (1916). *Dermatol. Wochschr.* **63**, 931.

Gelfand, H. H. (1936). *J. Allergy* **1**, 590.

Glatt, O. S. (1961). *Med. Times* **89**, 1293.

Goldberg, G. (1964). *Food Cosmet. Toxicol.* **2**, 669.

Goodman, H. (1931). *Brit. J. Dermatol.* **43**, 177.

Gross, P., and Robinson, L. B. (1930). *Arch. Dermatol. Syphilol.* **21**, 637.

Guba, E. F. (1950). *Mass., Agr. Expt. Sta., Bull.* **457**.

Haan, D., and Tilsner, V. (1965). *Angiology* **16**, 49.

Hagan, E. C., Hansen, W. H., Fitzhugh, O. G., Jenner, P. M., Jones, W. I., Taylor, J. M., Long, E. L., Nelson, A. A., and Brown, J. B. (1967). *Food Cosmet. Toxicol.* **5**, 141.

Harvald, B., Hilden, T., and Lund, E. (1962). *Lancet* **II**, 626.

Head, F. S. H., and Robertson, A. (1931). *J. Chem. Soc.* p. 1241.

Henry, S. A. (1933). *Brit. J. Dermatol.* **45**, 301.

Henry, S. A. (1938). *Brit. J. Dermatol.* **50**, 342.

Hirschberger, A., and Fuchs, H. (1936). *Muench. Med. Wochschr.* **83**, 1965.

Huebner, C. F., and Link, K. P. (1941). *J. Biol. Chem.* **138**, 529.

Hughes, J. C., and Swain, T. (1960). *Phytopathology* **50**, 398.

Imbrie, D. J., Daniels, F., Jr., Bergeson, L., Hopkins, C. E., and Fitzpatrick, T. B. (1959). *J. Invest. Dermatol.* **32**, 331.

Jacobs, M. B. (1953). *Am. Perfumer* **62**, 53.

Jensen, T., and Hansen, K. G. (1939). *Arch. Dermatol. Syphilol.* **40**, 566.

Kanof, N. B. (1959). *J. Invest. Dermatol.* **32**, 343.

Kaufman, K. D. (1961). *J. Org. Chem.* **26**, 117.

Kellum, J. M. (1952). *J. Am. Med. Assoc.* **148**, 1443.

Kitchevatz, M. (1934). *Ann. Dermatol. Syphilig.* **5**, 203.

Kitchevatz, M. (1936). *Bull. Soc. Franc. Dermatol. Syphilig.* **43**, 581.

Klaber, R. (1942). *Brit. J. Dermatol.* **54**, 193.

Knypl, J. S. (1963). *Nature* **200**, 800.

Kuske, H. (1938). *Arch. Dermatol. Syphilis* **178**, 112.

Lederer, E. (1949). *J. Chem. Soc.* p. 2115.

Legge, R. T. (1921). *Calif. State J. Med.* 461.

Legrain, P. M., and Barthe, R. (1926). *Bull. Soc. Franc. Dermatol. Syphilig.* **33**, 663.

Lerner, A. B. (1955). *Trans. Pacific Dermatol. Assoc.* **7**, 25.

Lerner, A. B. (1959). *J. Invest. Dermatol.* **32**, 285.

Lerner, A. B. (1968). *J. Am. Med. Assoc.* **207**, 960.

Lerner, A. B., Denton, C. R., and Fitzpatrick, T. B. (1953). *J. Invest. Dermatol.* **20**, 299.

London, I. D. (1959). *J. Invest. Dermatol.* **32**, 315.

Mamalis, P. (1969). *Chem. Abstr.* **70**, 22920X.

Mead, J. A. R., Smith, J. N., and Williams, R. T. (1958). *Biochem. J.* **68**, 67.

Medical Research Council. (1959). *Brit. Med. J.* **I**, 803.

Meischer, G., and Burckhardt, W. (1937). *Schweiz. Med. Wochschr.* **18**, 82.

Musajo, L., Caporale, G., and Rodighiero, G. (1954). *Gazz. Chim. Ital.* **84**, 870.

Nesbitt, B. F., O'Kelly, J., Sargeant, K., and Sheridan, A. (1962). *Nature* **195**, 1062.

Oppenheim, M., and Fessler, A. (1928). *Dermatol. Wochschr.* **86**, 183.

Overman, R. S., Stahman, M. A., and Link, K. P. (1942). *J. Biol. Chem.* **145**, 155.

Owren, P. A. (1961). *In* "Symposium on Anticoagulant Therapy," p. 118. Harvey & Blythe, London.

Palumbo, J. F., and Lynn, E. V. (1953). *J. Am. Pharm. Assoc.* **42**, 57.

Partridge, L. K., Tansley, A. C., and Porter, A. S. (1969). *Electrochim. Acta* **14**, 223.

Pathak, M. A. (1961). *J. Invest. Dermatol.* **37**, 397.

Pathak, M. A., and Fitzpatrick, T. B. (1959). *J. Invest. Dermatol.* **32**, 509.

Perone, V. B., Scheel, L. D., and Meitus, R. J. (1964). *J. Invest. Dermatol.* **42**, 267.

Phyladelphy, A. (1931). *Dermatol. Wochschr.* **92**, 713.

Pinkus, H. (1959). *J. Invest. Dermatol.* **32**, 281.

Quick, A. J., Stanley-Brown, M., and Bancroft, F. W. (1935). *Am. J. Med. Sci.* **190**, 501.

Reid, W. W. (1958). *Chem. & Ind.* (*London*) p. 1439.

Reppel, L. (1954). *Pharmazie* **9**, 278.

Riggs, D. S. (1950). *New Engl. J. Med.* **242**, 179.

Roderick, L. M. (1929). *J. Am. Vet. Med. Assoc.* **74**, 314.

Roderick, L. M. (1931). *Am. J. Physiol.* **96**, 314.

Rosenthal, F. (1924). *Zentr. Haut-Geschlechtskr.* **13**, 322.

Rupprecht, W. E. (1969). *Analyst* **94**, 126.

Sams, W. M. (1941). *A.M.A. Arch. Dermatol.* **44**, 571.

Sargeant, K., Carnaghan, R. B. A., and Allcroft, R. (1963). *Chem. & Ind.* (*London*) **53**, 5.

Scheel, L. D. (1967). *In* "Biochemistry of Some Foodborne Microbial Toxins" (R. I. Mateles and G. N. Wogan, eds.), p. 109. M.I.T. Press, Cambridge, Massachusetts.

Scheel, L. D. (1970). Personal communication.

Scheel, L. D., Perone, V. B., Larkin, R. L., and Kupel, R. E. (1963). *Biochemistry* **2**, 1127.

Schofield, F. W. (1924). *J. Am. Vet. Med. Assoc.* **64**, 553.

Sevitt, S. (1962). *Am. J. Med.* **33**, 703.

Shapiro, S. (1953). *Angiology* **4**, 380.

Shapiro, S., and Ciferri, F. E. (1957). *J. Am. Med. Assoc.* **165**, 1377.

Shapiro, S., Redish, M. H., and Campbell, H. A. (1943). *Proc. Soc. Exptl. Biol. Med.* **52**, 12.

Sidi, E., and Bourgeois-Gavardin, J. (1952). *J. Invest. Dermatol.* **18**, 391.

Soine, T. O. (1964). *J. Pharm. Sci.* **53**, 231.

Spath, E. (1937). *Chem. Ber.* **70**, 83.

Spath, E., and Holzen, H. (1933). *Chem. Ber.* **66**, 1137.

Spath, E., and Kuffner, F. (1936). *Monatsh. Chem.* **69**, 5.

Spensley, P. C. (1963). *Endeavour* **22**, 75.

Stahman, M. A., Huebner, C. F., and Link, K. P. (1941). *J. Biol. Chem.* **138**, 513.

Stegmaier, O. C. (1959). *J. Invest. Dermatol.* **32**, 345.

Stowers, J. H. (1897). *Brit. J. Dermatol. Syphilis* **9**, 285.

Straton, C. R. (1912). *Brit. Med. J.* **II**, 1139.

Thoms, H. (1911). *Chem. Ber.* **44**, 3325.

Tucker, H. A. (1959a). *J. Invest. Dermatol.* **32**, 277.

Tucker, H. A. (1959b). *J. Am. Med. Assoc.* **169**, 1698.

Ulhmann, E. (1927). *Med. Klin.* (*Munich*) **23**, 279.

Uritani, I., and Hoshiya, I. (1953). *J. Agri. Chem. Soc.* **27**, 161.

Uritani, I. (1967). *J. Assoc. Offic. Agr. Chemists* **50**, 105.

Van Sumere, C. F., Van Sumere, de P. C., Vining, I. C., and Ledingham, G. A. (1957). *Can. J. Microbiol.* **3**, 847.

Vogel, A. (1820). *Ann. Physik* **64**, 161.

Wender, S. H., Yang, C. H., and Nakagawa, Y. (1958). *J. Org. Chem.* **23**, 204.

White, J. C. (1897). *Boston Med. Surg. J.*

Wiswell, J. G., Irwin, J. W., Guba, E. F., Rackemann, F. M., and Neri, L. L. (1948). *J. Allergy* **19**, 396.

Wood, P. H. (1961). *In* "Symposium on Anticoagulant Therapy," p. 138. Harvey & Blythe, London.

Zimmerman, M. C. (1959). *J. Invest. Dermatol.* **32**, 269.

Zweidler, R., Hausermann, H., and Geigy, J. R. (1964). "Encyclopedia of Chemical Technology," 2nd ed., Vol. 3, p. 737. Wiley, New York.

SECTION B

CHAPTER 4

Stachybotryotoxicosis

JOSEPH FORGACS

I. History

During the latter part of 1931, a unique disease of unknown etiology occurred in horses in certain villages of the Ukrainian S.S.R. Subsequently, the malady attained enzootic proportions. This peculiar disease, which was characterized by an unusual clinical and anatomic complex, had a high mortality rate. Since the etiologic agent was not known initially, the terms "unknown" or "new" or the abbreviated form "NZ" (*neizvestnoe zabolevanie* – illness of unknown etiology) were used to designate the condition. Later, this abbreviation was supplanted by some by the notation "MZ" (*massovie zabolivanie* – massive illness) to signify its extensive nature.

Early studies conducted by Drobotko and co-workers (1946) indicated that the disease was neither contagious nor infectious. Healthy horses stabled adjacent to sick ones, but fed a different diet, did not develop the disease. Infectivity tests involving nasal smears taken from sick horses and applied to the throats of healthy animals failed to transmit the disease. Likewise, blood of ill horses or macerated tissues taken at necropsy from horses that died from the toxicosis did not transmit the disease upon injection into healthy horses. In addition, there was no evidence of transfer of the disease by ectoparasites or insects.

Epizootiological observations suggested a relationship between the toxicosis and straw used as roughage and bedding. A search for agricultural poisons, toxic plants, toxic insects, recognized poisonous fungi such as ergot, and other ustilagenous fungi in the suspect straw as well as in hay and other fodder was to no avail.

Although outbreaks of this mysterious disease continued for approximately 7 years, the etiology remained unknown. In 1938, the cause of this disease was found to be associated with the fungus referred to in early Soviet literature as *Stachybotrys alternans* and later reclassified as *Stachybotrys alternans* var. *jateli,* the latter used to designate toxic strains of this species. The solution to this toxicosis was accomplished within 1 year by the combined efforts of a number of research workers from diverse disciplines. Sarkisov (1947) indicated that a brigade from the Institute of Microbiology of the Academy of Sciences of the Ukrainian S.S.R., headed by Professor V. G. Drobotko, and a brigade from the Main Administration of Veterinary Medicine under the direction of Professor K. I. Vertinskii made an extensive study of various aspects of this toxicosis. Other Soviet scientists contributed to various phases of this disease. According to Sarkisov (1947), Professor N. A. Naumov in cooperation with M. I. Salikov and V. G. Panashenko introduced new material on the problem of the nature of the fungus *S. alternans*. Detailed clinical description of the disease as well as methods of therapy were investigated by V. V. Spasski, N. I. Olenik, and A. H. Maiborod. Professor K. I. Vertinskii and F. M. Panomarenko made a detailed study of the pathology of this complex toxicosis. Methods of treating toxic fodder and safe animal husbandry practices were studied by Professor V. P. Dobrynin and co-workers.

Examination by Drobotko *et al.* (1946) of a straw stack on a farm where the disease was enzootic revealed a large black sooty layer of fungal growth, which was located beneath a depression where water had seeped. Feeding the fungus-infected straw to a horse caused initial symptoms to appear within 36 hours, and after approximately 3 weeks of interrupted feeding, the horse developed other clinical manifestations of the toxicosis and perished. Necropsy findings, likewise, were typical of the toxicosis. Sarkisov (1947) indicated that P. Iatel (also spelled Jatel) first isolated and called attention of the expedition to the fungus *Stachybotrys alternans* as the etiologic factor of the toxicosis in horses. Subsequently, Drobotko and co-workers (1946) found that nontoxic damp straw artificially inoculated with the fungus, after 12 days incubation, formed sufficient toxin to produce typical clinical and pathological manifestations of the toxicosis in horses that were fed the cultured straw. More conclusive evidence of the toxicity of this fungus was established by feeding horses pure cultures of the fungus. The mold was cultured on wort agar in Petri plates until good sporulation occurred, and then mixed, primarily, with oats and fed to horses. Six animals were fed the contents of 30, 20, 10, 5, 3, and 1 plates of the fungus daily. Although all became ill, the intensity of the toxicosis and variations in clinical symptoms varied with dosage. The first horse that consumed the fungal contents of 30 Petri plates died on the fifth day; the second horse, after eating daily the fungal contents from 20 plates died on the ninth day; the third animal perished on the thirty-second day; the remaining three animals, although ill, remained alive, as did the control horses that received noninoculated agar, until the end of the test at 3 months. In horses that consumed 30 and 20 plates of fungal substratum, intensity of the toxicosis exceeded that observed in horses in field outbreaks, and, in addition, on histopathological examination, the liver showed marked fatty infiltration. The animal that received 10 plates of fungal substratum daily, developed the most typical manifestations of the field cases. Typical clinical manifestations with leukopenia but without an elevated body temperature were observed in the fourth horse. An inflammatory response on the oral mucosa and edema of the lips only occurred in the fifth and sixth animals.

Drobotko and associates (1946) also injected a horse parenterally (presumably intraperitoneally) with an ether extract of the fungus at a dosage equivalent to 10 Petri dish cultures of the fungus. The animal developed a granulocytic leukopenia and perished. Necropsy findings were typical of those observed in field cases.

Examination of available Soviet literature indicates that Drobotko (1942, 1945) and Drobotko and associates (1946) assigned the name stachybotryotoxicosis to this disease to correspond with the generic name of the causal fungus *Stachybotrys*.

Although the first outbreaks of stachybotryotoxicosis were reported in the Ukrainian S.S.R., Vyshelesski (1948) stated that a disease with similar characteristics was observed earlier in Hungary, although the etiologic agent was not established. Vyshelesski indicated that stachybotryotoxicosis prevailed not only in the Ukrainian S.S.R., but in Crimea, Moldavian S.S.R., and in many other oblasts, and stated that Sarkisov described extensive outbreaks of this equine disease in Bessarabia in 1936 and 1937. According to Sarkisov (1960), stachybotryotoxicosis has recurred more or less at a constant incidence in the Ukraine and sporadically in Moldavia.

Based on available literature, it appears that the toxicosis, shortly after its appearance in the Ukraine in about 1931, attained enzootic proportions for about a decade, and then apparently subsided, although sporadic cases recurred in previously enzootic areas. According to Shulyumov *et al.* (1960), who made a subsequent study of stachybotryotoxicosis in the southern Ukraine, outbreaks of this toxicosis in horses occurred in the

Odessa oblast in 1937 and 1938, were absent during the World War II years, and again reappeared during the winter of 1951 and 1952 on various farms, and, with the exception of 1954, outbreaks have occurred yearly at least until 1957.

Formerly, the horse was considered to be the only large farm animal susceptible to *Stachybotrys* toxin. In 1958, Forgacs and co-workers (1958a) in the United States reproduced experimentally the toxicosis in the horse, calf, sheep, and swine, using a toxic strain of *Stachybotrys atra*. Fortuskny *et al.* (1959) reported a natural outbreak of stachybotryotoxicosis among cattle that had consumed straw contaminated with *S. alternans*. This outbreak occurred on farms where the disease was known to be enzootic in horses. Other outbreaks in cattle have been reported (Nagornyi *et al.*, 1960; Matusevich *et al.*, 1960; Levenberg *et al.*, 1961; Kurmanov, 1961; Ismailov and Moroshkin, 1962), although Sarkisov as cited by Pidoplichko and Bilai (also spelled Bilay) in 1960 indicated that cattle are considerably more resistant to the toxin. Korneev (1948) produced various clinical signs of this toxicosis in white Swiss mice, guinea pigs, rabbits, and dogs, and Schumaier and co-workers (1963) found *S. atra* to be toxic to chicks. In addition, human beings exposed to aerosols of toxic strains of *S. alternans* or substrata infected with such strains have developed localized and systemic toxic manifestations (Drobotko, 1942, 1945; Drobotko *et al.*, 1946; Vyshelesski, 1948). Soviet scientists (Samsonov, 1960; Samsonov and Samsonov, 1960) have designated toxicoses due to inhalation of substrata infected not only with *S. alternans* but with other toxic fungi as well (*Aspergillus fumigatus, Aspergillus niger, Dendrodochium toxicum,* and others) as respiratory mycotoxicoses or pneumomycotoxicoses. It is reasonable to assume that other toxigenic fungi not yet studied will induce pneumomycotoxicoses.

Early Soviet studies focused on the etiology, clinical and pathological manifestations and on methods of controlling stachybotryotoxicosis. Thereafter, as outbreaks subsided, research activity on this toxicosis decreased. With the advent of more outbreaks in 1951 and 1952, however, interest in this disease again was renewed with emphasis on cultural, morphological, and physiological studies of various strains of *S. alternans* as related to toxicity, on rapid means of diagnosing stachybotryotoxicosis and detection of toxigenic strains of the causal fungus, and on newer methods of therapy and detoxification of infected fodder and feedstuffs.

II. Epizootiology

According to Vertinskii (1940), who, together with his colleagues, made a thorough study of the clinical and pathological aspects of stachybotryotoxicosis during the initial outbreaks of this toxicosis, the disease is generally seasonal. The first cases appear in the fall when the animals are stabled and consume straw, hay, or other fodder containing preformed *Stachybotrys* toxin. The number of affected animals tends to increase during the winter months, reaching a peak from about February through April, and the toxicosis subsides when the horses are turned onto pasture and stall feeding of toxic straw or other fodder is discontinued.

The toxicosis may be endemic. If the climatic and other conditions are favorable for fungal growth and toxin production, the toxicosis may appear in succeeding years, or may be repetitive in nature, recurring several times during a single season. Since the toxicosis can recur among the same horses during a single season or during succeeding years, the toxin apparently is nonantigenic, and consequently no immunity appears to develop to stachybotryotoxicosis.

The spread of the toxicosis does not have a consistent epizootiological pattern. On the contrary, the rate of spread of this toxicosis may vary widely. On certain farms, signs of the disease appear among a considerable number of animals within a group. Within 1 week, the number of afflicted horses may rise to 40 to 60% of the group, and within 2 weeks, virtually the entire group becomes ill. On other farms, however, the toxicosis may afflict the entire band within 6 to 8 hours following exposure. In other instances, cases of individual sickness and death occur sporadically over a period of 3 to 4 months, followed by a general massive outbreak. Cases have been reported where, on some farms, the toxicosis affected only one group of horses, while on others 70 to 100% of the herd suffered from the disease.

Since colts that nurse lactating mares affected with stachybotryotoxicosis do not develop the toxicosis, it has been concluded that the toxin is not excreted in the milk. Age, breed, sex, physical condition, level of nutrition, nature of work performed, and conditions under which the horse previously had been maintained appear to have no influence on susceptibility of the animal to the toxicosis.

Mortality among affected horses depends upon the form and stage of the toxicosis which in turn depend upon the amount and potency of toxic substratum consumed. Mortality appears to be highest among horses afflicted with the atypical or "shocking" (acute) form and also among those suffering from the third stage of the typical (chronic) form, particularly if methods of therapy are not initiated promptly. Horses afflicted with the first or second stage of the typical form rarely perish, providing feeding of toxic fodder is interrupted promptly and before symptoms of the third stage develop.

As indicated earlier, the horse formerly was considered to be the only large farm animal susceptible to the *Stachybotrys* toxin under field conditions. Subsequently, field outbreaks of this toxicosis have been reported in cattle (Fortuskny *et al.*, 1959; Nagornyi *et al.*, 1960; Matusevich *et al.*, 1960; Levenberg *et al.*, 1961; Kurmanov, 1961; Ismailov and Moroshkin, 1962). Tkachenko (1960) reported an outbreak of stachybotryotoxicosis in a hippopotamus and bison in a zoological park in the Soviet Union. A toxic strain of *S. alternans* was isolated from feed and litter and also from liver of the afflicted animals. In addition, histological examination of various organs taken at necropsy showed pathological changes typical of stachybotryotoxicosis.

III. Clinical Characteristics

Stachybotryotoxicosis may occur in either of two forms, the typical, which arbitrarily has been divided into three stages, and the atypical or shocking form, which exists in one stage.

A. TYPICAL FORM

According to most Soviet scientists who have contributed to stachybotryotoxicosis research, the typical form is observed most commonly under natural conditions. This form develops progressively while the horse consumes infected roughage or other feedstuffs that contain a low level of *Stachybotrys* toxin. The disease in this form progresses more or less sequentially through a series of clinical symptoms, manifestations of which later are reflected in a well-defined pathological process and changes in hematopoietic tissue, particularly in the bone marrow.

1. FIRST STAGE

The outstanding manifestation of the first stage is a stomatitis observed within 2 to 10 days following exposure to the toxin. The stomatitis is accompanied by appearance of fissures at the corners of the mouth. Initially, there appears a superficial and later a deep necrosis in the depths of the wrinkles of the skin at the mucocutaneous junction. The necrotic areas are dirty, gray, tough, exfoliating scales, or deep, bleeding grooves. The corners of the mouth become swollen, and the contours are slightly edematous. If the inflammatory process continues, edema of the underlying tissues results in swelling of the lower and upper lips, causing the mouth to assume a hippopotamus-like appearance. In some horses, excessive salivation occurs, and the submaxillary lymph nodes become enlarged and painful to the touch. Rhinitis and conjunctivitis may develop concurrently. In some horses, these local lesions are limited merely to a slight exfoliation of the oral and buccal mucosa. Consequently, in such animals, manifestations of the first stage of the toxicosis can easily be overlooked.

No particular abnormalities in the general condition of the horse are observed. Occasionally, however, an increase in body temperature of 1 to 1.5°C and a brief period of neutrophilia may be observed. The duration of the first stage may vary from 8 to 12 days, but it may extend over a period of 30 days or longer.

In spite of continued ingestion of sublethal amounts of toxic roughage, the typical well-developed local lesions may subside in many cases within 2 weeks of the initial onset, and then be followed within 5 days by a clinically quiescent period, which is the second stage.

2. SECOND STAGE

This stage is characterized by systemic manifestations. There appears a marked reduction in thrombocytes; instead of 40 to 60 thrombocytes per 3000 erythrocytes, there may be only 8–12, while toward the end of the second stage, there is a marked thrombocytopenia, the number of thrombocytes decreasing to 2 to 3 per 3000 red cells. Clot retraction time increases, and, in some cases, the blood fails to coagulate. Simultaneously, leukopenia and agranulocytosis develop. The number of leukocytes may drop to 2000 or less per cubic millimeter of whole blood. A nonreactive necrosis of the oral mucous membranes appears. Although respiration, cardiac function, and body temperature remain within normal limits in most cases, in rare instances it is possible to detect slight intestinal disturbances or atony. Duration of the second stage varies from 5 to 20 days.

3. THIRD STAGE

This stage is characterized by hyperthermia, the body temperature rising as high as 41.5°C. Changes in the peripheral blood picture observed in the second stage continue into the third stage. Thrombocytopenia and leukopenia radically intensify. Values as low as 400 to 100 leukocytes per cubic millimeter of whole blood have been reported by Vertinskii (1940). Complete inability of the blood to clot is observed in all cases. The afflicted horses usually are depressed, although some may remain active and contented. The pulse is weak, and in some cases, arrhythmic. Disturbances in the alimentary tract occur frequently and often are accompanied by colic. Fresh necrotic areas without distinct zones of demarcation develop on the mucous membranes of the cheeks, gums, frenum of the tongue, hard and soft palate, and on the lips. The necrotic areas frequently become infected by various species of bacteria. According to Vyshelesski

(1948), the blood glucose level decreases by 50% of normal; serum bilirubin levels increase; inorganic phosphorus diminishes five- to tenfold, and the chloride content of the serum and urine decreases.

This stage lasts from 1 to 6 days and, in most cases, terminates in death. The agonal period may be brief and quiescent, although it is occasionally prolonged and violent. Pregnant mares usually abort shortly before death.

B. Atypical Form

According to Soviet literature, the atypical or shocking form is not common. It develops after ingestion of large amounts of *Stachybotrys* toxin, but Forgacs and co-workers (1958a) in the United States, produced this form of the toxicosis in a horse after it consumed only 0.5 lb of highly toxic *S. atra*-infected straw over a period of 10 days. The atypical form may occur either in an individual case or among several members of a band of horses afflicted with the typical form of the toxicosis. Local signs of inflammation and necrosis of the lips and buccal cavity, generally observed in the first stage of the typical form, are absent in the atypical form. Nervous disorders comprise the main visible syndrome. Loss of reflex response, hyperesthesia, and hyperirritability, followed by loss of vision occur in the affected horse. The animal may become totally blind and stuporous. The horse avoids movement, stands with legs apart, or even crosses its forelegs and leans against an object for support. When induced to walk, it moves with an uncertain gait. There is anorexia, and although the horse drinks water avidly, it swallows with great difficulty. Body temperature increases 1.5°C above normal during the early stages of the toxicosis, and after 2 to 3 days, either returns to normal or remains elevated. In horses that recover, body temperature returns to normal on the second or third day.

Cardiac action at first is intensified and later greatly weakened. Atony of the gut, or, in rare cases, increased peristalsis and diarrhea, occur. The agonal period, as a rule, is prolonged and accompanied by incoordination and periodic clonic spasms and tremors. Death is due to respiratory failure.

In contrast to the typical form, animals afflicted with the shocking form have no blood dyscrasias. The symptoms just described appear within 72 hours following ingestion of the toxin. However, in some cases, there develop signs of shock, with a rapid rise in body temperature to over 41°C, weak pulse, shortness of breath, and cyanosis and hemorrhage of the visible mucosa. These signs develop within 10 to 12 hours following ingestion of the toxin, and practically all horses affected by this rapid form of the toxicosis perish.

IV. Pathological Findings

The pathology of equine stachybotryotoxicosis has been thoroughly described by Vertinskii (1940). According to this Soviet veterinarian, the pathological picture of this toxicosis is characterized by profuse hemorrhage and necrosis in many tissues and is essentially the same for both the typical and atypical forms; variations encountered are typically limited to intensity. The subcutis and skeletal musculature, the serous and mucous membranes, and certain parenchymatous organs are primarily affected. Hemorrhages resembling large bruises, blemishes, streaks, and ecchymoses are found on the costal pleura, diaphragm, mesentery, and in the capsule of the spleen. Numerous hemorrhages are observed in the large intestine, lymphatic nodes, and in the parenchyma of the lung and liver, the suprarenals, the brain, spinal cord, and in the cerebral meninges.

Necrotic processes on the mucous membranes appear along practically the entire length of the digestive tract and are quite characteristic of this toxicosis. In the buccal cavity, the necrotic foci of the lips, soft and hard palate, frenum of the tongue, and tonsils are circular or ovoid in shape, clearly outlined, and yellowish gray in color. The esophagus rarely is affected, but in the stomach, hemorrhages, necroses, and catarrhal inflammatory processes prevail. The small intestine is in a state of subacute inflammation. The most typical and constant changes are found in the large intestine, where, macroscopically, two types of lesions are observed. One type consists of small, yellowish gray papules on the mucous surface and the other, as large deep-seated lesions extending into the submucosa and muscularis. In the more acute phase of the disease, areas of trophic necroses predominate, alternating with hemorrhages. In some instances, focal inflammatory processes are of a serohemorrhagic type.

The lymph nodes are enlarged and hemorrhagic, the nodes of the pharynx showing the greatest involvement.

Changes in the liver, kidneys, and heart muscle are in the nature of circumscribed degenerative processes of the albuminous or adipose-dystrophic type. The lungs are intensely congested, edematous, emphysemic, and frequently contain hemorrhages. Incipient focal bronchial pneumonia is comparatively rare.

In the endocrine glands, changes in the suprarenals are the most constant and are expressed by hypoplasia of the cortex and depletion of the cortical lipids in the typical form and by hemorrhages in this area in the peracute form.

The pathological changes due to stachybotryotoxicosis are unique in that they have no zone of demarcation around the necrotic foci. The tissue appears to be in a nonreactive state. In the lymphatic nodes, the distended, serous-inflamed sinusoids are practically devoid of leukocytes. In the bone marrow, aside from necrotic foci, cellular forms are decreased, with an unusually small number of granulocytes.

Microscopically, the hemorrhagic areas contain nondescript pigmented, pleomorphic bodies which appear as aggregates in the epithelium and in the capillary walls of various organs.

Earlier attempts to identify the nature of these pigmented bodies by the use of various stains and reagents were fruitless, although some Soviet investigators were of the opinion that the pleomorphic bodies might be of fungal origin. They maintained that the fungal mycelia, while proliferating in the tissues, became pleomorphic and lost the morphological and staining characteristics shown by the fungus when cultured on Czapek's solution agar or other mycologic media. Sarkisov and Orshanskaiya (1944) described in great detail methods of isolating *S. alternans* from a wide variety of substrata, as well as from tissues taken at necropsy from horses that perished from the toxicosis. They indicated that during the initial phases of growth from tissues, the emerging aerial hyphae were quite different from those seen in repeated subcultures on ordinary mycologic media. These findings implied that at least some strains of *S. alternans* were pathogenic, some opportunistically so, and that depression of the hematopoietic tissue facilitated proliferation of the fungus within the host. Forgacs and co-workers (1958a) were unable to isolate *S. atra* from any tissues obtained from a calf which had leukopenia and died after feeding on straw infected with a toxic strain of their fungus.

V. Toxicity of *Stachybotrys* Fungi for Other Farm Animals

During the initial outbreaks of stachybotryotoxicosis, and for several years thereafter, this toxicosis was considered to affect only horses. In 1958, Forgacs *et al.* (1958a) observed that not only horses, but calves, sheep, and swine also were susceptible to the toxin produced by one of their strains of *S. atra*. The calf, however, was considerably more resistant to the toxin than the horse. Sterile moist wheat straw on which the fungus had been cultured for 60 days at room temperatures (about 24°C) was dried, pulverized in a Wiley mill, suspended in tap water, and force-fed by stomach tube to the three animal species.

A calf weighing 250 lb received 6 lb of the toxic straw over a period of 14 days. The animal was force-fed daily 0.5 lb of the toxic straw from the first to the seventh day, on the ninth and tenth days; 1 lb on the thirteenth day, and 0.5 lb on the fourteenth day. Four days after the initiation of force-feeding, the animal developed a slight depression and lachrymation, and after 6 to 8 days, anorexia, watery stools, and increased systemic depression also were observed. These abnormalities gradually intensified, and agranulocytosis, which became evident on about the ninth day, steadily progressed. The leukocyte count, which initially was 4500 per cubic millimeter of whole blood gradually decreased, and by the fifteenth day (2 days before death), the count was 150 per cubic millimeter (Table I). A pronounced increase was noticed in the blood clot retraction time. Blood clotting time changed from 9 minutes, observed on the first day, to 19 minutes on the fourteenth day, 3 days before the animal expired. Although body temperature remained within normal limits for 14 days, it rose rapidly to 41.8°C on the fifteenth day and remained elevated until shortly before death of the host. In the terminal stages, the calf experienced respiratory difficulty and dyspnea and died on the seventeenth day.

Gross necropsy findings corresponded, for the most part, to those observed in equine stachybotryotoxicosis. In addition, however, edema and a reddish amber-colored fluid in the submaxillary space, distention of the gall bladder, staining with bile pigment of adjacent mucosa and serosa, depletion of lipids in the adrenal cortex, atrophy and subcapsular petechiation in the spleen, and edema and injection of the intestinal mucosa accompanied by blood-tinted material in the lumen were seen.

Histological examination revealed areas of hemorrhage, congestion, and dilation in the collecting tubules of the kidneys; cytoplasmic vacuolation, areas of necrosis, and congestion of the sinusoids in the liver; sub-

TABLE I

BLOOD WHITE CELL COUNT OF CALF FED TOXIC *Stachybotrys atra* STRAW[a]

Day of test	Amount of infected straw fed[b] (lb)	Leuko-cytes/ml	Differential cell count							
							Lymphocytes			Monocytes and granulocytes
			Neutrophiles	Filaments	Bands	Juveniles	Large	Small	Total	
0	0	4500	50	47	2	1	14	32	46	4
1	0.5	3750	54	48	4	2	16	29	45	1
2	0.5	2800	49	46	2	1	24	21	45	6
3	0.5	2300	53	47	3	3	9	30	39	8
4	0.5	1350	55	28	15	12	25	12	37	8
5	0.5	650	65	37	8	20	10	20	30	5
6	0.5	1100	52	43	7	2	2	36	38	10
7	0.5	1050	57	49	2	6	5	37	42	1
8	0	550	61	41	15	6	7	30	37	2
9	0.5	650	63	38	15	10	14	20	34	3
10	0.5	700	65	40	13	12	20	22	42	1
11	0	750	67	42	11	14	16	14	30	3
12	0	850	68	44	18	6	18	12	30	2
13	1.0	650	74	51	14	9	15	10	25	1
14	0.5	400	72	43	21	7	19	17	36	2
15	0	150	2	2	0	0	76	22	98	0

[a]Forgacs *et al.* (1958a). Reproduced through the courtesy of The New York Academy of Sciences.
[b]The calf, in a previous experiment, was fed *S. atra* straw; it developed agranulocytosis, but recovered.

mucosal edema and red-staining refractile bodies of varying size in the gall bladder; and dystrophy and depletion of the lymphoid elements and presence of blast cells, in the spleen.

A 130-lb ewe was force-fed only 6 oz of the cultured straw in a single dose. The animal, when first observed 12 hours after feeding, was prostrate and breathing heavily. The sheep progressively became weaker, and died 25 hours after feeding. Gross necropsy and histological findings resembled to some degree those in the calf.

A pig weighing 125 lb was force-fed a single dose of approximately 3.5 oz of the toxic straw and died 12 hours later. Gross necropsy findings consisted primarily of hemorrhage and congestion in various tissues. Histological findings resembled those observed in the calf.

Fortuskny and associates (1959) reported outbreaks of stachybotryotoxicosis in stabled cattle that consumed straw contaminated with *S. alternans*. The cases occurred in 1958 and 1959 in the Ukrainian S.S.R. on farms where this toxicosis was concurrent in horses. Approximately 4000 head of cattle were affected, and mortality was high. The clinical, gross, and histopathological expressions of infection among cattle resembled to a great extent those observed in the atypical or shocking form of equine stachybotryotoxicosis. The toxicosis in cattle developed within 2 to 3 days, and death occurred on the fourth to the sixth day, while thereafter the surviving animals surprisingly recovered. Changes in the cellular blood picture of the peripheral blood indicated that, as the toxicosis progressed, a leukopenia followed by a leukocytosis developed (total white count dropped from 3500 to 500 or less per cubic millimeter), and within 3 to 4 days, the white count returned to normal, only to be followed by a panleukopenia which prevailed until death. Blood clot retraction time also increased, and, in some cases, the blood failed to clot.

Initial clinical characteristics of this toxicosis in affected cows were described as hyperthermia, decreased milk flow, diarrhea, conjunctivitis, and focal hyperemia of the mucosa of the oral cavity, nose, and vagina. In addition, as the general state of the animal deteriorated, a profuse nasal mucous discharge occurred, salivation increased, and anorexia developed. Hypotonia, followed by atony, occurred. Pulse rate was increased to 120 per minute, and body temperature rose and was maintained at 40 to 42.2°C for some time. Small necrotic lesions occurred on the mucosa of the lips, gums, and cheeks, and small petechiae were present at the base of the lesions. Edema was present in the lips. Thereafter, profuse bloody diarrhea and cachexia occurred. Near the terminal stages, clonic spasms occurred; a dull cough and grinding of the teeth was noted; lactation ceased; pregnant cows aborted, and there was edema in the submaxillary region. Death, which occurred usually on the fourth to sixth day, was violent. Necropsy findings consisted primarily of a hemorrhagic diathesis and degenerative changes in the liver, with inflammation and necrosis in the gastrointestinal tract. In contrast to equine stachybotryotoxicosis, the disease among cattle appeared to be less severe in younger cattle than in older ones.

Tkachenko (1960) diagnosed stachybotryotoxicosis in a hippopotamus and a bison that consumed fodder contaminated with a toxic strain of *S. alternans*. The hippopotamus, at first, developed snorting and nasal secretions, followed by profuse salivation. Intermittent gastrointestinal disturbances were observed during the toxicosis. The general health of the animal rapidly deteriorated, and the hippopotamus died 2 months after initial onset of toxic manifestations.

Toxic symptoms in the bison initially consisted of anorexia, cessation of rumination, and presence of a bloody diarrhea which continued for approximately 7 days, then subsided for approximately 2 weeks, but reappeared with much greater intensity. The animal perished 3 days after the second onset of toxicosis.

Although Tkachenko described in great detail the pathological findings in the hippopotamus and bison, the general necropsy findings were indicative of a hemorrhagic diathesis with various changes in the vital organs and particularly degenerative changes in the liver, inflammation and necrosis in the gastrointestinal tract, and a decrease in cellular forms in the bone marrow.

VI. Experimental Stachybotryotoxicosis in Small Animals

Korneev (1948) determined toxicity of *S. alternans* for mice, guinea pigs, rabbits, and dogs. He observed that when white Swiss mice were fed daily 2 gm of oats on which had been cultured a toxic strain of the fungus, they developed depression, cachexia, muscular incoordination, and difficulty in breathing within 2 days, followed by death on the third day. Gross necropsy findings consisted primarily of catarrhal-hemorrhagic inflammation of the stomach. Gray mice were equally susceptible. Autoclaving the infected oats did not destroy the toxin.

Guinea pigs that were fed 5 to 10 gm of unsterilized toxic oats *ad libitum* developed necrotic areas on the lips, diffuse necrosis of the buccal mucosa, and leukocytosis and died within 3 to 5 days. Gross necropsy findings included hemorrhage and necrosis of the mucosa of the mouth, tongue, and stomach.

In other experiments, guinea pigs were fed 1.5 to 2.0 gm of toxic oats daily and developed hyperemia of the mucous membranes of the mouth after 3 to 8 days, followed by edema and serous exudation and necrosis of the lips, particularly the lower lip. The local lesions disappeared by the fifth to eighth day, even though the animals were continually fed the toxic oats. Between the tenth and sixteenth day, leukopenia, agranulocytosis, and thrombocytopenia became evident. Leukopenia became progressively more pronounced, and in some cases, the white cell count dropped to 500 per cubic millimeter of whole blood. In the terminal stages, body temperature became elevated to 42°C, and death occurred within 25 to 40 days after the start of the experiment. Gross necropsy findings were characterized by a general hemorrhagic diathesis, changes being similar to those observed in the third stage of equine stachybotryotoxicosis.

Rabbits that consumed fungus-infected oats mixed with noninfected oats, at the rate of 10 to 20 gm daily on an *ad libitum* basis, did not develop oral lesions; however, within 25 to 30 days, they exhibited minute areas of necrosis and slight edema of the lips. Leukopenia persisted for 3 to 10 days, followed by leukocytosis for 2 to 5 days. The latter period was followed by a rapid rise in body temperature and ultimately by death. Autopsy findings corresponded, for the most part, to those observed in the third stage of the equine toxicosis with the exception that ulceration of the intestine was rarely observed.

Dogs that were fed a single dose of 1.0 ml of unpurified toxin mixed with raw meat developed anorexia and depression and died on the fourth to the sixth day. The chief lesion observed at necropsy was a hemorrhagic inflammation of the gastrointestinal tract. Dogs receiving 0.25 ml of the unpurified toxin daily developed leukocytosis in 10 to 20 days, followed by a leukopenia and thrombocytopenia by the 45th day, a rapid rise in body temperature, and death due to paralysis. Since the dogs bolted the toxic diets, no lesions developed on the lips or in the buccal cavity; however, lesions in various tissues, characteristic of the third stage of the equine toxicosis, were the principal changes observed at necropsy.

Forgacs and co-workers (1958a) fed white Swiss mice, *ad libitum*, oats on which a toxic strain of *S. atra* had been cultured for 3 weeks at room temperatures. Two days after the first day of feeding, the mice developed atony and slight depression. These symptoms and cachexia were observed from the fourth day until death, which occurred on the eighth day. Gross necropsy findings included mottling and congestion in the liver, hemorrhage and congestion in the lungs and stomach, together with edema in the small intestine. On histological examination, hemorrhage and congestion together with focal necrosis, and leukocytic infiltration were observed in the liver, hemorrhage and congestion in the heart and lungs, together with leukocytic infiltration in the stomach and small intestine, and slight congestion in the kidneys.

Schumaier *et al.* (1963) found that whole wheat grain on which a highly toxic strain of *S. atra* had been cultured and an extract prepared from such grain following 3 hours continuous extraction with diethyl ether in a Soxhlet apparatus were toxic to chicks. The toxic substrata were supplemented for noninfected wheat in a well-balanced basal diet. During a 13-day period, chicks fed either the mold substratum or the ether extract showed a profuse watery oral discharge, a generalized depression, paleness of combs, cachexia, and greatly reduced growth rates. Necropsy findings included necrotic lesions in the mouth and crop, but no gross changes in the remainder of the digestive tract or in other tissues. Histological examination of the crop lesions showed necrosis of the stratified squamous epithelium with an infiltration of neutrophiles and eosinophiles. These workers also observed that the fungus-cultured wheat, even after 3 hours continuous extraction with diethyl ether still was toxic to chicks. Attempts to recover *S. atra* from necrotic lesions following surface sterilization were fruitless.

Stachybotryotoxicosis primarily is considered an alimentary toxicosis in which various toxic manifestations (described earlier) develop, following ingestion of substrata contaminated with toxic strains of *S. alternans* or *S. atra*. It has been observed by various Soviet investigators that human beings working in areas where aerosols of such toxic fungi or their toxic substrata are present develop local and systemic effects following inhalation of the toxic substrata. However, as indicated by Samsonov (1960), although the local effects of toxic cultures of *S. alternans* have been tested dermally on animals and man, experiments demonstrating the pathological effects of introducing toxic cultures into the respiratory tract had been lacking. Therefore, Samsonov and Samsonov (1960) challenged a guinea pig by spraying spores of *S. alternans* into the lungs. Necropsy findings of such animals (no data were presented indicating the number of days the animal was on test prior to necropsy) showed that, in the lungs, a catarrhal-desquamative inflammation occurred in the tracheal and bronchial mucosa, and epitheloid degeneration, focal bronchopneumonia, areas of serous hemorrhagic interstitial inflammation, and atelectasis and compensatory emphysema were also seen. The heart showed degenerative changes of the muscle fibers and necrobiosis of the nuclei. Advanced degenerative changes were observed in the liver, with marked stasis in the capillaries. Degenerative changes also were observed in the kidneys. These workers used other fungi in similar studies and concluded that toxic fungal aerosols upon entering the respiratory tract settle on the mucous membranes where they exert irritating and caustic effects, resulting in a local serous inflammatory response, necrosis, and ulcerations. That no evidence of fungal proliferation was noted in the bronchopneumonic foci excluded these changes as being due to a mycosis. Pathological changes noted in other organs, particularly in the liver, spleen, kidneys, and heart, indicate that the toxic substance(s) present in the fungal morphological structures is absorbed in the lungs and acts systemically in the host.

VII. Stachybotryotoxicosis in Man

According to Vertinskii (1940), Drobotko (1942, 1945), and Drobotko and associates (1946), as well as Vyshelesski (1948), in areas where equine stachybotryotoxicosis is enzootic, people, particularly those who handled fodder, frequently developed a toxicosis. In rare cases, human beings who used straw for fuel or who slept on straw mattresses also became ill. Vertinskii (1940) indicated that affected humans, in the majority of cases, developed a dermatitis localized chiefly on the scrotum, in the axillary region, and less frequently on the hands and other parts of the body. The dermatitis initially was characterized by a hyperemia, subsequently by serum exudation and encrustations, followed by necrosis. Drobotko and co-workers (1946) indicated that, in addition to a dermatitis, affected people developed a catarrhal angina, a bloody rhinitis, and cough, and complained of pain in the throat, burning in the nasal passages, and tightness in the chest. On occasion, a slight rise in body temperature was recorded. In some individuals showing the afore-mentioned symptoms, leukocytosis followed by a leukopenia also developed. In some cases, the white cell count dropped to 500 per cubic millimeter. Mycologic examination of suspect straw resulted in isolation of strains of *S. alternans* which were toxic when tested by the rabbit dermal toxicity test. In addition, the aerial fructifications of the fungus, when rubbed directly onto the skin of the forearm of some members of Drobotko's brigade induced local and systemic toxic responses similar to those observed in naturally occurring cases. Some of the scientists who worked with naturally or artificially infected straw also developed the toxicosis.

Samsonov (1960) reviewed toxicoses in human beings associated with inhalation of dust aerosols heavily contaminated with mold spores of a variety of fungi to find that Ismailson in 1927 was the first Soviet scientist to call attention to this type of mycotoxicosis in people who were employed in a binder twine factory. Samsonov defined such mycotoxicoses as respiratory or pneumomycotoxicoses. This scientist cited subsequent works of several Soviet investigators who observed respiratory and systemic effects in humans in occupations where dust aerosols were heavily laden with a variety of mold spores. Occupations included those in cottonseed oil processing plants, grain elevators, plants used for reprocessing moldy grain, plants used for processing malt grains in breweries, and in textile mills, particularly those mills reprocessing materials containing various plant fibers. Samsonov characterized the general symptomology of the toxicoses in affected people as a general intoxication beginning with the nasal mucosa, ending in lung parenchyma, and involving the conjunctival mucosa. In some cases, a dermatitis occurred. Patients complained of irritation of the oral mucosa, severe malaise, nausea, fullness in the chest, shortness of breath, difficulty in exhaling, and evidenced sporadic cough accompanied by short periods of asphyxiation. A rise in body temperature of 1 to 1.5° frequently was noted, and temperature not infrequently rose to 39.5°C . On physical examination, only dry rales were detected in some cases, and X-rays of the lungs revealed only minor changes in the bronchii. Changes of a functional nature occurred in the cardiovascular system. Following interruption of exposure to the toxic aerosols, patients recovered quite rapidly; however, upon subsequent exposure, the disease recurred with more serious sequelae.

Mycologic examination of environmental aerosols, as well as of bronchial exudates of afflicted patients revealed identical flora, and comprised a variety of fungi, some of known toxigenic potentiality. Although there was no concerete evidence that some of the fungal isolates might have been responsible for the outbreaks of the toxicoses in humans, room for speculation certainly remained. Subsequently, Samsonov and Samsonov (1960) studied the clinical and pathological effects in guinea pigs of spores and mycelial fragments of some of these fungi, particularly *Dendrodochium toxicum, Stachybotrys alternans, Aspergillus fumigatus,* and *Aspergillus niger;* introduction was into the respiratory tract as aerosols. They observed local and systemic effects, as well as pathological changes in various tissues, particularly in the vital organs, indicating a systemic effect of the toxins following absorption in the lungs. The degree of clinical manifestations of toxicosis and the severity of lesions varied with the fungus, with *D. toxicum* being the most toxic, followed by *S. alternans.*

VIII. Biological and Ecological Characteristics of the Genus Stachybotrys and Its Toxin

In the taxonomic key of Clements and Shear (1931), the genus *Stachybotrys* is classified in the class Fungi Imperfecti, order Moniliales, family Dematiaceae, and subfamily Macronemiae. As indicated in this taxonomic key, the genus *Stachybotrys,* which is saprogenous, is characterized by dark, capitate conidia which are borne exogenously on simple conidiophores having terminal umbellate basidia.

The genus *Stachybotrys* was first defined by Corda in Prague in 1837 from a strain growing on wallpaper in a house. Since the time of Corda, many so-called new species and varieties of *Stachybotrys* have been described, but without critical study of morphological characteristics. Consequently, species limits within the genus are not clearly defined. Salikov (1940) recognized morphological differences, particularly between *Stachybotrys alternans* Bonorden and *Stachybotrys atra* Corda. Salikov characterized the genus *Stachybotrys* by the presence of hyaline to brown branching conidiophores at the upper portion of which are located cylindrical to clavate sterigmata on which are borne single, round to elongate, smooth or echinulate to spinulose conidia, ranging in color from dark brown to black. According to this mycologist, *S. alternans* Bonorden has hyaline to subhyaline conidiophores, sympodially branched, on which are borne seven dark red, clavate to ovoid sterigmata measuring $10 \times 4\text{-}5\ \mu$. The conidia are ellipsoidal, spinulose, and dark brown in color, and measure 8 to 12×5 to 7.5 μ. *Stachybotrys atra* Corda, as defined by Salikov, has forklike branching conidiophores on which are located spindle-shaped sterigmata (no measurements were indicated). Conidia are smooth, brown, ovoid to ellipsoidal, measuring 8 to 9 μ in length (width not indicated). White (1951) in the United States did not consider *S. alternans* a valid species, and, earlier, Bisby (1943) in England apparently did not recognize the species differences noted by Salikov (1940) and reduced the classification to *S. atra*.

Pidoplichko and Bilai (1960) made further taxonomic subdivisions of *S. alternans* based primarily on toxigenicity. They named the toxic strains *Stachybotrys alternans* var. *jateli* Pidoplichko. The cytoplasm of the sterigmata of this toxic variety stains pink when treated with hydrochloric acid containing resorcin. According to Bakai (1960) (also spelled Bakay), the Soviets now recognize two varieties of *S. alternans*, namely *S. alternans* var. *jateli* (apparently described in 1936) which is toxic and *S. alternans* var. *atoxica* which is nontoxic. According to Pidoplichko and Bilai (1960), toxic strains of *S. alternans* prevail over the southern half of Europe, whereas, the nontoxic strains predominate in the northern half. Earlier, Sarkisov and Orshanskaiya (1944) thoroughly described the morphological characteristics of *S. alternans,* but made no morphological differentiation between toxic and nontoxic strains. In efforts to develop a more rapid method for diagnosing stachybotryotoxicosis, Bakai (1960) studied the morphological and cultural characteristics of toxic and nontoxic strains of *S. alternans.* Using 9-day-old cultures of *S. alternans* var. *jateli* on Sabouraud's dextrose agar, he observed a more rapid rate of growth, less cottony trailing hyphae, more delicate appressed network of aerial hyphae and a wide marginal zone at the periphery of the colony, as well as pronounced pigmentation of the medium on reverse. Cultural characteristics of the weakly toxic strains of this fungus were strikingly similar to those of the nontoxic strains (*S. alternans* var. *atoxica*), but it was discovered that growth rate, macroscopic appearance of colonial growth, and degree of pigmentation were not consistent, but varied with cultural and environmental conditions. When examined in slide culture without a nutritive medium, in both the toxic and nontoxic strains, adjacent hyphae anastomosed, but subsequently the former gave rise to conidiophores and the latter did not. Furthermore, he observed that a minimum relative humidity of 98.5% was required for germination of spores of the toxic strains; for optimum growth, the temperature range was 22 to 26°C and the pH 6.2 to 6.9. The respective values for nontoxic strains were as follows: minimum relative humidity for germination of spores was 96.3%, temperature for optimum growth was 23 to 28°C, and pH was 6.0 to 6.6.

Forgacs and associates (1958a), while screening over 40 strains of *S. atra* for toxicity, found no correlation between cultural and morphological characteristics of toxic and nontoxic strains when growing on Czapek's solution agar and on an agar medium containing 2% milled wheat. These workers also noted no correlation between color and weight of an ether extract from straw on which their strains of *S. atra* had been cultured and dermal toxicity on rabbits.

Soviet scientists have indicated morphological differences, not only between *S. atra* and *S. alternans*, but between toxic and nontoxic strains of the latter; whereas, Bisby (1943) and White (1951) apparently have not recognized species differences. Since Soviet strains have not been available for critical examination, the author of this chapter makes reference to *S. alternans* and *S. atra* as separate species, at least until a critical comparison has been made of both species.

In the past, *Stachybotrys* fungi have been considered typical saprophytes that destroy cellulose. Although they have been considered good soil binders (Jensen, 1931; Meshustin and Gromyko, 1946), they are known to destroy books (Rybakova, 1955) and other materials containing cellulose. These fungi are not phytopathogenic, and although Sarkisov and Orshanskaiya (1944) and other Soviet scientists described procedures for isolating *S. alternans* from various organs of animals that succumbed to stachybotryotoxicosis, no concrete evidence was found in Soviet literature available to the author which would indicate that this fungus is truly pathogenic to normal animals. Certainly, futile attempts by Drobotko and co-workers (1946) to transmit this disease by parenteral injection into healthy horses of macerated tissues, which presumably contained *S. alternans* from horses that perished from stachybotryotoxicosis, would lend support to a lack of true pathogenicity of this fungus. Furthermore, Vertinskii (1940) stated that *S. alternans* cannot be considered a disease excitant possessing virulent properties. Parenteral injection of the fungus causes only a local response, but not systemic invasion. Therefore, it would appear that with suppression of the defense mechanisms, the host is subject to tissue invasion by a vast array of opportunistic microorganisms, including *S. alternans*. Forgacs and associates (1958a) found that in a calf that died following a leukopenic episode, histological examination revealed that various tissues were found to be invaded by *Bacillus subtilis*.

Members of the genus *Stachybotrys* are world wide in distribution and have been isolated from soil and a wide array of natural substrata rich in cellulose, such as various hays and straw; cereal grains; hemp; plant debris of all sorts; dead roots of various crops and weeds; dead stems and leaves of wild-growing plants of members of the families Compositae, Cruciferae, Leguminosae, Papaveraceae, and others; wood pulp; cotton; fabrics of plant origin; paper; glue in book binderies; plant fiber processing plants; and a host of other sources. According to Vyshelesski (1948), *Stachybotrys* fungi have been isolated not only in many areas of the U.S.S.R., but also in Austria, Belgium, Denmark, England, France, Hungary, and South America. White *et al.* (1949) in the United States isolated species of this fungus and its related genus *Memnoniella* from various natural substrata and items of clothing. In addition, White's collection of *S. atra* represent isolates from various parts of the world. The author of this chapter, in examining various hays and fodders collected from various parts of the United States, did not find *Stachybotrys* a common fungus, although, it was readily isolated from wet paper and clothing, primarily from dump heaps where the products were in an advanced stage of decomposition.

The range in temperature favorable for growth of this fungus is wide. Vyshelesski (1948) indicated that *S. alternans* will proliferate at temperatures ranging from 2 to 40°C, although the optimum is from 20 to 25°C. Spores of the fungus remain viable at temperatures as low as −30 to −40°C. These temperature ranges, for the most part, are in agreement with those of Salikov (1940), and Vertinskii (1940). Bakai (1960) indicated that the optimum temperature for growth of toxic strains of *S. alternans* which he used in his study ranged from 22 to 25°C. At temperatures above 33 to 37°C, there was marked reduction in toxin production. Various moisture optima have been indicated for growth of the fungus. The most reliable data indicate that a relative humidity of only 30% is required for initial growth, and the optimum conditions of moisture at which the fungus flourishes range from 60 to 100%.

Although the conidia of this fungus survive temperatures as low as −40°C, under conditions of high humidity they die within 1 hour at 60 to 65°C, within 10 minutes at 85 to 86°C, and within 5 minutes at 100°C.

Conidia retain their viability during passage through the gastrointestinal tract, but they are destroyed during the biothermal decomposition of manure. However, spores not subjected to such thermogenic fermentations remain viable for many years. When exposed to sunlight, actively growing cultures show a suppression of mycelial formation, but an increase in spore formation and pigmentation (Bakai, 1960). Viability of both conidia and mycelia of this fungus are destroyed readily by ordinary disinfectants. The cell wall of morphological structures, however, resists disintegration by various chemicals. Tkachenko (1960), in attempting to prepare allergens from *S. alternans* found that he could not disrupt the conidial wall with dilute or concentrated hydrochloric or sulfuric acid, and 60% aqueous solution of sodium hydroxide caused only slight disintegration after boiling for 40 minutes. Only concentrated nitric acid destroyed the conidial wall, following boiling for 40 minutes.

The fungus can overwinter in the soil, on plant stems and debris, wood shavings, paper, various kinds of forage and fodder, cereal grains, on fallen leaves, in old manure heaps, and on other substrata. On farms where the disease has occurred, the fungus has been isolated at least 20 years later. Forgacs and associates (1958a) observed that *S. atra* flourished and produced toxin on moist straw for at least 414 days (Table II). Shulyumov *et al.* (1960) maintained that, in growing on straw, *S. alternans* maintained the capacity to form toxin for 4 years.

TABLE II

EFFECT OF INCUBATION PERIOD OF STRAW INOCULATED WITH *Stachybotrys atra* ON DERMAL TOXICITY OF EXTRACTS[a]

Incubation period (days)	Weight of ether-extractable material (mg/0.125 ml)	Dermal reaction[b]
10	0.6	SE
19	1.3	N
25	1.6	N
33	1.6	N
47	1.7	N
74	1.6	N
109	1.0	N
125	0.9	N
160	0.8	N
220	0.8	N
414	0.4	N

[a]Forgacs *et al.* (1958a). Reproduced through the courtesy of The New York Academy of Sciences.

[b]Reaction observed 24 hours after three daily doses; SE = slight erythema; N = necrosis.

Stachybotrys fungi grow readily on natural substrata. Shulyumov and co-workers (1960) found that their strains of *S. alternans* developed best on oat and barley grains, on barley and wheat straw, and on corn stover. The fungi grew somewhat more slowly on wheat grain and oat straw, but very slowly on corn cobs. Schumaier *et al.* (1963), however, observed that their strains of *S. atra* grew quite readily and formed toxin on sterilized wheat grain. Although *Stachybotrys* fungi grow readily on such accepted mycological media as Czapek's solution agar, Sabouraud's dextrose agar, wort agar, and others, they do not proliferate as rapidly as many other fungi, including many *Aspergilli, Cladosporia, Fusaria,* mucoraceous molds, *Penicillia, Alternaria* species, and others that are common inhabitants of natural substrata such as hays and fodders. Consequently, members of the genus *Stachybotrys* frequently are overgrown by any of the afore-mentioned fungi, and thus overlooked during initial mycologic cultural examinations of suspect natural substrata.

As indicated earlier, toxic and nontoxic strains of both *S. alternans* and *S. atra* occur. Although several methods have been proposed by Soviet investigators for differentiating toxic from nontoxic strains, the most reliable to date is the rabbit dermal toxicity test. Ether extracts prepared from substrata on which toxic strains had grown produce a dermal inflammatory reaction when applied topically on the skin of a rabbit (Sarkisov and Orshanskaiya, 1944). According to Pidoplichko and Bilai (1960), P. D. Iatel was the first to make organic solvent extracts of *S. alternans* for use in the rabbit dermal toxicity test. Other Soviet literature indicates that development of the rabbit dermal toxicity test was the result of cooperative efforts of Sarkisov and Iatel. Evidently, Iatel worked under the capable direction of Professor Sarkisov, who has contributed immensely not only to research on stachybotryotoxicosis, but on other mycotoxicoses as well (Sarkisov, 1954). Forgacs *et al.* (1958a) found that a dermal inflammatory response, characterized initially by hyperemia and edema and subsequently by serum exudation and necrosis, can also be produced in horses and cattle. Following topical applications of a dermally toxic ether extract suspended in olive oil, both animal species also showed evidence of systemic toxicity which manifested itself as injection of the conjunctival mucous membranes, lacrymation, anorexia, and systemic depression. Respiration of the Cheyne-Stokes type increased to 120 per minute in the calf.

Forgacs and associates (1958a) tested over 40 strains of *S. atra* in detail and observed that the toxic strains varied in their ability to form the dermally toxic substance (Table III). These workers, using a special strain under laboratory conditions, found that the dermally toxic substance was

TABLE III
COLOR, WEIGHT, AND TOXICITY OF ETHER EXTRACTS FROM VARIOUS STRAINS OF
Stachybotrys atra INCUBATED 170 DAYS[a]

Strain tested	Color of extracts	Dose (mg/0.125 ml)	Dermal reaction[b]	
			Test strain	Control strain
816	Yellow-green	0.80	0	N
817	Yellow-green	0.92	0	N
820	Green	0.78	N	N
873	Amber	0.72	SE	N
875	Yellow	0.64	SE	N
926	Amber	1.21	SE	N
947	Dark green	0.75	SE	N
1006	Green	0.71	SE	N
1034	Green	0.98	SE	N
1041	Yellow	0.75	0	N
1065	Green	1.06	SE	N
1067	Amber	0.88	SE	N
1069	Amber	0.84	N	N

[a]Forgacs et al. (1958a). Reproduced through the courtesy of The New York Academy of Sciences.
[b]Reaction observed 24 hours after 3 daily doses; 0 = no reaction; SE = slight erythema; N = necrosis.

formed within 10 days, attained a high level in 19 days that was still present at 414 days, when the study was terminated (Table II).

Although the dermal toxin is soluble in various organic solvents, anhydrous ethyl ether appears to be the optimal solvent. The toxic substance can be extracted effectively from infected straw within 3 hours, the bulk of the material being removed during the first hour (Table IV). From observations of tissue changes in horses that died from acute stachybotryotoxicosis, Forgacs and co-workers (1958a) concluded that the toxin present in infected straw reacted with the gastric juice of the stomach and was absorbed in water-soluble form. This observation was confirmed by digestion of dermally toxic straw with pepsin at a low pH, which rendered the toxin water soluble and left the extracted straw nontoxic. Nonextracted, dermally toxic straw, when fed to a horse, produced an acute syndrome followed by death within 17 hours, whereas the pepsin-treated, water-extracted straw was nontoxic to a second horse.

According to Drobotko (1945), and Drobotko et al. (1946), Fialkov at the Academy of Sciences at Moscow isolated the crystalline toxin and determined its empirical formula. Pidoplichko and Bilai (1960) indicated that Fialkov and Serebriani in 1949 isolated the toxin from a wort agar

TABLE IV

INFLUENCE OF TIME OF EXTRACTION OF *Stachybotrys atra*-INOCULATED STRAW ON
REMOVAL OF DERMALLY TOXIC SUBSTANCE[a]

Extraction time (hours)	Ether solubles		
	Percent extracted	Weight (mg/0.125 ml)	Reaction[b]
1	1.28	1.750	PE
2	0.12	0.577	ME
3	0.08	0.117	ME
4	0.05	0.061	0
5	0.03	0.061	0
6	0.03	0.037	0
7	0.03	0.036	0

[a]Forgacs *et al.* (1958a). Reproduced through the courtesy of The New York Academy of Sciences.

[b]Dermal reaction on rabbits; PE = pronounced erythema; ME = moderate erythema; 0 = no reaction.

culture of *S. alternans,* which they named stachybotryotoxin, having an empirical formula of either $C_{25}H_{34}O_6$, or $C_{26}H_{38}O_6$. They indicated that the toxin is representative of a newly discovered group of naturally occurring cardiac toxins. The author has found no reference to the structure of the toxin, although presumably the chemical structure has been determined. It appears that, under favorable cultural conditions, *S. alternans* produces more than one toxin.

Apparently the toxin is extremely potent. It has been suggested that 1 mg of the pure toxin will cause death in a horse. Forgacs *et al.* (1958a) found that 0.00175 μg of a crude ether extract from toxic straw suspended in 0.125 ml of olive oil produced a hyperemia when applied topically to the skin of a rabbit. These authors purified the extract, and although they obtained some refined preparations showing strong dermal toxicity, they did not crystallize the toxin. Some Soviet workers maintain that there exist both dermally and orally toxic factors in toxic straw; however, the relationship between the two entities remains vague.

Stachybotrys alternans toxin is resistant to sunlight, ultraviolet light, and X-rays; it is thermostable, withstanding temperatures of 120°C for at least 1 hour; it is not inactivated by 2% concentrations of inorganic or organic acids, but is readily destroyed by alkalis (Vertinskii, 1940, Dobrynin, 1946; Vysheleski, 1948; Poliakov, 1948). Since the toxin is inactivated by various alkalis, various Soviet scientists have devised procedures using several alkaline materials, including gaseous ammonia, for detoxification of fodders and feedstuffs. Immunity to stachybotryotoxicosis does not develop, since the disease may reappear in a given band of horses within the same season or in succeeding years.

IX. Prophylaxis and Treatment

In areas where outbreaks of stachybotryotoxicosis have occurred, spores of *S. alternans* literally are omnipresent and are disseminated readily during normal agronomic practices, particularly during the summer and fall. Accordingly, essentially all forage, fodder, and feedstuffs become contaminated with conidia of *S. alternans*. Since the fungal spores will not germinate in substrata containing less that 20% moisture, the primary and most important step in prevention of stachybotryotoxicosis comprises prompt harvesting of roughage and other feedstuffs and storing under dry conditions. Straw, or hay, which has become wet in the field and subsequently dried, should not be stacked for feeding. When hay or straw is stored in the field, proper methods of stacking to prevent seepage of water have been recommended. In areas where the fungus is widespread, straw, or hay, from the upper portion of stacks is not used for feeding horses. In stacks showing a black fungal growth, it has been recommended that the moldy layer be discarded, and the inner layers which grossly appear mold free be used for feeding only after laboratory examination has indicated absence of toxin. Early studies concerned with prevention of this toxicosis recommended that toxic roughage be destroyed or fed to cattle, since it was considered nontoxic to such animals. With subsequent reports of field outbreaks of stachybotryotoxicosis in cattle, presumably this recommendation has been withdrawn.

Based on early information that *Stachybotrys* toxin is destroyed readily by alkalis, Dobrynin (1946) and Poliakov (1948) described methods for successfully detoxifying infected fodders using basic substances, including sodium, potassium, calcium, and ammonium hydroxide, and gaseous ammonia. Chlorine gas also was used. According to Pidoplichko and Bilai (1960), Reynfeld, Bakai, and others in 1944 developed a simple method for completely detoxifying roughage using gaseous ammonia. Bakai (1960) indicated further that treating toxic straw with ammonia in hot water vapor for 4 hours rendered the straw nontoxic, and, following evaporation of the gas, the processed straw had a pleasant odor and was readily consumed by farm animals with no ill consequences.

Prevention of endemic spread of the fungus dictates sound sanitary agricultural practices, particularly in reducing conditions favorable for proliferation and dissemination of the fungus. To this end, it has been recommended that stubble and debris left in the field either be quickly plowed under after harvest or burned. Straw or hay not used for feeding should not be allowed to accumulate, and since spores of *S. alternans* are not destroyed in passage through the gastrointestinal tract, corrective measures should be taken in handling manure both in stables and in the field.

Prevention of epizootics also lies in early diagnosis of stachybotryotoxicosis. With the appearance of symptoms of the first stage, particularly those involving the mucosa of the lips, feeding of suspected roughage promptly is interrupted and bedding material is removed. Generally, this results in prompt recovery of the animal.

Since the second stage usually is quiescent, clinically, it is more difficult to diagnose the disease at this time; however, it has been recommended that animals routinely be subjected to hematological examination, particularly clot retraction time. Horses showing blood dyscrasias are treated as those showing initial manifestations of the toxicosis. Earlier recommendations also suggested use of various medications for treatment of local lesions and for systemic therapy. Moseliani (1940) questioned the efficacy of the therapeutic measures, since none of the medications employed in his studies produced any beneficial effect. The current measures used in therapy of equine stachybotryotoxicosis include detoxification and stimulation of the host, regulation of the cardiovascular system and other organs, and prevention of sepsis, particularly in animals whose defense mechanisms have been affected (Shulyumov *et al.*, 1960). Combined parenteral administration of adrenalin, a tissue extract, and antibiotics apparently have resulted in a positive clinical response, particularly in horses in the first or second stage of the typical form of stachybotryotoxicosis, although results in horses in the third stage have not been entirely successful.

X. Isolating and Testing *Stachybotrys* Fungi for Toxicity

Sarkisov and Orshanskaiya (1944) described in great detail proper collection and handling of field specimens and laboratory methods for isolating *S. alternans* from various types of roughage, grain, excreta of sick or dead horses, and from tissues taken at necropsy. These workers also described methods for performing the rabbit dermal toxicity test. Forgacs (1962c), and Forgacs and Carll (1962) also described techniques which they and their associates used for isolating not only *S. atra*, but other fungi as well, from coarse fodder, grain, broiler mash, litter, and other substrata. For testing the fungal isolates for toxicity, these workers used the rabbit dermal toxicity test which they modified to meet the requirements of their respective studies. In addition, they employed other methods for testing their fungal isolates for toxicity. For detailed descriptions of the afore-mentioned procedures, the interested reader may refer to the publications indicated above.

XI. General Comments

The etiology, clinical manifestations, and pathogenesis of equine stachybotryotoxicosis was described first by Soviet scientists. The only other available published literature outside the Soviet Union on this toxicosis in large animals is that by Forgacs and co-workers (1958a) and by Forgacs (1962a,b,c, 1964, 1965a,b, 1966c). Attempts by Forgacs *et al.* (1958a) to produce the three stages of the typical form of equine stachybotryotoxicosis were not entirely successful, since horses refused to eat *S. atra*-infected straw, presumably due to local irritating effects of the substratum. Two horses previously starved for 48 hours were fed *ad libitum* 0.25 lb of the ground toxic straw mixed with 0.25 lb of cane sugar and 1 gallon of oats. The horses readily consumed only three to four mouthfuls of the material. Apparently the odor and taste were not objectionable, but after the fourth mouthful, some unpleasant sensation occurred, because the animals refused to eat more of the toxic mixture. Nevertheless, both horses developed local and systemic reactions typical of the first stage of the typical form of stachybotryotoxicosis. A third horse, after considerable inducement, consumed over a period of 10 days, approximately 0.5 lb of the toxic milled straw mixed with blackstrap molasses and oats. This animal also developed signs of the first stage of the typical form, but on the eleventh day, signs of the atypical form suddenly developed and the animal died. Similarly, two horses, force-fed by stomach tube 1.5 and 2.0 lb, respectively, of toxic milled straw suspended in water, developed the atypical form of the toxicosis and perished within 17 hours. A calf was force-fed 6 lb of a similar slurry in graded doses over a period of 14 days and developed local and systemic manifestations of a toxicosis, with blood dyscrasias typical of the third stage of the typical form of the equine toxicosis, it died on the seventeenth day. Gross necropsy findings in the horses and calf were similar to those described by Soviet scientists for equine stachybotryotoxicosis.

Early studies by Soviet workers on stachybotryotoxicosis indicated that the horse was the only large farm animal subject to this toxicosis. In fact, it was considered safe to feed toxic straw to cattle, since they were considered resistant to the toxin (Vyshelesski, 1948). Pidoplichko and Bilai (1960) maintained that fodder contaminated with *S. alternans* is considered practically harmless to cattle and stated that Sarkisov indicated the cow to be approximately 100 times more resistant to *Stachybotrys* toxin than the horse. Resistance of the former was postulated to be due to the detoxifying effect of the alkaline reaction of the saliva of the ruminant. However, Forgacs and associates (1958a) in the United States found that not only the horse, but calf, sheep, and swine also are susceptible to the toxin. Later, field outbreaks of stachybotryotoxicosis were reported in cattle in the Soviet Union (Fortuskny *et al.*, 1959; Nagornyi *et al.*, 1960; Matusevich *et al.*, 1960; Levenberg *et al.*, 1961; Kurmanov, 1961; Ismailov and Moroshkin, 1962). Perhaps, the strains of *S. alternans* previously toxic to the horse and not cattle mutated when proliferating on natural substrata to give rise to more toxic strains. Forgacs and associates (1958a) observed that one of their nontoxic strains of *S. atra* (Strain 1002), upon repeated subculturing on straw, became extremely toxic, and straw on which this altered culture had grown not only was lethal to the horse, but to the calf and sheep as well.

As to why equine stachybotryotoxicosis suddenly appeared in the Ukraine about 1931 as a new disease in enzootic form remains vague. Certainly, it long had been established that *Stachybotrys* fungi are distributed over many parts of the world and are present in the natural mycologic population of soil and various decomposing natural substrata rich in cellulose. In fact, Corda, as early as 1837 in Prague, first described the genus from a fungus which he isolated from wallpaper. Perhaps sporadic isolated cases occurred prior to 1931 but remained undiagnosed for various reasons, such as lack of economic importance, an initial dearth of qualified diagnosticians familiar with this group of diseases, or other reasons. Outbreaks of various mycotoxicoses have occurred in other parts of the world in livestock and poultry prior and subsequent to stachybotryotoxicosis, but having somewhat the same epizootiologic pattern. These include sweet clover poisoning in cattle in the United States and Canada (Schofield, 1924), cornstalk disease in horses in the midwestern United States (Schwarte *et al.*, 1937), moldy corn toxicosis in swine and cattle in the southeastern United States (Sippel *et al.*, 1953), certain cases of bovine hyperkeratosis in the United States (Carll *et al.*, 1954, Forgacs *et al.*, 1954; Carll and Forgacs, 1954), poultry hemorrhagic syndrome in the United States (Forgacs *et al.*, 1955, 1958b, 1963; Forgacs, 1962a,b,c, 1964, 1966a,b) and other parts of the world, as well as other mycotoxicoses. The history, clinical findings, and other aspects of these and other mycotoxicoses have been reviewed in the English language earlier by Forgacs and Carll (1962), and by Forgacs (1962a,b,c); the interested reader may refer to these reports. It can be postulated, perhaps, that the causal toxic fungi of the afore-mentioned mycotoxicoses could have resulted from naturally induced mutations of parent strains and changes in agronomic practices, together with unusual climatic conditions conducive to growth of the respective fungi. These factors could have contributed to the initial outbreaks of the toxicoses and as a result introduced a preponderance of these fungi into the natural mycologic population, with subsequent enzootic outbreaks of a mycotoxicosis.

Members of the genus *Stachybotrys*, particularly *S. atra* which also contains both toxic and nontoxic strains, have been isolated from diverse parts of the world. In examining his *Stachybotrys* accession list of his stock culture collection, the author found that toxic and nontoxic strains of this fungus have been isolated from various substrata from as far north of the equator as Finland, and from New Guinea and the Solomon Islands to the south. Thus geographic distribution of toxic strains of this fungus is wide indeed.

To the best knowledge of the author, no authenticated outbreaks of equine stachybotryotoxicosis have been reported in the United States. Perhaps cases have occurred in the past but remained undiagnosed. The author was informed of a suspect case that occurred in North Carolina in 1960 in a band of horses that consumed a shipment of baled hay that was heavily molded with a black fungus (Ryan, 1969). Although the clinical and pathological findings in the afflicted horses were typical of the shocking form of equine stachybotryotoxicosis, it is regrettable that no efforts were made to isolate the causal fungus. Whether authenticated cases may occur in the future in the United States remains to be determined, but with a reduction in draft horses on farms, and a lack of diagnosticians familiar with stachybotryotoxicosis, the probability is poor. To complicate the picture even more, *Stachybotrys* fungi are difficult to isolate from a natural substratum by ordinary mycologic techniques, particularly in the presence of other less fastidious fungi which readily overgrow culture plates, leaving the more fastidious fungi undetected. However, current changes in agronomic practices, particularly in allowing baled hay to remain in the field unprotected from the elements from fall till spring, may pose a potential health hazard, particularly to cattle that are allowed to forage such hay, not only from possible mycotoxins elaborated by *S. atra*, but by a host of other fungi such as the cryophilic molds, whose optima of growth and toxin formation lies near or slightly above the freezing point. The author has examined a considerable number of such hay samples over the past 5 years, and to date has not isolated *S. atra*, but found most samples to be heavily laden with a variety of fungi, some which represent known toxigenic groups, particularly of the genera *Aspergillus*, *Fusarium*, *Penicillium*, and especially members of the family Dematiaceae. Even though sufficient toxin may not be elaborated to cause serious illness or even death among cattle that consume such molded forage, consumption of sublethal amounts of toxin over prolonged periods of time may depress the hematopoietic centers and thereby leave the host subject to opportunist infection by various microorganisms of low initial viru-lence. Perhaps of equal importance is that animals that consume sublethal quantities of mycotoxins show poor feed conversion and thus net gain is low. This has been demonstrated well by various workers in the case of moldy feed toxicosis in chickens (poultry hemorrhagic syndrome).

Although some aspects of stachybotryotoxicosis have been studied thoroughly, there are other areas in need of further investigation. There appears a need for reevaluating the taxonomic status of *S. alternans* and *S. atra*. It appears expedient that strains of *S. alternans* used by Soviet scientists, as well as those of *S. atra* used by investigators in the western countries, be subjected to arbitrary critical morphological and taxonomic studies to determine whether indeed they both represent two valid spe-cies.

The ecological significance and potential health hazards of *Stachybo-trys* fungi is in need of further study, particularly in the Western world. Efforts should be made to determine the potential health hazard that *Stachybotrys* fungi may pose to animal health in outbreaks of unknown etiology in farm animals in which the clinical and pathological findings resemble those described for stachybotryotoxicosis. Accordingly, fod-ders and other feedstuffs should be examined mycologically for *Stachybo-trys* fungi. Concurrently, the role of other toxigenic fungi also should be determined.

As indicated by Sarkisov (1960), one of the major problems in control-ling stachybotryotoxicosis is timely detection of infection of fodder and feedstuffs with toxic strains of *S. alternans*. He stressed that the best available and most dependable diagnostic method presently in use by So-viet scientists, which consists of isolating the fungus and subsequently establishing its toxicity, must be considered inadequate and must be su-perceded by a more rapid method of identifying the toxin in the natural or suspect substratum. Thus, it would appear that studies should be initiated on determining the chemical structure of *Stachybotrys* toxin with the ulti-mate aim of using such information for devising a rapid chemical method for both qualitative and quantitative measurement of the toxin not only in a variety of animal feedstuffs, but also in food consumed by human beings. Once chemically characterized, the crystalline toxin should be studied pharmacologically and toxicologically to determine its tissue dis-tribution, excretion, mode of action, detoxification in the body, and means of counteracting the effect once the toxin has entered the body. Although the resorcin test has been suggested for differentiating toxic from non-toxic strains of *S. alternans*, the method apparently has not been accepted widely for field conditions. Tkachenko (1960), in preliminary studies, employed allergenic methods for diagnosing this toxicosis in horses, but it appears that results so far are only in the exploratory stage.

Additional studies are needed on the biochemical activity of *Stachybotrys* fungi. Whether in nature there exist physiological races of these fungi which produce more than one toxin should be determined. It appears plausible that, depending on precursors, environmental conditions, strain variation, and other factors, *Stachybotrys* fungi could produce more than one toxin, each having somewhat different physiological and toxicologic properties.

Regarding the toxicologic properties of *Stachybotrys* toxin, no data have been found indicating the chronic effect in animals of sublethal graded doses of the toxin. Certainly, the potential carcinogenic properties of the toxin should be studied, since this would have a bearing upon the welfare and health of humans who are exposed to sublethal aerosols of the toxic substrata. Although Samsonov and Samsonov (1960) described clinical and pathological findings in guinea pigs experimentally challenged with *Stachybotrys* pneumomycotoxicosis, following a single respiratory dose, no data are available on tissue changes in animals on long-term studies.

Preventing outbreaks of stachybotryotoxicosis in animals lies chiefly in sound agronomic and sanitary practices, such as prompt harvesting of fodders and storing them under dry conditions, plowing under straw and other plant refuse, and general sanitary procedures associated with the feeding of animals and disposal of manure. Perhaps use of antifungal compounds for treating harvested fodders should be investigated. Such agents could either prevent fungal growth or alter metabolism of mycotoxins, as has been demonstrated by Forgacs and co-workers (1962b, 1963) in their investigations on moldy feed toxicosis in poultry. One method would be a systemic fungicide that could be applied to the growing crops in the field and that would alter metabolism of not only parasitic fungi growing on the plant but of saprophytic toxic fungi as well that subsequently infect harvested fodder or feedstuffs.

Finally, as has been indicated earlier, stachybotryotoxicosis affects not only animals, but also human beings exposed to aerosols from *Stachybotrys*-infected substrates (Vertinskii, 1940; Drobotko, 1942, 1945; Drobotko *et al.*, 1946; Vyshelesski, 1948). Toxicity in humans is characterized primarily by a systemic effect. Samsonov (1960) named such mycotoxicoses as pneumomycotoxicoses, and Samsonov and Samsonov (1960) reproduced such toxicoses in guinea pigs by introducing into the respiratory tract spores not only of *S. alternans,* but of other toxic fungi as well. Furthermore, there is abundant evidence in published Soviet literature that leaves many other fungi suspect as agents in pneumomycotoxicoses. Many of these fungi have been studied in the Western world and have been found toxic to animals when infected substrata were fed *ad libitum* or administered orally. Some fungi, because of their xerophytic nature, are dispersed readily by air currents and thus contaminate adjacent areas. This is true particularly of many *Aspergilli* and *Penicillia,* some of which are toxic. Accordingly, the investigator who works with toxic fungi should take the necessary precautions to avoid inhalation of fungal spores and aerosols. This applies not only to isolation techniques, which should be carried out under an exhaust-equipped hood, but particularly to drying and grinding of substrata, the manipulations of which always should be done in an area adequately equipped with a filtering exhaust system. The individual performing the drying, grinding, and extraction of toxic substrates should wear rubber gloves, protective clothing, and a well-fitting efficient gas mask and should change clothing and shower before reentering a noncontaminated area. When extracting toxic straw or other substrates with organic solvents, particular care must be exercised to prevent the extract from contacting the skin, since the toxins are readily absorbed and elicit a systemically toxic effect, as can be verified by the author and some of his colleagues. This applies particularly to *Stachybotrys* toxin, which is very toxic to man when dissolved in an organic solvent.

REFERENCES

Bakai, S. M. (1960). *In* "Mycotoxicoses of Man and Agricultural Animals" (V. I. Bilai, ed.), pp. 163-167. Izd. Akad. Nauk Ukr. S.S.R., Kiev.
Bisby, G. R. (1943). *Brit. Mycol. Soc. Trans.* **26,** 133.
Carll, W. T., and Forgacs, J. (1954). *Military Surgeon* **115,** 187.
Carll, W. T., Forgacs, J., and Herring, A. S. (1954). *Am. J. Hyg.* **60,** 8.
Clements, F. E., and Shear, C. L. (1931). "The Genera of Fungi." Hafner, New York.
Dobrynin, V. P. (1946). *Konevadstvo* **16,** 27.
Drobotko, V. G. (1942). "Stachybotryotoxicosis, a New Disease of Horses and Humans." Report presented at the Academy of Science, U.S.S.R.
Drobotko, V. G. (1945). *Am. Rev. Soviet Med.* **2,** 238.
Drobotko, V. G., Marushenko, P. E., Aizeman, B. E., Kolesnik, N. G., Iatel, P. D., and Melnichenko, V. D. (1946). *Vrachebnoe Delo* **26,** 125.
Forgacs, J. (1962a). *Proc. Maryland Nutr. Conf. Feed Mfrs., Washington, D.C. 1962* pp. 19-31. Dept. of Poultry Husbandry, Univ. Maryland.
Forgacs, J. (1962b). *Feedstuffs* **34,** 124.
Forgacs, J. (1962c). *Proc. 66th Ann. Meeting, U. S. Livestock San. Assoc., Washington, D.C.* pp. 426-448.
Forgacs, J. (1964). *New Hampshire Poultry Health Conference, Durham, N. H., 1964* pp. 16-26. Poultry Dept., Univ. New Hampshire.

Forgacs, J. (1965a). *In* "Mycotoxins in Foodstuffs" (G. N. Wogan, ed.), pp. 87-104. M.I.T. Press, Cambridge, Massachusetts.

Forgacs, J. (1965b). *25th Ann. Meeting, Inst. Food Technologists, Kansas City, Mo., 1965* Abstr. No. 58, pp. 59-60.

Forgacs, J. (1966a). *Feedstuffs*, March 12, **38**, 18, 66, 71.

Forgacs, J. (1966b). *Feedstuffs*, March 19, **38**, 26, 28, 30, 71.

Forgacs, J. (1966c). *Food Technol.* **20**, 46.

Forgacs, J., and Carll, W. T. (1962). *Advan. Vet. Sci.* **7**, 273.

Forgacs, J., Carll, W. T., Herring, A. S., and Mahlandt, B. G. (1954). *Am. J. Hyg.* **60**, 15.

Forgacs, J., Koch, H., and Carll, W. T. (1955). *Poultry Sci.* **34**, 1194.

Forgacs, J., Carll, W. T., Herring, A. S., and Hinshaw, W. R. (1958a). *Trans. N. Y. Acad. Sci.* **20**, 787.

Forgacs, J., Koch, H., Carll, W. T., and White-Stevens, R. H. (1958b). *Am. J. Vet. Res.* **19**, 744.

Forgacs, J., Koch, H., Carll, W. T., and White-Stevens, R. H. (1962a). *Avian Diseases* **6**, 363.

Forgacs, J., Koch, H., and White-Stevens, R. H. (1962b). *Avian Diseases* **6**, 420.

Forgacs, J., Koch, H., and White-Stevens, R. H. (1963). *Avian Diseases* **7**, 56.

Fortuskny, V. A., Govrov, A. M., Tebybenko, I. Z., Biochenko, A. S., and Kalitenko, E. T. (1959). *Veterinariya* **36**, 67.

Ismailov, I. A., and Moroshkin, B. F. (1962). *Veterinariya* **4**, 27.

Jensen, H. L. (1931). *J. Agr. Sci.* **21**, 81.

Korneev, N. E. (1948). *Veterinariya* **25**, 36.

Kurmanov, I. A. (1961). *Veterinariya* **38**, 41.

Levenberg, I. G., Ivantsvov, L. I., and Prostakov, M. P. (1961). *Veterinariya* **38**, 38.

Matusevich, V. G., Fekilstov, M. H., and Rozhdestvenskii, V. A. (1960). *Veterinariya* **37**, 71.

Meshustin, E. H., and Gromyko, E. P. (1946). *Mikrobiologiya* **15**, 169.

Moseliani, D. V. (1940). *Veterinariya* **17**, 42.

Nagornyi, I. S., Possiginskii, M. M., Govzdov, A. V., and Rybka, N. V. (1960). *Veterinariya* **37**, 69.

Pidoplichko, N. M., and Bilai, V. I. (1960). *In* "Mycotoxicoses of Man and Agricultural Animals" (V. I. Bilai, ed.), pp. 3-37. Izd. Akad. Nauk Ukr. S.S.R., Kiev.

Poliakov, A. A. (1948). *In* "Rukovodstvo po Veterinarnoi Dezinfeksii," pp. 154-158. Ogiz-Selskhozgiz, Moscow.

Ryan, T. B. (1969). Personal communication to J. Forgacs.

Rybakova, S. G. (1955). *Mikrobiologiya* **24**, 608.

Salikov, M. I. (1940). *Sov. Botan.* **6**, 53.

Samsonov, P. F. (1960). *In* "Mycotoxicoses of Man and Agricultural Animals" (V. I. Bilai, ed.), pp. 131-140. Izd. Akad. Nauk Ukr. S.S.R., Kiev.

Samsonov, P. F., and Samsonov, A. P. (1960). *In* "Mycotoxicoses of Man and Agricultural Animals" (V. I. Bilai, ed.), pp. 140-151. Izd. Akad. Nauk Ukr. S.S.R., Kiev.

Sarkisov, A. Kh. (1947). *Veterinariya* **24**, 25.

Sarkisov, A. Kh. (1954). "Mycotoxicoses." *Goz. Izd.* Selskohoziastvennoi Literatury, Moscow.

Sarkisov, A. Kh. (1960). *In* "Mycotoxicoses of Man and Agricultural Animals" (V. I. Bilai, ed.), pp. 155-163. Izd. Akad. Nauk Ukr. S.S.R., Kiev.

Sarkisov, A. Kh., and Orshanskaiya, V. N. (1944). *Veterinariya* **21**, 38.

Schofield, F. W. (1924). *J. Am. Vet. Med. Assoc.* **64**, 553.

Schumaier, G., DeVolt, H. M., Laffer, N. C., and Creek, R. D. (1963). *Poultry Sci.* **62**, 70.

Schwarte, L. H., Biester, H. E., and Murray, C. (1937). *J. Am. Vet. Med. Assoc.* **90,** 76.

Shulyumov, Ye. S., Kuzmin, A. F., and Fedko, P. A. (1960). *In* "Mycotoxicoses of Man and Agricultural Animals" (V. I. Bilai, ed.), pp. 167-180. Izd. Akad. Nauk Ukr. S.S.R., Kiev.

Sippel, W. L., Burnside, J. E., and Atwood, M. B. (1953). *Proc. 90th Ann. Meeting, Am. Vet. Med. Assoc., Toronto, Canada* pp. 174-181.

Tkachenko, A. F. (1960). *In* "Mycotoxicoses of Man and Agricultural Animals" (V. I. Bilai, ed.), pp. 187-195. Izd. Akad. Nauk Ukr. S.S.R., Kiev.

Vertinskii, K. I. (1940). *Veterinariya* **17,** 61.

Vyshelesski, S. N. (1948). "Special Epizootiology," pp. 374-382. Ozig-Selskozgiz, Moscow.

White, W. L. (1951). Personal communication to J. Forgacs.

White, W. L., Yeager, C. C., and Shotts, H. (1949). *Farlowia* **3,** 399.

SECTION C

PHYTOPATHOGENIC TOXINS

CHAPTER 5

Phytopathogenic Toxins*

H. H. LUKE AND V. E. GRACEN, JR.

I. Introduction

Basic concepts and terminology of the toxin theory of plant disease should be clearly defined before a discussion of specific examples of pathogenic toxins is presented. This chapter, therefore, is devoted to history, concepts, and terminology. Other chapters in this volume discuss toxins produced by species of *Helminthosporium, Alternaria,* and *Didymella.* These genera were selected because many of the phytopathogenic toxins are produced by them. These examples should be sufficient to give the reader a thorough insight into the history, problems, and progress of the toxin theory of plant disease. Additional details have been published in recent reviews (Wheeler and Hanchey, 1968; Wright, 1968; Owens, 1969).

In some cases toxins are produced *in vivo* by the host as well as by the pathogen. Consequently, Chapter 9 has been included to discuss compounds that accumulate in plants after fungal infection.

II. History

The concept that plant pathogens produce pathogenic toxins originated about a century ago (de Bary, 1886), but almost three decades elapsed before experimental evidence was obtained to support this concept (Hutchinson, 1913). Moreover, proof that phytopathogens induce disease through toxigenic action was not obtained until the 1960's.

Acceptable experimental evidence for toxigenic action of human pathogens was established by medical pathologists as early as 1888 (Roux and Yersin, 1888). Why had so long a period elapsed before it was established

*U.S. Department of Agriculture, ARS, Plant Science Research Division, Plant Pathology Department, University of Florida, Gainesville, Florida. Florida Agricultural Experiment Station Journal Series Paper No. 3441.

with phytopathogens? This question is particularly perplexing because the germ theory of disease was first demonstrated by plant pathologists (Prevost, 1807) about a half century before it was established in medical pathology. Differences between the anatomy and physiology of the two suscepts, as well as differences in the toxins produced by animal and plant pathogens probably contributed to the delay in demonstrating toxigenic activity in plant pathogens. Human pathogens, which induce diseases through toxic components, usually produce large proteinaceous toxins that move freely through the circulatory system. In contrast, phytopathogens produce a wide variety of toxic compounds. Some produce two or more toxins of widely different natures. The phytopathological problem becomes more complex when the host as well as the pathogen produces toxic components. Since plants do not have an efficient circulatory system, most phytotoxigenic compounds are not readily translocated. Therefore, injury some distance from the point of infection does not occur rapidly. This observation is basic to the idea that microbes produce toxins that induce disease. Nevertheless, these obstacles have been surmounted, and a few cases showing that plant pathogens produce toxins have been reported.

The long and arduous path to these discoveries was opened by an excellent work published in 1913 (Hutchinson, 1913). For three decades after Hutchinson's report, many papers were published that indicated that phytopathogens produce toxins injurious to plants. For the most part, results were obtained by growing the pathogens in artificial culture media and applying the culture filtrate to the suscept. Workers did not appear to realize that microorganisms produce numerous toxigenic compounds in enriched culture media that are not produced in the host. Reports during this era (1913-1943) led to the erroneous conclusion that many phytopathogens induced disease through toxigenic action. These conclusions persisted, even though the toxic metabolites from culture fluid affected plants that were not parasitized by the pathogen. The pendulum began to swing in the opposite direction when attempts to extract toxin from infected plants met with considerable difficulty (Gottlieb, 1943, 1944). The turning point came with the idea that one could not prove that a given pathogen produced a disease-inducing toxin unless the toxin could be produced in the infected host and extracted from it (Dimond and Waggoner, 1953).

III. Concepts

The idea that the toxigenic potential of phytopathogens could be evaluated by extracting the toxin from the infected host led to the vivotoxin postulation. A vivotoxin was defined "as a substance produced in the infected host by the pathogen and/or its host, which functions in the production of disease, but is not itself the initial inciting agent of disease" (Dimond and Waggoner, 1953). The latter part of this definition implies that the pathogen and not the toxin is the "initial" inciting agent of disease. A vivotoxin was designated a pathogenic agent and three criteria were established to prove the pathogenicity of vivotoxins. These criteria demanded: (1) reproducible separation from the diseased plant, (2) purification, and (3) reproduction of at least some of the disease symptoms by the toxin after it is placed in a healthy plant (Dimond and Waggoner, 1953).

Although the vivotoxin concept does not establish a satisfactory set of criteria by which the toxin theory can be tested, it is useful in bringing about a critical examination of the *in vitro* system. In practice, the vivotoxin idea is insufficient because many toxins produced *in vivo* are labile and are produced in low quantities. Thus it is virtually impossible to extract and identify a disease-inducing toxin from the host. Another reason for the failure of the vivotoxin idea is that many toxic substances produced as a result of host–pathogen interaction are not directly involved in the disease syndrome. Nevertheless, the vivotoxin concept marked a turning point in the development of the toxin theory because it provoked a more critical analysis of experimental criteria to be fulfilled before a toxin could be considered a disease-inducing agent.

Another decade elapsed before a set of criteria was formulated that would adequately test the toxin theory of plant disease (Wheeler and Luke, 1963). In brief, these criteria demanded: (1) that a reasonable concentration of the toxin induce in the suscept all of the symptoms characteristic of the disease, (2) that the pathogen and the toxin exhibit similar suscept specificity, and (3) that a positive correlation exist between toxin production and pathogenicity. When these conditions are met, a single toxin can be said to induce disease. A toxin fulfilling these requirements is described as a pathotoxin, meaning a toxin that causes disease. It was realized that some toxins may play a minor role in the disease syndrome. Therefore, the term phytotoxin was proposed to denote a toxin produced by a phytopathogen that is nonspecific, expresses a low order of activity, or incites few or none of the symptoms incited by the pathogen (Wheeler and Luke, 1963).

The set of criteria required to satisfy the pathotoxin idea was another milestone in the development of the toxin theory. These criteria, however, have not been as useful as they might have been because of the general interpretation that if all the rules are not satisfied, the toxin in question is not involved in the disease syndrome. This was not the objective of the pathotoxin proposal. Instead, the objective was to propose a set of rules that if satisfied would prove that a single toxin produced by a phyto-pathogen was indeed a disease-inducing entity. Obviously, all the rules would be difficult to satisfy if two or more toxins are involved in the same test system.

The misinterpretation of the pathotoxin proposal was further com-pounded by the proposal of the term host-specific plant toxin. "A host-specific plant toxin is defined as a metabolic product of a pathogenic microorganism which is toxic to the host of that pathogen" (Pringle and Scheffer, 1964). In essence, this definition is similar to the second crite-rion (that the pathogen and the toxin exhibit similar suscept specificity) proposed for pathotoxicity (Wheeler and Luke, 1963). Undue emphasis placed on the term host-specific plant toxin gives the impression that host-specific toxins are the only ones that are important in plant pathology. Since there are only six known host-specific plant toxins, emphasis placed on specificity has left the impression that the induction of plant disease through a toxic component is a rare phenomenon. The importance of non-specific phytopathogenic toxins, which may play some role in disease development, should not be ignored.

A recent report on the status of the toxin theory of plant disease sug-gested that the vivotoxin idea has not and cannot help identify a true toxin (Kalyanasundaram and Charudattan, 1966). Kalyanasundaram and Cha-rudattan did not define a true toxin, but stated that in the case of human diseases the involvement of specific toxins in specific diseases has been demonstrated not by *in vivo* detection but by applying the following cri-teria: (1) the organism is known to produce the toxin, (2) the virulent vari-ants produce the toxin and the avirulents do not, (3) infection by the toxin separately from the organism produces symptoms that mimic the disease, (4) the infecting organisms produce the disease without spreading exten-sively, and organs far removed from the seat of infection are affected, and (5) the disease can be prevented by immunization against the toxin.

Before these five criteria can be applied to phytopathogens the fifth cri-terion (prevention of disease by immunization) should be deleted. Dele-tion is necessary because phytopathogenic toxins are usually small mole-cules that are not antigenic, and plants are not known to have immunological systems comparable to those of animals. Further, the first part of the fourth criterion (the infecting organism produces the disease without spreading extensively) should be more clearly defined. This is suggested because some microorganisms that produce toxins spread ex-tensively through host tissues, but still induce symptoms somewhat re-moved from the area of infection. Moreover, in leafspot diseases, only a few millimeters separate the point of infection from the periphery of the affected cells. This may be because most leafspot pathogens do not produce toxins, or the toxins produced are not easily translocated.

Most investigators who have made an effort to establish a set of criteria to test the toxigenic potential of phytopathogens have maintained that the toxin should reproduce the symptoms incited by the pathogen. Although many plant pathologists have accepted this principle, it is not sound unless tempered with some judgment. Failure of the toxin to reproduce the disease syndrome certainly leaves doubt about the role of the toxin in the disease, but reproduction of the disease syndrome is not conclusive evidence for pathotoxicity. The latter is true because many symptoms caused by a wide variety of irritants are similar. For example, the characteristic symptoms of wilting are similar regardless of the mechanism that initiates the process.

IV. Terminology

When the toxin theory of plant disease is considered, two divergent points of view emerge. One maintains that few if any phytopathogens incite disease through toxigenic action, whereas the other claims that a phytopathogen must be toxigenic before it can be pathogenic. The views differ in the criteria with which one chooses to test the toxin theory and the manner in which one chooses to define the term toxin. When the matter of semantics has been resolved, the truth will be found to lie between these two extremes.

It would appear that the development of adequate criteria to test different cases in which toxins are suspected would be the paramount issue, but such is not the case. Instead, the most recent controversy concerns the lack of agreement on the meaning of the term toxin. The term, as used in plant pathology, has taken on many different connotations; therefore, its meaning has become ambiguous. Such terms as phytotoxin, vivotoxin, pathotoxin, and host-specific toxin have been coined. Also, the term mycotoxin was recently used to designate all toxins produced by fungi (Wright, 1968). All of these terms are somewhat synonymous because they strive to denote a toxin produced by a phytopathogen which is involved in plant disease. The following quotation illustrates our inability to develop absolute terminology. "In recent years many new words have been coined, such as phytotoxin, vivotoxin, pathotoxin and host-specific toxin, and in our opinion they are terms coined to support the importance of particular toxins rather than as absolute and final definitions" (Kalyanasundaram and Charudattan, 1966). Since no single term gives an absolute definition of a phytopathogenic toxin, this viewpoint has some validity. For example, a "phytotoxin" could be a toxic compound produced by a plant or one toxic to a plant. A "vivotoxin" is obviously produced *in vivo*, but no indication of the type of organism producing the toxin is given, nor does the term identify the organism affected. A "pathotoxin" denotes a toxin that incites a disease, but again the organism that produces the toxin and the organisms affected remain unknown. A "host-specific plant toxin" is a cumbersome term indicating that the toxin has the same host range as the pathogen, but gives no idea as to the organism producing it. A "mycotoxin" denotes a toxin produced by a fungus but does not signify the organism affected. Clearly none of these terms concisely define the desired idea. Moreover, it would be difficult to construct a grammatically correct latinized term to do so. Perhaps an absolute and final definition is somewhat analogous to the proverbial absolute truth.

When considered in their original text, most of the terms mentioned are quite acceptable. Thus, new terminology is not proposed; instead, "toxin" is used as a general term to denote a metabolite of fungal origin that is involved in plant disease. Although enzymes, auxins, and cytokinins exhibit toxigenic activity, they are not included.

V. Discussion

When the enormous population of phytopathogenic organisms is contrasted with the small number of infectious diseases normally occurring in a given plant population, the power of plants to resist disease is truly amazing. In fact, it might be said that microbial infection of a plant which leads to a disease situation is a rare phenomenon. The question thus arises, what mechanism or mechanisms are the dynamic forces in the defense system?

When the anatomy of higher plants is considered, it appears that each cell of a given tissue acts as an individual that possesses two mechanisms of defense. One is a general and nonspecific mechanism that responds to any physical or pathological injury. In many cases, the nonspecific mechanism appears to involve the polyphenol synthetic system of the host. Thus, the individual cells in a tissue are capable of rapidly defending themselves against a large number of microbes or other irritants.

The second line of defense is a more specific mechanism that may be largely conditioned by the physiology of the invading pathogen. Some toxins produced by fungi seem to affect the specific mechanism. This does not imply that all microbial toxins are specific, but it perhaps helps to explain why it took so long to establish the toxin theory of plant disease. The concept that toxins are the causal agents of disease was not completely accepted until a specific toxin which caused a disease was discovered and used as a model system. This discovery, along with the development of an adequate set of test criteria, has placed the toxin theory of plant disease on firm ground.

Finally, it must be made clear that both specific and nonspecific microbial toxins are involved in plant disease. But when one is working with a specific and a nonspecific toxin in a single system, the problem becomes so complex that it is most difficult to obtain a satisfactory solution. Therefore, the most conclusive evidence that toxins induce plant disease has been obtained from systems that employ a single specific toxin.

REFERENCES

de Bary, A. (1886). *Botan. Zt.* **44**, 337.
Dimond, A. E., and Waggoner, P. E. (1953). *Phytopathology* **43**, 229.
Gottlieb, D. (1943). *Phytopathology* **33**, 126.
Gottlieb, D. (1944). *Phytopathology* **34**, 41.
Hutchinson, C. M. (1913). *India, Dept. Agr., Mem., Bact. Ser.* **1**, 67.
Kalyanasundaram, R., and Charudattan, R. (1966). *J. Sci. Ind. Res. (India)* **25C**, 63.
Owens, L. D. (1969). *Science* **165**, 18.
Prevost, A. (1807). *In* "Phytopathological Classic No. 6." Am. Phytopatholo. Soc., St. Paul, Minnesota (reprint, 1939).
Pringle, R. B., and Scheffer, R. P. (1964). *Ann. Rev. Phytopathol.* **2**, 133.
Roux, E., and Yersin, A. (1888). *Ann. Inst. Pasteur* **5**, 17.
Wheeler, H., and Luke, H. H. (1963). *Ann. Rev. Microbiol.* **17**, 223.
Wheeler, H., and Hanchey, P. (1968). *Ann. Rev. Phytopathol.* **6**, 331.
Wright, D. E. (1968). *Ann. Rev. Microbiol.* **22**, 269.

CHAPTER 6

Helminthosporium Toxins*

H. H. LUKE AND V. E. GRACEN, JR.

I. Introduction

An unusual phenomenon is the fact that pathogenic species of *Helminthosporium* rarely parasitize plants other than members of the grass family (Gramineae). Further, members of this genus cause many of the most devastating diseases of grasses. In this chapter, we will discuss the role of pathogenic toxins produced by six different species of *Helminthosporium*.

Six species of *Helminthosporium* are known to produce phytopathogenic compounds. Some of these are major disease-inducing agents, while others play minor roles in the disease syndrome. We will discuss in detail one case in which the toxin is known to incite disease. Other toxins produced by different species of *Helminthosporium* are also presented. In most instances, toxin production, purification, and molecular structure will be considered. The physiological interaction of the toxin with the suscept and mode of action will also be discussed. Basic data concerning the toxins that will be discussed are presented in Table I.

Because several reviews are available (Wheeler and Luke, 1963;

*U.S. Department of Agriculture, ARS, Plant Science Research Division, Plant Pathology Department, University of Florida, Gainesville, Florida. Florida Agricultural Experiment Station Journal Series Paper No. 3442.

139

TABLE I

GENERAL MOLECULAR STRUCTURE, TOXICITY, SPECIFICITY, AND MODE OF ACTION OF PHYTOPATHOGENIC TOXINS PRODUCED BY *Helminthosporium* SPECIES

Toxin	Producing organism	Type of compound	Toxicity (µg/ml)	Specificity[a]	Mode of action
Victorin	*H. victoriae*	Peptide and secondary amine	0.01	Yes	Disrupts permeability?
Victoxinine	*H. victoriae*	Secondary amine	75	No	—
Helminthosporal	*H. sativum*	Cyclic terpenoid	50–100	No	Dual effect[b]
Helminthosporol	*H. sativum*	Cyclic terpenoid	30–50	No	Growth regulator
Ophiobolins	*H. oryzae*	Cyclic terpenoid	3–10	No	Enzyme inhibitor?
H. carbonum toxin	*H. carbonum*	Polypeptide	1	Yes	—
Carbtoxinine	*H. carbonum*	Polypeptide	25	No	—
H. maydis toxin	*H. maydis*	—	—	Yes?	—
H. sacchari toxin	*H. sacchari*	Peptide?	—	Yes?	—

[a] "Specific" denotes that the toxin exhibits the same host range as the pathogen.
[b] Helminthosporal inhibits the energy transfer system and uncouples oxidative phosphorylation.

Pringle and Scheffer, 1964; Kalyanasundaram and Charudattan, 1966; Wright, 1968; Wheeler and Hanchey, 1968; Owens, 1969), we will not attempt an exhaustive presentation of the literature, but will instead concentrate on an objective evaluation of current information pertinent to the toxin theory of plant disease. Finally, a summary of our present knowledge about the toxigenic potential of phytopathogens is presented.

II. *Helminthosporium victoriae* Toxins

"The season was right for Iowa to have a record oat crop in 1946, but it was only average and many of you probably know the cause—the new blight and root rot disease—Helminthosporium. It cut the yields of many fields, some in half" (Murphy, 1946). The above is a quotation from the first report of the fungus *Helminthosporium victoriae* Meehan & Murphy, later shown to produce a toxin (victorin) causing disease symptoms on certain oat cultivars. This quotation is pertinent because the toxin produced by *H. victoriae* is the first toxin produced by a phytopathogen shown conclusively to be a disease-inducing agent. Discovery of the toxin produced by *H. victoriae* not only renewed interest in the idea that phytopathogens induce disease through toxigenic action, but ideas evolving from the use of this system have led to new approaches concerning the basic nature of phytopathogenesis. Victorin, therefore, is a prime example to be used in a discussion of the toxin theory of plant disease.

Early papers reported a description of the fungus and the disease syndrome, which occurs only on the oat cultivar Victoria or hybrids derived from it (Murphy and Meehan, 1946; Meehan and Murphy, 1946a). Infected seedlings exhibit necrosis at the base of the stem and striping or reddening of the leaves. Leaf striping and discoloration progress upward from the lower leaves. Since the fungus could not be isolated from discolored leaves, it was suggested that the symptom may be the result of a toxin which originated at the site of infection (Meehan and Murphy, 1946b). Investigation of this observation revealed that culture filtrates diluted 1:45 and sterilized by filtration were very toxic to a susceptible cultivar (Meehan and Murphy, 1947). Cultivars of oats highly resistant or immune to the fungus were not injured by the toxin. Thus, the toxin was shown to be highly specific. In this chapter, specific denotes that the toxin exhibits the same host range as the pathogen. It was implied that pathogenicity of the fungus may depend largely upon the injurious effects of a toxic substance. These observations were soon confirmed and extended (Litzenberger, 1949). That the *H. victoriae* toxin is indeed a disease-inducing agent was determined by other investigators (Luke and Wheeler, 1955). This conclusion was based on three lines of evidence: (1) nonpathogenic cultures of the fungus did not produce toxin, (2) differences in pathogenicity among pathogenic cultures were positively correlated with differences in toxin production, and (3) toxin production was directly related to the growth rate of pathogenic isolates (Luke and Wheeler, 1955). Hence the toxin produced by *H. victoriae,* which was given the trivial name victorin (Wheeler and Luke, 1954), was established as a model system useful in studying the toxin theory of plant disease.

A. TOXIN PRODUCTION

The toxin theory of plant disease evolved from two basic observations: (1) toxins are produced by phytopathogenic microorganisms when grown on synthetic media, and (2) these toxins incite a part or all of the symptoms characteristic of the disease initiated by the pathogen. The validity of these observations has been the topic of much discussion and debate. Consequently, at least one case which would positively substantiate the toxin theory was critically needed. Victorin was one of the first toxins that fully supported the idea that plant pathogens produce a pathogenic toxin *in vitro* (Luke and Wheeler, 1955).

The first attempts to produce victorin *in vitro* indicated that fungal-free filtrates from a liquid culture medium (Richard's solution) caused mild symptoms on susceptible plants when diluted 1:90 (Meehan and Murphy, 1947). Extracts from oatmeal agar on which the fungus had been grown were toxic at a concentration of 0.5% (Litzenberger, 1949). Only a few isolates of the pathogen were tested by these workers; consequently, it is not known if the low toxicity of the toxin was the result of the culture medium used or the inability of the fungus to produce large quantities of the toxin.

Other workers tested several different media and found that high toxin production was obtained on a modified Fries medium (Luke and Wheeler, 1955). Toxicity of culture filtrates from the Fries medium was determined by a bioassay based on the inhibition of root growth of seedlings. This bioassay has been used to determine the relative amounts of toxin in culture filtrates. The quantity of toxin in 1 ml required to reduce root elongation 50% was designated as 1 milliunit. Levels of toxic activity were determined by assaying serial dilutions of toxic culture filtrates, plotting seedling root lengths against the various dilution factors, and interpolating from the curve obtained the dilution factor required to retard root growth 50% as compared to a distilled water control (Luke and Wheeler, 1955).

Differences in pathogenicity have been positively correlated with differences in toxin production. Toxin production is directly related to the growth rates of various isolates. Growth rates of different isolates, however, vary inversely with the pH of the culture medium, with peak toxin production occurring at the lowest pH values (3.0 to 3.5). When the pH of the culture medium increases to 6.0 or above, toxicity of the culture fluid drops sharply. The decrease in toxin remaining in culture fluids at pH values above 6.0 is thought to be due to toxin breakdown (Luke and Wheeler, 1955).

Only six toxins produced *in vitro* by phytopathogens have been shown to incite disease. Two explanations for this appear likely. First, pathogenesis mediated through toxigenic action of a phytopathogen may be the exceptional case. Second, toxin production may be common *in vivo* but not *in vitro*. This is not surprising because vast nutritional, biochemical, and biophysical differences occur between the *in vitro* and *in vivo* environments. Although it has been suggested that most disease-inducing toxins are produced *in vivo* (Dimond and Waggoner, 1953), proof to substantiate this idea has been difficult to obtain. In many instances, toxins produced in the suscept are labile compounds produced in micro quantities and are difficult to extract and identify. Since only a few toxins produced either *in vitro* or *in vivo* have been shown to cause disease, we are confronted with the question, is pathogenesis mediated through toxigenic action the exceptional case? The authors, as well as others, think not (Wheeler and Luke, 1963; Owens, 1969). Before this assumption can be seriously considered, some explanation for the fact that only a few toxins have been shown to incite disease should be offered.

In this connection, pathogens that produce phytopathogenic toxins usually exhibit a narrow host range, while many toxins produced in artificial cultures exhibit a wider host range than the pathogen. Therefore, it appears that most toxins responsible for inciting disease are not produced *in vitro*. The premise that most toxins are produced *in vivo* has not been supported by good evidence because it is difficult to distinguish between toxins produced by the pathogen and those produced by the host.

In many cases, the plant disease syndrome is conditioned by the physiology of the pathogen. Consequently, an assessment of the toxigenic potential of phytopathogenesis requires two approaches: (1) accumulate more knowledge concerning nutritional and physical requirements of the pathogen and from this information attempt to closely duplicate *in vitro* the biochemical and biophysical environment of the suscept, and (2) use cytochemical techniques and radioisotopes to detect microbial toxins in the suscept (Wheeler, 1953). It is desirable to produce phytopathogenic toxins *in vitro* because of the extreme difficulty of extracting them from infected plants.

B. Toxin Purification

Several attempts have been made to purify and characterize victorin. Early experiments revealed that the toxin is relatively stable at pH values below 4.0 and labile at pH values above 10.0 (Luke, 1954). Initial attempts to purify victorin showed that it was adsorbed from acidified culture fluid by activated charcoal. Fractions eluted from charcoal by acetone or butanol were further purified on silicic acid, celite, and alumina adsorption columns. The highest toxin activity was obtained from celite columns, but attempts to crystallize the eluted toxin were unsuccessful. Purified toxin from adsorption columns was soluble in methanol, ether, petroleum ether, and carbon tetrachloride (Luke, 1954; Luke and Wheeler, 1955).

Further efforts to isolate victorin revealed that chromatography of concentrated culture fluid on an acid alumina column resulted in the recovery of material toxic at 0.02 μg/ml (Pringle and Braun, 1957). Instability of the partially purified toxin made it difficult to further purify fractions from the alumina column. Additional purification on a starch column gave fractions that were active at 0.01 μg/ml. When fractions from the starch column were dried and precipitated with ethanol and acetone, a material was obtained that was toxic at 0.0002 μg/ml (Scheffer and Pringle, 1963).

Although freshly purified toxin is ninhydrin negative, two ninhydrin positive fractions are observed when toxin is treated with saturated sodium bicarbonate and chromatographed on paper strips (Pringle and Braun, 1958). One of these spots was reported to be a tricyclic secondary amine with an empirical formula of $C_{17}H_{29}NO$ (molecular weight 264.4). This compound was given the trivial name victoxinine. The other compound was reported to be a peptide containing five amino acids. A quantitative analysis of the peptide has not been reported. Further observations indicated that the tricyclic secondary amine contained one double bond close to its single nitrogen atom and that its single oxygen atom was in some ether linkage other than methoxyl (Braun and Pringle, 1959).

A yield of 14% victoxinine was obtained by sodium bicarbonate cleavage of the toxin molecule. From this it was concluded that the molecular weight of the intact toxin was less than 2000, the weight proposed for the toxin from molecular sieving data (Pringle and Scheffer, 1964). It was assumed that the peptide contains five amino acids (Pringle and Scheffer, 1964), but the 14% yield of victoxinine could be an indication that the peptide actually contains more than five amino acids. If the toxin which yielded 14% victoxinine was composed of victoxinine and peptide, 86% of the intact toxin would be peptide. If 86% of the toxin is peptide, then the peptide would contain about 15 amino acids. Since five amino acids were identified when the peptide was hydrolyzed, it is likely that some of these amino acids are repeated in the peptide. When it is assumed that the peptide contains fifteen amino acids instead of five, the molecular weight of the toxin, based on the average molecular weight of the amino acids, is about 1872. This weight is in agreement with molecular sieving data. However, the fraction from which the determinations were made could have contained contaminants, which would decrease the number of amino acids indicated by these calculations. We do not propose that these assumptions are valid. The chemical character of the molecule is un-known and must be determined experimentally, but it should be pointed out that the peptide may be larger than indicated in previous reports (Pringle and Braun, 1957; Pringle and Scheffer, 1964).

Because the intact toxin does not react with ninhydrin, it was suggested that linkage of the peptide and the secondary amine involved the amino groups of both entities (Pringle and Braun, 1958). This interpretation is questionable because a covalent bonding of the two amino groups of the peptide and victoxinine would be extremely labile, and under normal con-ditions the toxin is stable. Moreover, covalent bonding between nitrogen atoms is rarely produced in nature. The nitrogen atoms of victoxinine and the peptide may not share a common bond, but they could be linked by another system which is hydrolyzed by bicarbonate. Perhaps the peptide and victoxinine are linked by a bond between the carboxyl groups of the peptide and the nitrogen of the base, similar in nature to a peptide bond. The victorin molecule may be ninhydrin negative because of the length of the peptide, the cyclic nature of the peptide moiety, or chelation involving the alpha amino groups of the two moieties. These suggestions are not conclusive, but represent alternative explanations of the original data.

Victoxinine completely inhibited the growth of roots of both toxin-re-sistant and toxin-susceptible seedlings at a concentration of 75 μg/ml. Thus, victorin was shown to be about 7500 times more toxic than victoxi-nine (Pringle and Braun, 1958). It was suggested that victoxinine may be responsible for the toxicity of victorin, and that specificity may be a func-tion of the peptide moiety. The conclusion that victoxinine is the toxic moiety appears questionable because victorin is much more toxic than victoxinine. Moreover, victoxinine inhibits respiration of susceptible tis-sue, but victorin causes a sharp increase in respiration. It appears that the intact toxin molecule is needed for high toxic activity and specificity. Vic-toxinine may contribute to toxicity only as a component of the intact toxin molecule, and the low order of toxicity of victoxinine is probably due to its structure. One could speculate that the structure of victorin is inte-grally related to its mechanism of action. Dissociation of the peptide from victoxinine is probably sufficient to partially destroy the activity of the toxin. For example, dissociation could transform the toxin from a compound capable of chelating into one that can not chelate. One can visualize how the dissociation of victorin could lower its activity. The fact that one of the dissociation products of victorin, namely victoxinine, happens to be a nonspecific toxin is perhaps coincidental. It has not been definitely determined that victoxinine is the toxic portion of the victorin molecule. If victoxinine is the toxic moiety, information concerning its origin and function is needed.

No attempt has been made to determine the origin of victoxinine in cultures of *H. victoriae*. It has been suggested that victoxinine is a breakdown product of victorin (Pringle and Braun, 1958). Victoxinine may be a breakdown product of a contaminant occurring in the partially purified fraction from which victoxinine was identified. Toxicity of victorin at the time victoxinine was identified was stated to be 0.01 μg/ml (Pringle and Braun, 1957). However, a later report disclosed that victorin was 100-fold more toxic (Scheffer and Pringle, 1967). Thus, most of the fraction from which victoxinine was identified could have been some compound other than victorin.

Early experiments indicated that highly purified victorin was labile, yet the unpurified toxin is rather stable. Recent observations may resolve this conflict. That is, several lines of evidence suggest that two or more species of the toxin molecule may exist (Luke and Gracen, 1970). Highly active species appear to be labile, but less active species are stable. The highly toxic types do not commonly occur in raw culture fluids, but are observed after initial purification stages. This indicates that culture filtrates may contain a toxin inhibitor or purification procedures may result in changes in the structure of the toxin molecule. The latter appears more plausible. Although the origin of the highly active species is not known, it may be related to chelation or to a rearrangement of the peptide moiety.

Since the peptide is known to contain both aspartic and glutamic acids, the toxin molecule should have at least two potentially free carboxyl groups. If the base and peptide are linked by a nitrogen-nitrogen bond as has been proposed, there should be a third potentially free group in the carboxyl end of the peptide. If the carboxyl groups are not bound, they should dissociate at pH 3.0 to 4.0, resulting in two or three negative charges counteracted by a positive charge around the nitrogen-nitrogen bond. This arrangement would yield a negatively charged molecule. The molecule is positively charged up to pH values between 8.0 and 9.0, indicating that the carboxyls are not free to dissociate. Two possibilities are offered as to how the carboxyls are bound: (1) the peptide portion of the toxin is cyclic in nature or (2) the toxin forms a chelation complex in which the carboxyls are bound to a metal ion. If the peptide were bound to victoxinine across its terminal carboxyl group, it would be possible to produce a molecule with two nitrogen atoms and at least two carboxyl groups which could participate in a chelation complex. The nature of the bond between the peptide and the base is unknown. The chelation idea is speculation, but the possibility should be investigated.

C. CONCEPTS OF RESISTANCE AND SUSCEPTIBILITY

The mechanisms of resistance and susceptibility are not known, but some progress has been made toward elucidating their nature. Studies have resulted in several working hypotheses that are being tested. Although specific names have not been assigned to these hypotheses, they will be referred to as: (1) the exclusion hypothesis, (2) the inactivation hypothesis, (3) the plasma membrane hypothesis, (4) the adsorption site hypothesis, and (5) the self-repair hypothesis.

1. THE EXCLUSION HYPOTHESIS

The exclusion hypothesis represents the first attempt to explain the mechanism of resistance to victorin. It is based on the premise that toxin enters cells of susceptible but not of resistant plants (Braun and Pringle, 1959). Although this idea is not clearly stated, it is implied that victorin enters the cytoplasm of susceptible cells, but is excluded from the cytoplasm of resistant cells.

The exclusion hypothesis evolved from the observation that the toxin is comprised of two components, a tricyclic secondary amine and a peptide. The secondary amine (victoxinine) is toxic to both resistant and susceptible plants, indicating that victoxinine may be the toxic moiety and that the peptide conveys the specificity to the toxin molecule (Braun and Pringle, 1959). This suggests that the intact toxin molecule does not enter the cytoplasm of resistant cells. The idea became less attractive when it was reported that victoxinine may not be the toxic portion of the victorin molecule (Scheffer and Pringle, 1964). Nevertheless, the concept that toxin does not enter the cytoplasm of resistant cells needs to be tested. The exclusion concept should be tested because it is difficult to demonstrate physiological effects on resistant tissue with reasonable concentrations (0.1-1 unit/ml) of the toxin. In fact, it is virtually impossible to cause wilting of leaves of resistant cultivars with victorin. Moreover, toxin does not cause increased respiration of resistant tissue at concentrations that drastically increase respiration of susceptible tissue. Thus, it appears that toxin may not enter the cytoplasm of resistant tissue. Two alternate explanations (inactivation, lack of receptor sites) for this lack of physiological reaction of resistant tissue to victorin have a degree of credibility and will be discussed later (Section II, C, 2 and 4). Alternative explanations to the exclusion hypothesis do not render it invalid. It is an attractive hypothesis needing further investigation.

A direct approach to the detection of victorin in oat tissue has been considered by most who have worked with this toxin. The technique, which was first suggested in 1953 (Wheeler, 1953), consisted of tagging the toxin with a radioactive label and determining its absorption by resistant and susceptible cells. Experiments of this kind have not been successful because of the instability of purified toxin and the difficulty in obtaining sufficient quantities of pure toxin. A carbon-14 label results in a diffuse autoradiograph; therefore, a solution awaits improved methods of toxin purification, a stable form of the toxin, and a tritium labeling system. Conjugation of victorin with a fluorescent compound or with electron-dense ferritin might permit detection of the toxin in oat tissue. These types of experiment are needed before the site of action of victorin can be definitely established.

The exclusion hypothesis is an opinion not supported by data, but it is based partly on the observation that resistant tissue is virtually insensitive to the toxin. An alternative explanation is that victorin does invade the protoplast of resistant cells but is rapidly inactivated.

2. THE INACTIVATION HYPOTHESIS

This hypothesis represents another attempt to explain the mechanism of resistance. It states that victorin is inactivated by intact resistant tissue (Romanko, 1959). When victorin was added to homogenated tissue, toxin was recovered from both resistant and susceptible tissue. It was concluded that inactivation may not result from a simple one-step reaction, but was the result of an interaction occurring in intact cells (Romanko, 1959). In early experiments, small quantities of toxin were recovered from susceptible tissue (Romanko, 1959). A recent report indicating that susceptible tissue either firmly binds or inactivates victorin until cells are severely damaged may explain why small quantities of toxin were recovered (Wheeler, 1969). These results (Wheeler, 1969), however, are contrary to previous observations which indicated that maximum recovery of toxin occurred when exposure times were relatively short (Romanko, 1959).

The inactivation hypothesis was criticized by another group of workers (Scheffer and Pringle, 1964), who after many attempts to repeat previous results (Romanko, 1959), were unable to extract detectable quantities of toxin from either resistant or susceptible tissue. This report (Scheffer and Pringle, 1964) is contrary to observations of several independent workers (Litzenberger, 1949; Gracen, 1970; Wheeler, 1969). In fact, the first attempt to extract victorin from susceptible oat plants indicated that a specific toxin was recovered (Litzenberger, 1949). Unfortunately, extracts from resistant plants inoculated with *H. victoriae* were not used in these experiments. More recently, victorin was sealed in coleoptile sheaths of resistant and susceptible oat cultivars (Wheeler, 1969). When the toxin recovered from coleoptile sheaths was measured, resistant coleoptile segments contained 30-fold less toxin than susceptible coleoptiles and 80-fold less than that recovered from toxin sealed in glass tubes. It was concluded that victorin is inactivated at about the same rate by both susceptible and resistant tissue until susceptible cells are severely damaged. Toxin inactivation then ceases in susceptible tissue, but continues in resistant tissue (Wheeler, 1969). Recent results (Gracen, 1970) disagree with others (Scheffer and Pringle, 1964; Romanko, 1959). When root tissues were allowed to take up victorin and were then washed in a desorption solution for 30 minutes, toxin was easily extracted from both resistant and susceptible oat cultivars. Increasing the length of the desorption period resulted in decreased amounts of toxin recovered (Gracen, 1970). Thus, it appears that most toxin extracted from roots came from intercellular spaces. At any rate, extraction of victorin from either leaf or root tissues gives no indication that the toxin enters the cytoplasm of resistant or susceptible cells.

3. THE PLASMA MEMBRANE HYPOTHESIS

The plasma membrane hypothesis, which evolved from several lines of experimental evidence, represents the first attempt to explain the reaction of susceptible tissue to victorin. Originally, the hypothesis stated that victorin disrupts membrane systems that control water balance (Luke *et al.*, 1966). Additional evidence (Luke *et al.*, 1969) supports the view that victorin causes a malfunction of the plasmalemma. In its current form the plasma membrane hypothesis contends that victorin disrupts the physiochemical function of the plasma membrane, resulting in the loss of intracellular components and cell turgor (Luke *et al.*, 1969).

The current hypothesis was tested by investigating the effects of the toxin on membrane systems. Results revealed that the first effects of victorin are exerted on either the outer surface of the plasma membrane, the inner surface of the cell wall, or the gelatinous-like material occurring between the two areas. The earliest effect observed in electron micrographs was a partial separation of the plasma membrane from the cell wall. Blister-like structures were formed because the plasma membrane retained its physical integrity. The blister-like structures are illustrated in Fig. 1. The membrane appeared swollen. Further observations indicated that this was not the case; instead, a dark-staining material (electron

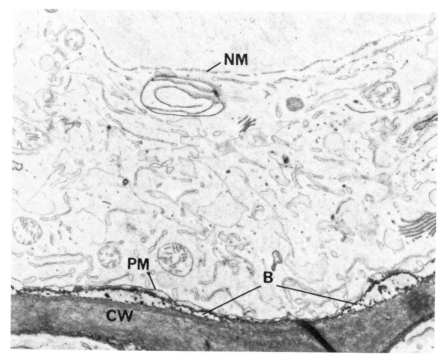

Fig. 1. A portion of a root cell treated with victorin showing blister formation (× 25,000); blister-like body (b), cell wall (cw), plasma membrane (pm), and nuclear membrane (nm). From Luke *et al.* (1966).

dense) located between the cell wall and the plasma membrane gave the swollen appearance because it adhered to the outer surface of the membrane. It was inferred that the dark-staining material did not originate in the cytoplasm, but was a reaction product between victorin and the cell wall or between victorin and the outer surface of the plasma membrane. This led to the conclusion that the separation of the plasma membrane from the cell wall disrupts the osmotic properties of the membrane, which in turn results in the loss of cellular components and cell turgor. Electron micrographs did not reveal any effect of toxin on the tonoplast of susceptible cells, nor did toxin cause discernable effects on membrane systems of resistant tissue (Luke *et al.*, 1966).

These results were soon confirmed and extended by others who indicated that the blister-like structures originate in the cell wall (Hanchey *et al.*, 1968). The origin of lesions within the cell wall suggests that victorin treatment results in an activation of enzymes associated with the cell wall.

Further investigation (Hanchey and Wheeler, 1969) led to the conclusion that the plasmalemma is either more resistant to toxin disruption than other membranes or more capable of self-repair. However, alternative explanations appear plausible: (1) victorin may not cause physical damage to the plasma membrane, or (2) physical damage to the membrane may be below the resolution level of electron micrographs made to date.

The plasma membrane hypothesis has gained additional support from evidence that victorin causes bursting of susceptible protoplasts free of cell walls, but causes little damage to resistant protoplasts (Samaddar and Scheffer, 1968). These observations led to the suggestion "that toxin combines with or affects an unknown component in the susceptible cell, resulting in disorganization of the surface" (Samaddar and Scheffer, 1968). Although the above statement is not clear, the term "cell surface" may refer to the plasma membrane.

Results from these studies (Luke *et al.*, 1966; Hanchey *et al.*, 1968; Samaddar and Scheffer, 1968) preclude any firm conclusion concerning the direct effect of toxin on the plasma membrane. A different approach, however, revealed that when oat leaf tissue was allowed to absorb and phosphorylate mannose-^{14}C, toxin treatment causes phosphorylated hexoses to leak from susceptible tissue but does not affect the permeability of resistant cells. The rate of loss of phosphorylated hexoses is directly proportional to the concentration of victorin (Luke *et al.*, 1969). These results show that toxin causes a malfunction of the plasma membrane. This conclusion was reached because phosphorylated hexoses will not pass through plasma membranes that function normally (Humphreys and Garrard, 1964). One hour of victorin treatment causes susceptible tissue to leak about 50% of the radioactive mannose absorbed. The toxin thus causes abrupt and massive damage to the plasma membrane (Luke *et al.*, 1969).

Victorin affects the uptake and retention of rubidium in susceptible but not in resistant roots (Gracen, 1970). Such effects can be detected in 2 minutes, indicating that victorin indeed exerts a rapid effect on membrane permeability. The exact site(s) of the rubidium uptake mechanism are not known (Epstein, 1966), but the mechanism for uptake of this ion at the concentration used (0.1 mM) is believed to be located in the plasma membrane. These observations lend additional support to the idea that the plasma membrane is drastically affected by victorin.

Five lines of evidence (electrolyte leakage, ultrastructural changes, protoplast bursting, leakage of phosphorylated hexoses, and retardation of uptake) imply that victorin causes a malfunction of the plasma membrane. The mechanism of action of this malfunction is not known, nor is it known if the effect on the plasma membrane is a primary or a secondary one.

Nevertheless, when the mechanism of action of victorin is uncovered, it would appear that the plasma membrane will be involved.

4. The Adsorption Site Hypothesis

In its original form, this hypothesis states that "toxin uptake by susceptible tissue is considered to involve adsorption to a receptor that is lacking in resistant tissue" (Scheffer and Pringle, 1964). The idea was later modified to state "that resistant tissue simply lacks a toxin receptor, or toxin-sensitive point" (Pringle and Scheffer, 1964). To account for intermediate reactions to the toxin (Luke and Wheeler, 1964) it was stated that "plants with intermediate types of resistance could have fewer receptors, or sites with less affinity for the toxin" (Scheffer and Pringle, 1967).

The receptor site hypothesis is not supported by experimental data, but it is a projection of the observation that resistant tissue is virtually insensitive to the toxin. If one considers the report that toxin uptake appears to be a simple one-step process in which the toxin binds to a receptor (Scheffer and Pringle, 1964), then the adsorption site hypothesis is simply an amendment to the exclusion hypothesis.

The hypothesis is based on several lines of circumstantial evidence. The first observations indicate that toxin uptake is not affected by temperature, osmotic pressure, or metabolic inhibitors; suggesting that cellular energy is not required for adsorption. This view is in conflict with the suggestion that a specific transport enzyme may occur on the surface of susceptible cells but not on the surface of resistant cells (Kalyanasundaram and Charudattan, 1966).

Recent evidence suggesting that victorin is inactivated by resistant tissue (Wheeler, 1969), plus the observation that victorin causes physiological changes in resistant tissue (Wheeler and Doupnik, 1969) is contrary to the receptor site hypothesis. If these observations are correct, portions of the receptor site hypothesis concerning resistance to the toxin must be rejected or modified. If modified, it must be assumed that resistant plants contain abundant receptor sites of very low binding potential or that they contain binding sites that are rare and relatively inaccessible (Wheeler, 1969). The proposed modification hinges on the validity of the observation that victorin truly triggers the same physiological disruptions in both resistant and susceptible tissue.

However, the portion of the receptor site hypothesis indicating that uptake by susceptible tissue involves a specific receptor is acceptable. This is neither a new nor a novel idea, since it has been concluded for some time that a compound must have a binding site to be active. Identification of the receptor is essential before the hypothesis can be considered of significant value. Finally, several lines of evidence indicate that the receptor site may be in the plasma membrane (Luke *et al.*, 1966; Hanchey *et al.*, 1968; Samaddar and Scheffer, 1968), but its exact nature is not known and should be determined.

5. THE INDUCED SELF-REPAIR HYPOTHESIS

The induced self-repair hypothesis is based on the premise that victorin incites similar physiological effects in both susceptible and resistant tissue. It states that "in resistant tissue these initial physiological changes result in activation of a general defense mechanism in the plant which repairs the damage" (Wheeler and Hanchey, 1968; Wheeler, 1969). This idea was based on the observation that toxin at high concentrations (100 units/ml) increases respiration, reduces transpiration, and causes electrolyte loss from resistant tissue (Wheeler and Doupnik, 1969). Thus, it was concluded that victorin induces physiological disruptions in resistant tissue and that this damage is repaired before symptoms that can be seen occur. Although these physiological symptoms are common to susceptible tissue treated with toxin, they also occur in many host–pathogen interactions. Moreover, it is not known if physiological disturbances that occur in resistant tissue are definitely caused by the action of victorin.

Culture filtrates with average activity (500 units/ml) do not incite physiological symptoms in resistant oat cultivars when diluted 20-fold, indicating that victorin preparations with high activity are required before resistant tissue is affected (Wheeler and Doupnik, 1969). Recent evidence may explain the requirement for high activity. For example, preliminary results imply that lyophilized culture filtrates extracted with methanol contain two or more species of the toxin molecule (Luke and Gracen, 1970). One of these species appears to be highly active and relatively labile. Consequently, it may be that only the highly active species of the toxin molecule are capable of inducing physiological symptoms in resistant tissue.

Three additional lines of evidence support the induced self-repair hypothesis. The first indicates that senescing leaves that are presumably less capable of self-repair are more sensitive to victorin than are young leaves (Wheeler, 1969). This observation may support the induced self-repair hypothesis if both old and young leaves take up the same quantity of toxin. No indication of the amount of toxin absorbed by old and young tissue was given. The second line of evidence indicates that puromycin increases sensitivity of resistant oat leaves to victorin, suggesting that self-repair depends upon protein synthesis. The third line of evidence shows that resistant tissue treated with high concentrations of victorin exhibits the same changes in peroxidase isoenzymes as does susceptible tissue (Novacky and Wheeler, 1969).

The induced self-repair hypothesis appears to be an attractive idea in its formative stages. Several lines of direct evidence are needed before it can be placed in final form. For example, the mechanism or site of action should be supported with conclusive evidence that observed physiological disturbances in resistant tissue are indeed caused by victorin.

D. Speculation on Mode of Action

When the hypothetical explanations concerning the nature of reaction to victorin are considered, it becomes apparent that all five have some merit. While a discussion of the interrelationships among these hypotheses would be of little value, it appears pertinent to mention that the adsorption site, the self-repair, and the plasma membrane hypotheses are closely interrelated. In the final analysis, major evidence indicates that the plasma membrane plays a vital role in the disease induced by victorin. Thus, speculation on the mode of action of the toxin can be divided into two components, one based on the assumption that the plasma membrane is the primary site of action of victorin and the other maintaining that some other cellular component is the primary site of action.

The first speculation assumes that victorin directly affects the plasma membrane, causing a disruption of semipermeability. When this line of thought is pursued, two options become obvious. The plasma membranes of resistant and susceptible cells are either functionally similar or functionally dissimilar. If they are similar, then toxin could induce the same type of damage to membranes of both types of cells. It follows that resistant cells have the capacity to either inactivate the toxin or repair the damage to the membrane, and that susceptible cells do not. Conversely, if the plasma membranes of resistant and susceptible cells are functionally different, then membranes of susceptible cells are sensitive to victorin, but those of resistant cells are not.

Although the mechanism by which victorin causes loss of semipermeability is not known, several lines of evidence allow us to consider this problem. Calcium, strontium, and uranyl ions prevent victorin from disrupting the semipermeability of susceptible tissue, but other divalent cations apparently are not involved in this reaction. Ultrastructural changes in susceptible tissue treated with victorin are very similar to those observed in calcium-deficient barley tissue. Additional evidence indicates that susceptible cultivars are more sensitive to calcium deficiency than resistant cultivars. Thus, it is tempting to theorize that victorin causes susceptible cells to lose their ability to maintain an adequate calcium balance in the plasma membrane where it is vitally needed. In other words, victorin may disrupt the calcium binding sites. Because calcium is said to be essential for the physiological integrity of the plasma membrane, any disruption of the calcium function in the membrane could result in an abrupt loss of semipermeability. Loss of semipermeability is indeed the most striking physiological change brought about by victorin. The assumption is that toxin disrupts the function of calcium in the plasma membrane of susceptible cells but has no effect on its function in resistant cells. No indication of the mechanism involved is given. Recent observations may further support this idea. That is, at physiological pH values (6-8) victorin carries a positive charge (Luke and Gracen, 1970). Thus victorin may bind to negative sites (phosphoryl or carboxyl groups) on the membrane surface. The fact that calcium and uranyl ions suppress the activity of victorin lends support to this view. Moreover, uranyl not only has a higher affinity for negatively charged sites than does calcium, but it is also more effective than calcium in suppressing the effects of victorin. These two cations may bind to negative charges on the membrane and thus prevent adsorption of victorin. The structure of the plasma membrane has not been determined; therefore, speculation on the relationship between structure and function would be of little value.

Another possibility is that some organelle other than the plasma membrane is the primary site of action of victorin. This idea is supported by the observation that puromycin increases the sensitivity of resistant tissue to victorin, thus suggesting that new or increased protein synthesis is required to repair damage caused by toxin treatment.

While the above observation is only preliminary, it is interesting to speculate on the self-repair concept. To do so, it is necessary to assume that the plasma membrane exists in a dynamic state of physiological change. If maintenance of the membrane is dependent upon protein synthesis, we may theorize that toxin affects transfer RNA, which partially prevents the repair of the plasma membrane. Obviously, such a speculation is but one part of a complex interaction. Puromycin does not transpose resistant tissue to the completely susceptible state. If self-repair of membranes in resistant tissue is dependent upon protein synthesis, then puromycin would render resistant tissue as sensitive to victorin as susceptible tissue. It does not. This suggests that puromycin is not effectively absorbed by oat tissue, or that metabolic forces other than protein synthesis are also involved in the self-repair phenomenon. Moreover, the inhibition of the synthesis of specific proteins (i.e., lipoproteins) would have to be positively correlated with increased sensitivity of resistant tissue for this idea to be valid.

III. Helminthosporium sorokinianum Toxins

Reports that *Helminthosporium sorokinianum* Sacc. in Sorok. (*H. sativum* Pam., King and Bakke) initiates disease by toxigenic action are based on the observation that sterilized culture filtrates predispose barley seedlings to infection by this pathogen (Ludwig *et al.*, 1956; Ludwig, 1957). Other reports indicate that different isolates of *H. sativum* produce a metabolite with a low order of toxicity (Earhart, 1959). Treatment of soil with sterilized cultures of *H. sativum* not only predisposes barley seedlings to invasion by other fungi, but also predisposes wheat and oats to infection (Ludwig, 1957). Since this culture of *H. sativum* does not parasitize oats under field conditions, the observed result may have been due to general injury and not to a specific disease-inducing agent. It also appears unlikely that enough fungal growth would occur in a natural soil environment to produce sufficient quantities of the compound(s) toxic to higher plants. Nevertheless, it was concluded that a toxin(s) produced by *H. sativum* predisposes barley seedlings to invasion by inducing premature senescence (Ludwig, 1957).

Although the validity of this conclusion has not been tested, considerable information has been compiled concerning the identification and mode of action of two compounds isolated from this fungus. Both compounds are structurally similar. One has been considered a toxin (helminthosporal) and the other a growth regulator (helminthosporol).

A. HELMINTHOSPORAL

Helminthosporal (H-al) is a nonspecific toxin produced by *H. sativum*. This toxin has a low order of activity and its role as a disease-inducing agent is dubious. At best, if involved at all, it perhaps plays a minor role in the disease inducing property of the parasite. Nonetheless, considerable information concerning its purification and structure is available.

Initial attempts to identify the toxin showed that it is a sesquinterpenoid with an empirical formula of $C_{15}H_{22}O_2$. The structure was shown to be a bicyclic dialdehyde; one aldehyde being saturated and the other unsaturated (De Mayo *et al.*, 1961; De Mayo and Spencer, 1962). Additional observations concerning the stereochemistry and total synthesis of H-al showed that this toxin is indeed a bicyclic dialdehyde (De Mayo *et al.*, 1963; Corey and Nozoe, 1965).

It was suggested that while H-al may not be the primary disease-inducing agent, it is essential for disease development (Spencer *et al.*, 1966). Further observations on the mode of action of H-al showed that the toxin inhibits respiration in barley and wheat root tissue (Taniguchi and White, 1967). Results also indicate that the site of action of the toxin is between flavoprotein dehydrogenase(s) and cytochrome c. In order to determine the precise site of action of H-al, low temperature spectra of mitochondria oxidizing succinate with and without toxin were taken. In the presence of the toxin, absorption maxima were greatly diminished, demonstrating that H-al prevents the reduction of cytochromes b, c, and a + a_3. Moreover, H-al does not act strictly as an uncoupler or as an energy transfer inhibitor, but it has a dual effect on both electron transfer and oxidative phosphorylation.

The site and mode of action of H-al was determined using isolated mitochondria from sweet potato and mouse liver. In fact, mouse liver mitochondria are more sensitive to the toxin than sweet potato mitochondria; *H. sativum* does not incite disease symptoms in either sweet potato or mice. The data show, however, that H-al has a wide spectrum of toxicity when energy-transfer systems are directly exposed to the toxin. The question then arises as to why the pathogen exhibits a narrow host range, yet one of its metabolic products has a very wide spectrum of toxicity. While the answer to this question remains unknown, one may speculate that toxin is not produced in effective quantities in nonhosts, or it does not enter the cytoplasm of nonhost organisms.

B. HELMINTHOSPOROL

Although helminthosporol (H-ol) was isolated from culture filtrates of *H. sativum* and has a molecular structure similar to helminthosporal (H-al), it has generally been referred to as a growth regulator. This is not surprising, because the gibberellin and cytokinin-type growth regulators are toxic at certain concentrations. We will not discuss H-ol in detail, but instead will use it to illustrate the close relationship between growth regulators and toxins. In some cases, the differences in physiological function of growth regulators and toxins are very subtle. Thus, a distinction between growth regulators and toxins will not be attempted.

H-ol, which was first isolated from a Japanese strain of *H. sativum,* proved to be very similar to H-al (Tamura *et al.*, 1963). The latter is a bicyclic dialdehyde and H-ol is a bicyclic aldehyde alcohol. Even though the molecular structure of these compounds is very similar, the Japanese strain of *H. sativum* does not produce H-al. Attempts to convert H-ol to H-al were unsuccessful (Tamura *et al.*, 1963).

The growth-promoting properties of H-ol are more similar to gibberellin than to cytokinins or auxins (Mori *et al.*, 1966). That is, H-ol increases the elongation of leaf sheaths of certain grasses and enhances the germination of lettuce and tobacco seed (Hashimoto and Tamura, 1967).

However, gibberellin is about 100-fold more active than H-ol (Katsumi *et al.*, 1967), and H-ol does not affect as many plant species as gibberellin (Hashimoto *et al.*, 1967). It was suggested that H-ol does not affect as many plant species as gibberellin because the active groups of the two compounds are different (Hashimoto and Tamura, 1967).

The maximum growth promoting activity of H-ol on shoots of rice occurred at about 50 ppm, but this concentration inhibited root growth of rice seedlings. However, H-ol had no effect on shoots of wheat seedlings, but inhibited root growth of wheat seedlings at 30 ppm (Tamura *et al.*, 1963). Thus, it appears that when rice shoots are used in the assay system, H-ol is a growth regulator of low activity, but if it is used to treat wheat roots it is toxigenic. In this case, it should be noted that the difference in growth promotion and growth retardation is not conditioned by the concentration of H-ol used in the experiments. Thus, the distinction between growth regulators and toxins is indeed a narrow one. In some instances, this distinction depends upon the concentration of toxin or the tissue used in the assay.

Although the differences between growth regulators and toxins are sometimes rather subtle, some phytopathogens are thought to produce growth regulators which in turn initiate disease. The authors believe that further investigation will reveal that disruption of the growth regulatory system of the host by the pathogen will add additional examples to the present list. A disruption of the growth regulatory system affects many physiological functions of the host. This, coupled with the fact that higher plants maintain a delicate balance of growth regulators in different tissues, leads to the supposition that only a small quantity of the growth regulator produced by the pathogen would be required to cause severe physiological malfunction of the host. Therefore, a phytopathogen could easily disrupt the physiology of the host by causing an imbalance of the growth regulatory systems. After all, disruption of the physiology of the host is the primary function of the pathogen in the initiation of disease.

IV. *Helminthosporium oryzae* Toxins

Helminthosporium oryzae Breda de Haan, which causes a leafspot of rice, was first reported to initiate disease symptoms through toxigenic action in 1939 (Yoshii, 1939). The report was based on the observation that host cells were killed prior to hyphal invasion. Since that time, six species of *Helminthosporium* have been reported to produce a toxin or toxins with a similar molecular structure (Oku, 1968a).

The toxin, thought to play a role in the disease syndrome, was isolated and named cochliobolin (Orsenigo, 1957). This compound was also isolated from *H. oryzae* by another group of workers and named ophiobolin (Nakamura and Ishibashi, 1958). More recently, both groups collaborated in characterizing this group of toxins, for which they proposed the trivial names ophiobolin A, B, C, and D (Tsuda *et al.*; Canonica *et al.*, 1967). Since the structures and functions of ophiobolin A, B, C, and D are very similar, we will refer to the toxin(s) produced by *H. oryzae* as ophiobolin.

An early report concerning the molecular structure of ophiobolin indicated that it is a tetracyclic terpenoid containing 25 carbon atoms (Nozoe *et al.*, 1965). Since C_{25} terpenoids are rarely found in nature (Robinson, 1963), it may be that the early work involved a sterol. A later report (Tsuda *et al.*; Canonica *et al.*, 1967) suggested that ophiobolin is a tricyclic terpenoid.

Although much is known about the structure of ophiobolin, its role in disease development is uncertain. Nevertheless, some information concerning the mode of action of this toxin is available. At the onset of infection, it appears that the pathogen produces two or more toxins (zizanin A and B) that are structurally very similar to ophiobolin (Oku, 1968a). These toxins kill host cells prior to hyphal development, suggesting that analogs of ophiobolin (zizanin) are responsible for the penetration of the host by the pathogen. Following the early phases of infection, ophiobolin is produced in the host (Oku, 1967) and is thought to be responsible for the malfunction of polyphenol metabolism of the host.

Ophiobolin was described as a vivotoxin because a compound showing a spectral absorption identical to that of ophiobolin was found in fractions extracted from infected leaves (Oku, 1967). Obviously, the validity of this circumstantial evidence depends largely upon the purity of extracts from inoculated plants.

One of the hypothetical explanations for the toxigenic action of the fungus indicates that the fungus invades host cells and produces ophiobolin. Toxigenic injury stimulates the host to produce various polyphenols, and these in turn are oxidized to quinones by fungal polyphenol oxidase. The quinones which inhibit additional fungal invasion of the host are polymerized. These polymers initiate small necrotic brown lesions which cause the characteristic symptoms of the disease (Oku, 1964).

Physical injury or environmental stress stimulate polyphenol synthesis in many plants. It does not seem reasonable that the pathogen would trigger a system (polyphenol oxidase) resulting in the synthesis of compounds (quinones) that would inhibit its own growth. The polyphenol oxidase produced by *H. oryzae* may not play a key role in disease development. Instead, it may be involved in the development of symptoms which occur rather late in the disease syndrome. Assuming that polyphenol oxidase of the pathogen is not a key to disease development, the question arises as to the mode of action of the fungus and its toxin, ophiobolin. The answer is not clear, but two assumptions are plausible. One of these was presented in a recent paper indicating that the activity of phloroglucinol oxidase was stimulated either by infection with *H. oryzae* or by ophiobolin treatment. It was assumed that ophiobolin depresses regulatory factors maintaining the normal enzyme synthesis in healthy plants (Oku, 1968b). The second assumption concerns the possibility that ophiobolin is not the toxin that triggers a series of physiological disturbances culminating in the symptoms observed on infected plants. Moreover, ophiobolin is not specific to rice and appears to be involved in a symptom occurring late in disease development. The second assumption does not imply that ophiobolin is not involved in some way in disease development. It suggests that ophiobolin may not be the specific toxin which triggers the initial malfunction culminating in the disease syndrome. Although the toxicity of ophiobolin is relatively low (10 ppm), it occurs in high concentrations (100 μg/gm fresh weight) in infected plants (Oku, 1968b). Concentrations of this order must have some effect on the plant.

V. *Helminthosporium carbonum* Toxins

When *Helminthosporium carbonum* Ullstrup (Ullstrup, 1944) is cultured on a modified Fries medium, culture filtrates are found to contain a toxic substance (Pringle and Scheffer, 1967). The *H. carbonum* toxin inhibits root growth of susceptible corn seedlings at 1:400 dilutions; however, 1:50 dilutions are required to completely inhibit growth. Purified toxin kills roots of susceptible seedlings at concentrations of 1 μg/ml (Pringle and Scheffer, 1967). Roots of resistant corn cultivars or roots of nonhost plants are not inhibited by high toxin concentrations (1:50 dilutions) (Scheffer and Ullstrup, 1965). The toxin and the pathogen exhibit the same host range; thus, the *H. carbonum* toxin is shown to be specific.

The *H. carbonum* toxin, which was isolated by chloroform extraction and countercurrent distribution, has been crystallized from ethyl ether. It is soluble in water, lower alcohols, acetone, and chloroform; insoluble in petroleum ether, diethyl ether, and carbon tetrachloride (Pringle and Scheffer, 1967). The toxin is inactivated under mild alkaline conditions and does not react with ninhydrin before acid hydrolysis. However, it yields ninhydrin positive products upon hydrolysis, indicating a polypeptide component (Pringle and Scheffer, 1967).

A secondary toxin produced by *H. carbonum* was isolated from culture filtrates by chloroform extraction. The secondary toxin, called carbtoxinine, has been crystallized (Pringle and Scheffer, 1967). It inhibits the root growth of both resistant and susceptible corn cultivars at a concentration of 25 μg/ml. Carbtoxinine is very soluble in water and methanol; less soluble in ethanol, butanol, and acetone; and insoluble in chloroform, ether, and benzene (Pringle and Scheffer, 1967). Carbtoxinine yields amino acids upon hydrolysis, indicating a polypeptide component.

Carbtoxinine, although not detected in raw culture filtrates, was isolated from partly deproteinized filtrates extracted with chloroform. It has been suggested that the failure to detect carbtoxinine in raw filtrates is either because of its low order of toxicity in relation to the primary toxin or because the concentration of carbtoxinine in culture filtrates is not great enough to permit its detection. The former explanation appears contrary to the fact that carbtoxinine inhibits growth of resistant roots at a concentration of 25 μg/ml, while the *H. carbonum* toxin inhibits the growth of resistant roots to the same extent at 50 μg/ml. Thus, carbtoxinine should be detected in resistant tissues at lower concentration than that required to detect *H. carbonum* toxin. Another explanation may be that carbtoxinine does not exist in substantial quantities in raw filtrates, but may be produced by the breakdown of some component of culture filtrates. This possibility should be investigated.

Helminthosporium carbonum toxin is thought to be more stable than victorin, but less active on a dry weight basis. Also, *H. carbonum* toxin appears to be produced in larger quantities in culture filtrates than victorin. The above comparisons are based on purified toxin solutions and do not necessarily apply to the toxins in crude culture filtrates. It is difficult to compare relative amounts of the two toxins because they differ in stability and toxicity.

The perfect stage of *H. carbonum* (*Cochiobolus carbonum* Nelson) has been produced by pairing compatible conidial isolates in culture (Nelson, 1959). Conidial cultures of *H. victoriae* and *H. carbonum* are sexually compatible (Nelson and Kline, 1963). Progeny from crosses between the two species segregate 1:1:1:1 for pathogenicity to oats, to corn, to both, and to neither (Scheffer *et al.*, 1967). Thus, a different single gene pair controls toxin production in *C. carbonum* and *C. victoriae*. Crosses of different isolates of *C. carbonum* indicate that the relative level of toxin produced is controlled quantitatively (Scheffer *et al.*, 1967). Race 1 and race 2 of *C. carbonum*, although morphologically identical and compatible, differ in pathogenicity by a single gene (Nelson and Ullstrup, 1961).

Until recently, little has been reported about the biochemical and physiological effects of *H. carbonum* toxin. *Helminthosporium carbonum* infection reduces the level of malic acid in susceptible but not in resistant plants and causes a 40% reduction in $^{14}CO_2$ fixation in infected susceptible tissue in the dark (Malca *et al.*, 1964; Malca and Zscheile, 1963, 1964). These results appear to disagree with a recent report indicating that the *H. carbonum* toxin causes an increase of carbon dioxide fixation in the dark (Kuo and Scheffer, 1968). It is difficult to evaluate this disagreement because one case involved the use of infected plants and the other involved toxin treated tissue.

Additional observations indicate that *H. carbonum* infection and toxin do not stimulate oxygen uptake. Neither the toxin nor the pathogen inhibits incorporation of ^{14}C-labeled amino acids or uridine, nor do they immediately affect the permeability of susceptible corn cultivars. Sulfite does not suppress the action of *H. carbonum* toxin (Kuo and Scheaffer, 1967). All of the reactions listed above are stimulated by victorin, but are not affected by *H. carbonum* toxin. Thus, it appears that physiological disruptions incited by victorin and *H. carbonum* toxin are widely different. These differences are unexpected because the toxins are thought to be chemically similar.

The uptake of *H. carbonum* toxin is temperature sensitive and is dependent on toxin concentration and exposure time (Kuo and Scheffer, 1968). Toxin apparently is not taken up under anaerobic conditions and uptake is decreased by 2,4-dinitrophenol, sodium azide, and potassium cyanide. The uptake of *H. carbonum* toxin seems to be an energy requiring process.

Although the *H. carbonum* toxin is not extremely toxic, it is stable and should be obtainable in adequate quantities in purified form. Therefore, further study on the molecular structure and site of action of the toxin is needed to elucidate the role of the toxin in disease development.

VI. Other *Helminthosporium* Toxins

In addition to the toxins already mentioned, two other species of *Helminthosporium* have been reported to produce toxins involved in plant disease. Because only a few reports about them have been published, toxins produced by these two species have been placed under one heading. This does not imply that these toxins do not play major roles in the induction of disease. In fact, both appear to be specific; thus it is likely that additional work may show that they are vitally involved in the induction of plant disease.

A. *Helminthosporium maydis* TOXINS

Helminthosporium maydis Nisik. and Miyake, which causes a leaf blight on corn, has been shown to be pathogenic to a wide variety of gramineous species (Nelson and Kline, 1966). The blight is characterized by a diffuse water-soaking and chlorosis of leaves, followed rapidly by necrosis (Orillo, 1952). Some isolates of *H. maydis* induce blight on a single cultivar, while others are capable of producing blight on several hosts (Nelson and Kline, 1968).

Two groups of investigators working with different isolates of *H. maydis* reported that the fungus produces a toxin of low activity (Orsenigo and Sina, 1961; Quimio and Quimio, 1966). When leaves were soaked in culture filtrates, it was observed that nonhost plants were affected by the culture fluid. Furthermore, wilting symptoms produced by the toxic filtrate differed from leaf blight symptoms caused by the pathogen. Thus, early reports indicated that the toxin produced by *H. maydis* is nonspecific and does not produce the symptoms caused by the pathogen.

A recent report, however, presents evidence that a specific toxin is produced by certain isolates of *H. maydis* (Smedegar-Petersen and Nelson, 1969). Comparisons are not available; thus, it is not known if conflicting results obtained by the different investigators were the result of cultural methods used, or because of metabolic differences of the isolates. The latter appears more likely, because a modified Fries medium was used by two groups who obtained conflicting results.

An assay system based on root inhibition was found to be unsatisfactory for detecting low quantities of toxin; therefore, another system was devised (Smedegar-Petersen and Nelson, 1969). The new assay technique, which consists of floating leaf tissue on diluted culture filtrate, permits the detection of low quantities of toxin. Moreover, when the leaf assay system is used, symptoms similar to those incited by the pathogen are obtained. These observations indicate that the detection of low quantities of toxin is dependent upon the use of the proper host tissue.

Although the activity is very low (1:160 dilution), it appears that the toxin is specific. This conclusion was based on the fact that eight host species were resistant to both the pathogen and the toxin. Two other species were not only susceptible to the pathogen and to the toxin, but they exhibited typical disease symptoms to both. The fact that all plant species tested exhibit the same pattern of resistance and susceptibility to two isolates and their toxigenic products indicates that the toxin produced by *H. maydis* is specific (Smedegar-Petersen and Nelson, 1969).

Efforts to determine the characteristics of the *H. maydis* toxin revealed that the toxin is dialyzable and heat stable (Quimio and Quimio, 1966; Smedegar-Petersen and Nelson, 1969). It is adsorbed to charcoal and to acid cation exchange columns. Adsorption by a cation exchange column indicates that the toxin molecule carries a positive charge. It was implied that the toxin may be a low molecular weight peptide, but no evidence was presented to support this suggestion (Smedegar-Petersen and Nelson, 1969).

The pathogen was reported to produce a nonspecific toxin of low activity (Quimio and Quimio, 1966). Recent observations, however, show that *H. maydis* does indeed produce a specific toxin *in vitro* and *in vivo* (Smedegar-Petersen and Nelson, 1969). Although the activity of *H. maydis* toxin is low, it appears that pathogenicity may be mediated through toxigenic action. Further work on this toxin and its effect on the physiology of the suscept is needed before it can be assigned a definite role in pathogenesis.

B. *Helminthosporium sacchari* TOXIN

Two reports indicating that *Helminthosporium sacchari* (Breda de Haan) Butl. produces a toxin have been found (Lee, 1929; Steiner and Byther, 1969). Although the earlier paper (Lee, 1929) presents a thorough report on a toxin produced by *H. sacchari*, it was not determined whether the toxin is specific. The undiluted culture filtrate used in many experiments caused wilting of both resistant and susceptible sugar cane cultivars. It was observed, however, that leaves of a resistant cultivar reacted more slowly to the toxin than leaves of a susceptible type. Old leaves of susceptible cultivars were more sensitive to the toxin than young leaves. In addition, it was observed that the lowest concentration (25%) of the culture filtrate caused more typical disease symptoms than undiluted material.

The toxic principle, which was heat stable, was considered to be some form of nitrite. The author (Lee, 1929) concluded that the fungus has a strong capacity to reduce some of the lower nitrogen compounds to nitrites and that the nitrites were toxic to leaf tissue of the host plant. It was observed that dilute solutions of potassium nitrite and undiluted culture filtrates induced similar symptoms on susceptible leaves. Unfortunately, resistant cultivars were not tested with potassium nitrite. It appears unlikely that one of the nitrites was the toxic principle because these compounds are general biological poisons and would probably have been as toxic to the pathogen as to the host. Moreover, recent observations fail to confirm nitrite accumulation in the culture medium (Steiner, 1969).

A recent report indicates that *H. sacchari* produces a specific, low molecular weight (about 1500) toxin (Steiner and Byther, 1969). Resistant cultivars of sugar cane, as well as a number of monocotyledonous and dicotyledonous plant species, were unaffected by both the fungus and the toxin. When 0.01 μg of the partially purified toxin was injected into susceptible sugar cane cultivars, typical disease symptoms were observed. Further experimentation could yield evidence that this toxin is a disease-inducing entity.

VII. Discussion

When the molecular structure and mode of action of toxins produced by *Helminthosporium* species are considered, two trends become evident. The first is that *Helminthosporium* species produce two general classes of toxigenic compounds — peptides and terpenoids. The second is that the peptides are specific and the terpenoids are nonspecific. The peptides appear to trigger the primary physiological disruption in the disease syndrome. Conversely, the terpenoids usually exert their effects late in the disease cycle. Therefore, the terpenoids may be one of several toxins required to induce a given set of symptoms.

The terpenoids adversely effect electron transfer and oxidative phosphorylation. The peptides also disrupt oxidative mechanisms, but this seems to be a secondary effect. One of the primary effects of terpenoids is the stimulation of phenol synthesis in the suscept. In some cases, the terpenoids predispose the host to pathogenic attack; however, the mechanism of predisposition is not known.

Peptide toxins have the same host range as the pathogen that produces them, but terpenoid toxins have a much wider host range than the organism that produces them. Constructive speculation on this observation awaits more knowledge concerning the site and mode of action of these two different types of toxins.

Another unresolved paradox is that some toxins mimic growth regulators and some growth regulators are toxigenic. Thus it is difficult to distinguish between growth regulators and toxins. The distinction has sometimes been conditioned by the viewpoint of the investigator. For example, gibberellin which was first reported to cause a systemic disease of rice later became commonly accepted as a growth regulator. In certain systems, the growth-promoting or toxigenic properties of a given fungal metabolite are conditioned by its concentration or by the tissue used in the assay.

A comparison of the toxigenic compounds produced by *Helminthosporium* species may give an indirect indication that production of toxigenic compounds play a significant role in phytopathology. Within one genus, toxins are produced which vary greatly in chemical characteristics, stability, and specificity. These toxins are shown to affect respiration, oxidative phosphorylation, carbon dioxide fixation, semipermeability, and even to interfere with hormonal control mechanisms. There is no reason to assume that the production of toxigenic compounds is limited to *Helminthosporium* species. In fact, several other genera are known to produce toxins which play a role in disease development. It seems plausible to conclude that many other plant pathogens will be shown to produce toxins when techniques of toxin isolation, chemical characterization, and adequate assay methods are developed.

REFERENCES

Braun, A. C., and Pringle, R. B. (1959). *In* "Plant Pathology Problems and Progress 1908-1958" (C. S. Holton, ed.), pp. 88-99. Univ. Wisc. Press, Madison, Wisconsin.
Canonica, L., Fiecchi, A., Kienle, M. G., and Scala, A. (1967). *Tetrahedron Letters* **35**, 3369.
Corey, E. J., and Nozoe, S. (1965). *J. Am. Chem. Soc.* **87**, 5728.
De Mayo, P., and Spencer, E. Y. (1962). *J. Am. Chem. Soc.* **84**, 494.
De Mayo, P., Spencer, E. Y., and White, R. W. (1961). *Can. J. Chem.* **39**, 1608.
De Mayo, P., Spencer, E. Y., and White, R. W. (1963). *Can. J. Chem.* **41**, 2996.
Dimond, A. E., and Waggoner, P. E. (1953). *Phytopathology* **43**, 229.
Earhart, R. W. (1959). *Plant Disease Reptr.* **43**, 1184.
Epstein, E. (1966). *Nature* **212**, 1324.
Gracen, V. E. (1970). Ph.D. Dissertation, University of Florida.
Hanchey, P., and Wheeler, H. (1969). *Can. J. Botany* **47**, 675.
Hanchey, P., Wheeler, H., and Luke, H. H. (1968). *Am. J. Botany* **55**, 53.
Hashimoto, T., and Tamura, S. (1967). *Plant Cell Physiol. (Tokyo)* **8**, 197.
Hashimoto, T., Kakurai, A., and Tamura, S. (1967). *Plant Cell Physiol. (Tokyo)* **8**, 23.
Humphreys, T. E., and Garrard, L. A. (1964). *Phytochemistry* **3**, 647.
Kalyanasundaram, R., and Charudattan, R. (1966). *J. Sci. Ind. Res. (India)* **25**, 63.
Katsumi, M., Tamura, S., and Sakurai, A. (1967). *Plant Cell Physiol. (Tokyo)* **8**, 399.
Kuo, M. S., and Scheffer, R. P. (1967). *Phytopathology* **57**, 817 (abstr.).
Kuo, M. S., and Scheffer, R. P. (1968). *Phytopathology* **58**, 1056 (abstr.).
Lee, A. (1929). *Plant Physiol.* **7**, 193.
Litzenberger, C. S. (1949). *Phytopathology* **39**, 300.
Ludwig, R. A. (1957). *Can. J. Botany* **35**, 291.
Ludwig, R. A., Clark, R. V., Julien, J. B., and Robinson, D. B. (1956). *Can. J. Botany* **43**, 653.
Luke, H. H. (1954). Ph.D. Dissertation, Louisiana State University.
Luke, H. H., and Gracen, V. E. (1970). Unpublished observations.
Luke, H. H., and Wheeler, H. E. (1955). *Phytopathology* **45**, 453.
Luke, H. H., and Wheeler, H. (1964). *Phytopathology* **54**, 1492.
Luke, H. H., Warmke, H. E., and Hanchey, P. (1966). *Phytopathology* **56**, 1178.
Luke, H. H., Freeman, T. E., Garrard, L. A., and Humphreys, T. E. (1969). *Phytopathology* **59**, 1002.
Malca, I., and Zscheile, F. P. (1963). *Phytopathology* **53**, 341.
Malca, I., and Zscheile, F. P. (1964). *Phytopathology* **54**, 1281.
Malca, I., Huffaker, R. C., and Zscheile, F. P. (1964). *Phytopathology* **54**, 663.
Meehan, F., and Murphy, H. C. (1946a). *Phytopathology* **36**, 406 (abstr.).

Meehan, F., and Murphy, H. C. (1946b). *Science* 104, 413.
Meehan, F., and Murphy, H. C. (1947). *Science* 106, 270.
Mori, S., Inoue, Y., and Mitso, K. (1966). *Plant Cell Physiol.* (*Tokyo*) 7, 503.
Murphy, H. C. (1946). *Iowa Farm Sci.* 1, 3.
Murphy, H. C., and Meehan, F. (1946). *Phytopathology* 36, 407 (abstr.).
Nakamura, M., and Ishibashi, K. (1958). *J. Agr. Chem. Soc. Japan* 32, 732.
Nelson, R. R. (1959). *Phytopathology* 49, 807.
Nelson, R. R., and Kline, D. M. (1963). *Phytopathology* 53, 101.
Nelson, R. R., and Kline, D. M. (1966). *Plant Disease Reptr.* 50, 382.
Nelson, R. R., and Kline, D. M. (1968). *Plant Disease Reptr.* 52, 620.
Nelson, R. R., and Ullstrup, A. J. (1961). *Phytopathology* 51, 1.
Novacky, A., and Wheeler, H. (1969). *Phytopathology* 59, 116 (abstr.).
Nozoe, S., Morisaki, M., Tsuda, K., Iitaka, Y., Takahashi, N., Tamura, K., Ishibashi, K., and Shirasaka, M. (1965). *J. Am. Chem. Soc.* 87, 4968.
Oku, H. (1964). *In* "Host Parasite Relations in Plant Pathology" (Z. Kiraly and G. Ubrizsy, eds.), pp. 183–191. Res. Inst. Plant Protect., Budapest, Hungary.
Oku, H. (1967). *In* "The Dynamic Role of Molecular Constituents in Plant-Parasite Interaction" (C. J. Mirocha and I. Uritani, eds.), pp. 237–255. Am. Phytopathol. Soc., St. Paul, Minnesota.
Oku, H. (1968a). Jubilee publication in commemoration of 60th birthday of Professor M. Sakamato.
Oku, H. (1968b). *In* "Biochemical Regulation in Diseased Plants or Injury," pp. 253–260. Phytopathol. Soc. Tokyo, Japan.
Orillo, F. T. (1952). *Philippine Agriculturist* 36, 327.
Orsenigo, M. (1957). *Phytopathol. Z.* 29, 189.
Orsenigo, M., and Sina, P. (1961). *Nuovo Giorn. Botan. Ital.* [N.S.] 68, 64.
Owens, L. D. (1969). *Science* 165, 18.
Pringle, R. B., and Braun, A. C. (1957). *Phytopathology* 47, 369.
Pringle, R. B., and Braun, A. C. (1958). *Nature* 181, 1205.
Pringle, R. B., and Scheffer, R. P. (1964). *Ann. Rev. Phytopathol.* 2, 133.
Pringle, R. B., and Scheffer, R. P. (1967). *Phytopathology* 57, 1169.
Quimio, T. H., and Quimio, A. J. (1966). *Philippine Agriculturist* 49, 778.
Robinson, T. (1963). "The Organic Constituents of Higher Plants." Burgess, Minneapolis, Minnesota.
Romanko, R. R. (1959). *Phytopathology* 49, 32.
Samaddar, K. R. (1968). *Phytopathology* 58, 1065 (abstr.).
Samaddar, K. R., and Scheffer, R. P. (1968). *Plant Physiol.* 43, 21.
Scheffer, R. P., and Pringle, R. B. (1963). *Phytopathology* 53, 465.
Scheffer, R. P., and Pringle, R. B. (1964). *Phytopathology* 54, 832.
Scheffer, R. P., and Pringle, R. B. (1967). *In* "The Dynamic Role of Molecular Constituents in Plant-Parasite Interaction" (C. J. Mirocha and I. Uritani, eds.), pp. 217–236. Am. Phytopathol. Soc., St. Paul, Minnesota.
Scheffer, R. P., and Ullstrup, A. J. (1965). *Phytopathology* 55, 1037.
Scheffer, R. P., Nelson, R. R., and Ullstrup, A. J. (1967). *Phytopathology* 57, 1288.
Smedegar-Petersen, U., and Nelson, R. R. (1969). *Can. J. Botany* 47, 951.
Spencer, E. Y., Ludwig, R. A., De Mayo, P., White, R. W., and Williams, R. E. (1966). *In* "Advances in Chemistry Series" (R. F. Gould, ed.), p. 106. Amer. Chem. Soc., Washington, D.C.
Steiner, G. W. (1969). Personal communication.
Steiner, G. W., and Byther, R. S. (1969). *Phytopathology* 59, 1051 (abstr.).

Tamura, S., Sakurai, A., Kainuma, K., and Takai, M. (1963). *Agr. Biol. Chem.* (*Tokyo*) 27, 738.
Taniguchi, T., and White, G. A. (1967). *Biochem. Biophys. Res. Commun.* 28, 879.
Tsuda, K., Nozoe, S., Morisaki, M., Hirai, K., Itai, A., and Okuda, S. (1967). *Tetrahedron Letters* 35, 3369.
Ullstrup, A. J. (1944). *Phytopathology* 34, 214.
Wheeler, H. E. (1953). *Phytopathology* 43, 663.
Wheeler, H. (1969). *Phytopathology* 59, 1093.
Wheeler, H., and Doupnik, B. (1969). *Phytopathology* 59, 1460.
Wheeler, H. E., and Luke, H. H. (1954). *Phytopathology* 44, 334. (Abstr.).
Wheeler, H., and Luke, H. H. (1963). *Ann. Rev. Microbiol.* 17, 223.
Wheeler, H., and Hanchey, P. (1968). *Ann. Rev. Phytopathol.* 6, 331.
Wright, D. E. (1968). *Ann. Rev. Microbiol.* 22, 269.
Yoshii, H. (1939). *Ann. Phytopathol. Soc. Japan* 9, 170.

CHAPTER 7

Alternaria Toxins Related to Pathogenesis in Plants

G. E. TEMPLETON

I. Introduction

Several plant toxins have been isolated from culture filtrates produced by species of the fungus genus *Alternaria* (Brian *et al.*, 1949; Fulton *et al.*, 1960; Grove, 1952; Hiroe and Aoe, 1954; Pound and Stahmann, 1951; Starratt, 1968; Sugiyama *et al.*, 1966a; Templeton *et al.*, 1967c). They are all relatively low molecular weight, nonenzymatic compounds ranging from small peptides to simple phenols. Their roles, if any, during pathogenesis are yet to be proven, but there are convincing arguments that some of them participate in symptom development at certain times during infection, incubation, or sporulation of the fungus on the host plant.

That there is a dynamic interchange between host and parasite is a concept that has been acknowledged since the establishment of the germ theory of disease; yet there is no rigorously proved, well-defined, completely satisfactory explanation of the biochemical mode of action of any plant toxins. The arguments for their participation in pathogenesis center on two main points: (1) Some are host specific, i.e., they affect only the host that the fungus will infect, and loss of ability to produce the toxin in culture is accompanied by loss of pathogenicity; or (2) the pathological effects of the toxin are so distinctly similar to all or part of the parasitic disease syndrome that there can be little doubt that the toxin is involved. These two points have been the basis for categorizing pathogenic metabolites of plant parasitic fungi as either primary or secondary determinants of pathogenicity. The former is considered essential for pathogenicity of the parasite, and the latter is thought to influence virulence but not control pathogenicity (Pringle and Scheffer, 1964). Both primary and secondary determinants of pathogenicity are found among the toxic substances produced by species of *Alternaria* in artificial culture.

Unfortunately, plants are limited in the number or type of distinctive responses, such as chlorosis, necrosis, and wilting, exhibited upon treatment, either externally or internally, with toxin preparations. It is not surprising, therefore, that most of the *Alternaria* toxins studied in detail produce chlorosis at some stage in the ontogeny of the lesion, just as most of the diseases caused by *Alternaria* species, or most other plant pathogens for that matter, produce chlorosis at some time during symptom development.

In addition to the primary and secondary determinants of pathogenicity produced by *Alternaria* species, there have been many compounds isolated from mycelial mats and culture filtrates of this genus that have not been examined for biological activity of this type (Miller, 1961; Raistrick *et al.*, 1953; Rosett *et al.*, 1957; Stickings, 1958; Thomas, 1961). This would seem a fruitful area of research, since there is the distinct possibility that many compounds are released by fungal parasites in plants that have quite subtle effects, imperceptible in ordinary bioassays, yet of distinct importance in metabolic interactions of host and parasite. Perhaps assays representing a spectrum of biochemical reactions with compounds isolated from metabolized media or mycelial mats would be an indirect but possibly fruitful area of research. Thus, the *Alternaria* toxins present some promising systems for solution of two major biological problems: (1) the mechanisms that govern specificity of parasites for particular hosts and (2) the relationships between host and parasite at the molecular level.

II. Black Spot Disease of Japanese Pear — *Alternaria kikuchiana*

A. PHYTOALTERNARINS A, B, AND C

These compounds have been isolated from culture filtrates and mycelial mats of *Alternaria kikuchiana* Tanaka, the causal organism of black spot disease of Japanese pear (Hiroe, 1952; Tanaka, 1933). The disease is of considerable interest because of its exceptional degree of specificity for certain high quality varieties of Japanese pears. European and North American pear varieties are highly resistant or immune, as are most Japanese varieties, but the variety Nijisseiki and several other high quality varieties are extremely susceptible. The symptoms occur on leaves and fruits as small black spots that become necrotic and are frequently surrounded by a yellow halo (Hiroe *et al.*, 1958; Mori, 1962).

Isolates of the organism vary considerably in their virulence on susceptible varieties. Also, virulence is positively correlated with the ability of isolates to produce metabolites in culture media that initiate black spot symptoms when applied in a water droplet to mechanically injured points on the epidermis of detached leaves.

Hiroe and Aoe (1954) isolated phytoalternarin A, B, and C from combined culture filtrates and mycelial mats by using the injured leaf bioassay with the Nijisseiki variety. The size of spots around the injured area after 24 hours of incubation at 28°C was estimated as an assay for tracing these toxic metabolites through the purification steps. The filtrate and mycelial mats were obtained from 30-day-old still cultures on Richards solution at 28°C. The fresh mycelial mat was extracted with acetone, which was then removed, and the extract was concentrated *in vacuo* and added to the culture filtrate. The combined filtrate and acetone extract was adjusted to pH 3.5 with hydrochloric acid and then extracted with ether. The ether-soluble portion was dried with anhydrous sodium sulfate, and taken to dryness leaving a brown residue. This brown residue was extracted with hot carbon tetrachloride, the extract was applied to a column of activated alumina for chromatography, and the residue was taken up in cold ethanol from which phytoaltenarin C crystallized as colorless needles. Phytoaltenarin A and B were eluted from the column with ethanol, B emerging first and remaining as a yellow fluorescent liquid after separation of nontoxic crystals that developed when the ethanolic eluate was held at a cold temperature. Finally, phytoaltenarin A emerged from the column as a white fluorescent band and was crystallized from methanol. The physical properties and biological activity reported by Hiroe and Aoe (1954) for these compounds may be seen in Table I.

The phytoalternarins give positive reactions to ninhydrin and the Liebermann and Xanthoproteic test, but are negative to the Fehling and Biuret reactions, indicating that they are low molecular weight proteins or peptides (after Mohri *et al.*, 1967).

In a biological activity test, phytoalternarin A was demonstrated to have the same host specificity as the fungus on three susceptible and fourteen resistant pear varieties. Its effect further resembled symptoms of the fungus infection in tissues of different age (increasing tissue age correlates with decreasing spot size) and in that the optimum temperature for disease development coincided with the optimum temperature for symptom expression when the pure toxin was applied.

The apparent host specificity and peptide nature of the phytoalternarins certainly warrant their consideration as possible primary determinants of pathogenicity. Further studies to confirm these investigations and characterize these toxins more completely, both biologically and chemically, should add substantially to our understanding of the nature of host resistance and the physiology of parasitism.

TABLE I

PROPERTIES OF PHYTOALTERNARINS AND RELATED COMPOUNDS

| Compound | Melting point (°C) | Form | Ultraviolet absorption[a] | | Biological activity |
			Major peak	Minor peak	
Phytoalternarin A	—	Colorless needles	256–258	280	Host specific
Phytoalternarin B	—	Yellow fluorescent liquid	255–260	—	Toxic on susceptible varieties
Phytoalternarin C	235	Colorless needles	257	335	Toxic on susceptible varieties
Nontoxic crystals	146–147	Colorless needles	—	—	Nontoxic on susceptible varieties

[a] In 95% ethanol.

B. ALTENIN

Sugiyama and associates (Sugiyama *et al.*, 1966a,b, 1967a,b,c) used different culture conditions and a slightly different bioassay to isolate altenin, another metabolite from *A. kikuchiana*, that produces black spots on the susceptible pear variety Nijisseiki. Altenin, a yellow liquid, $C_9H_{14}O_6$, is ethylhydroxy-5-(1-hydroxyethyl)-4-oxotetrahydrofuroate.

The bioassay to guide purification was conducted with detached young leaves of the Nijisseiki variety to which was applied drops of test solution in ammonium acetate buffer at pH 8.0. The treated leaves were held at 23°C for 17 hours and then the area of each black spot was measured. A linear plot of spot size against concentration yielded a straight line with ether extracts of the culture filtrates between concentration of 0.7 mg/ml and 85 mg/ml. Activity of the pure toxin in aqueous solution was detectable at a concentration of 2×10^{-5} mg/ml.

Altenin was isolated from culture filtrates of *A. kikuchiana* grown for 8 to 9 days at 30°C in a 200-liter fermenter (120 liters/batch). The medium contained sucrose (4%), dibasic ammonium phosphate (0.2%), monobasic potassium phosphate (1.1%), magnesium sulfate (0.05%), and potassium chloride (0.05%). Aeration was continuous at a rate of 20 liters/minute.

Two procedures for purification of altenin were developed. One involved adsorption of activity from culture filtrates on charcoal followed by acetone elution, ether extraction from acid solution, and then chromatography on silica gel and alumina. The other procedure involved ethyl acetate extraction of the culture filtrate and then chromatography on silica gel. The former yielded purest preparations of altenin plus succinic acid, and the latter yielded somewhat less pure preparations of altenin plus diheptylphthalate, myristic acid, 2,8-dihydroxy-1-hydroxymethyl-9,10-anthraquinone, and two incompletely identified compounds thought to be a C_{15} carboxylic acid and a C_{19} steroid. Altenin was the only active metabolite in the pear leaf bioassay when the components were tested individually.

In the charcoal adsorption procedure for purification of altenin, 10 liters of culture filtrate was adjusted to pH 3.0, mixed with 100 gm active charcoal, and stirred for 5 hours. The charcoal was filtered and air dried and the active element was eluted with 1 liter of acetone, which was then concentrated to 200 ml *in vacuo* at 35-40°C in a nitrogen atmosphere. The acetone solution was then adjusted to pH 9.0 with aqueous ammonia and extracted with ethyl ether which was discarded. The aqueous layer was adjusted to pH 3.0 with dilute hydrochloric acid and extracted with three 70-ml portions of ethyl ether. The ether extracts were combined, dried over anhydrous sodium sulfate, and the ether evaporated (35-40°C), which left a residue of 170 mg. The residue was dissolved in a cold mixture of benzene and acetone (2:1 v/v) and filtered. The precipitate was recrystallized from ethyl ether and identified as succinic acid, mp 183°C. The benzene-acetone filtrate was passed through a silica gel column with a 2:1 v/v benzene-acetone mixture, and fractions containing a blue fluorescent band combined and passed through an alumina column (6.7% water) with methanol and finally through a silica gel column with a benzene-acetone (4:1 v/v) mixture. The yield was 2.5 mg of altenin as a yellow liquid.

Thin-layer chromatography on silica gel (Wakogel B-5) with benzene-acetone (1:1 v/v) moved the altenin to R_f 0.76, while that of methyl red dye was 0.37. On alumina (Merck 1090) plates with methanol:water (5:1 v/v), the R_f was 0.87, while that of methyl red was 0.41. The spots were detected either by charring with concentrated sulfuric acid at 80°C for 1 hour or by developing a yellow color by spraying with an o-dianisidene solution and heating.

The structure of altenin, as illustrated in Fig. 1, was determined by Sugiyama and associates (1966b) from the elemental analysis and spectral data summarized in Table II. The structure was confirmed by synthesis, and the biological activity of the synthesized material in the pear leaf test was practically equal to that of altenin from culture filtrates (Sugiyama et al., 1967a). By comparison of altenin with its various isomers and certain analogous compounds, Sugiyama et al. (1967b,c) concluded that the active portion of the molecule is in the endiol carbonyl grouping.

Further characterization of the biological activity of this compound would be an especially intriguing area of pursuit. Does this compound affect black spot resistant pears or is it host specific? Also of considerable interest is the relationship between this compound and the phytoalternarins, which are apparently host specific. Does the phytoalternarin molecule have an endiol carbonyl function? With pure altenin now available, it will be possible to examine its effect on plant respiratory metabolism, permeability, and ultrastructure. If this is possible, it could add more meaningful evidence to the arguments for or against its role in parasitism as well as contribute to our understanding of these basic biological processes, whether or not it plays a role during parasitism by A. kikuchiana.

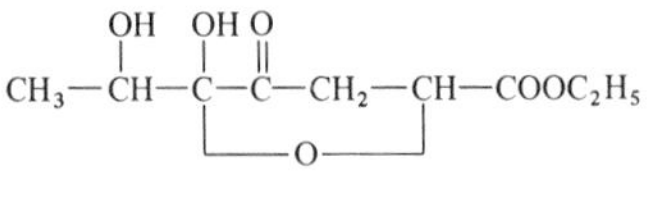

$$CH_3-CH-C-C-CH_2-CH-COOC_2H_5$$

FIG. 1. Altenin.

TABLE II

SUMMARY OF SPECTRAL DATA AND PHYSICAL PROPERTIES OF ALTENIN

Physical state at standard condition	Yellow liquid; heat labile (10 minutes, 80°C); pH 7.5
Elemental analysis	$C_9H_{14}O_6$; molecular weight 218 (mass spectroscopy)
Ultraviolet peak	274 μ; ϵ = 228 (in ethanol)
Infrared peaks	3400 cm^{-1}, 1730 cm^{-1}
Optical rotatory dispersion	Specific rotation at 589μ = 5.1;
	inversion from levo to dextro at 245 mμ
Nuclear magnetic resonance	

τ (ppm)	Intensity	Coupling	J (Hz)
4.83	1 H	Quartet	7.0
5.80	2 H	Quartet	7.0
5.83	1 H	Triplet	—
6.9-8.0	2 H	Multiplet	—
7.98	0.5 H	Singlet	—
9.46	2 H	Doublet	7.0
8.45	2 H	Doublet	7.0
8.51	3 H	Triplet	7.0

III. Early Blight Disease of Tomato and Potato—*Alternaria solani*

ALTERNARIC ACID

Alternaric acid was isolated (almost simultaneously in Wisconsin and England) from culture filtrates of *Alternaria solani* (Ell. and Mart.) Jones and Grout, the causal organism of early blight disease of tomato and potato (Brian *et al.*, 1951; Pound and Stahmann, 1951). This disease is a common foliage blight of these crops and is of appreciable economic importance in the United States. It appears first on the leaflets as circular to angular, dark brown to black spots that range in size from a pin point to about 4 mm in diameter. Lesions may occur on stem and fruit. Usually a narrow chlorotic zone is present around the necrotic spot, and it has been observed that when lesions are localized on one side of a petiole, leaflets on the affected side are often chlorotic or necrotic. Leaflets with numerous spots prematurely wither, droop, and, more commonly, drop off (Walker, 1952).

Brian and associates (Brian *et al.*, 1949, 1951) isolated alternaric acid from culture filtrates of *A. solani* using an assay based on its inhibitory activity during the germination of *Botrytis allii* Munn. Germination of conidia is not inhibited by alternaric acid at concentrations as high as 100 mg/ml, but elongation of spore germ tubes is markedly retarded shortly after germination by concentrations down to 0.01 μg/ml. The procedure was to assay a series of twofold dilutions to determine the greatest dilution at which at least 90% of the germ tubes would not exceed 200 μ in length after incubation for 16 to 18 hours at 25°C in a nutrient medium. Normal germ tubes under the same conditions are at least 400 μ in length. The activity index was recorded as the number of twofold dilutions made before exceeding the 200 μ germ tube length.

The culture filtrates from which the metabolite was purified were obtained by growing the fungus in modified Czapek-Dox medium in earthenware culture vessels containing 0.5 to 1.0 liters of solution at 25°C until maximum mycelial growth was attained. The optimal medium contained a high concentration (7-10% w/v) of sucrose, and nitrogen was supplied as nitrate or casein hydrolyzate. Nitrogen supplied in the ammonium form in conjunction with 0.25% w/v acetic acid was particularly favorable for good yields of toxin.

To purify the toxin the highly active filtrates were adjusted to pH 3.5 and extracted twice with 1/10 volume of chloroform. The chloroform was evaporated under reduced pressure, leaving a brown gummy material that was taken up in hot benzene. Alternaric acid crystallized from the benzene upon cooling and yielded on the order of 100 mg/liter.

Alternaric acid, $C_{21}H_{30}O_8$ (mp 138°C), was characterized and its structure determined by Bartels-Keith and Grove (1959; Grove, 1952). It is 12-(5,6-dihydro-4-hydroxy-6-methyl-2-oxopyran-3-yl)-4,5-dehydroxy-3-methyl-9-methylene-12-oxododec-6-ene-5-carboxylic acid, a hemiquinone derivative, as illustrated in Fig. 2.

During the course of characterization and structural determination, many derivatives of alternaric acid were prepared. The physical properties of two of them, the monohydrate and the methyl ester, are included in Table III, which summarizes the properties of alternaric acid.

Alternaric acid either inhibits germination, causes wilting or chlorosis and necrosis in certain higher plants, and inhibits germination of certain fungi while only retarding the germination rate of others. It inhibited germination of conidia of *Absidia glauca, Myrothecium veruccaria,* and *Stachybotrys atra* at concentrations ranging from 0.1 to 1.0 μg/ml, while concentrations as high as 200 μg/ml only retarded the rate of hyphal extension from spores of *Botrytis allii, Fusarium caeruleum,* and *Penicillium digitatum.* In higher plants, it caused a severe wilt followed by death

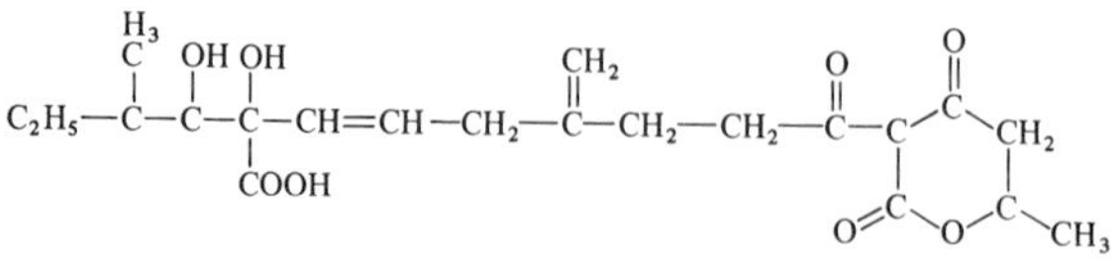

FIG. 2. Alternaric acid.

TABLE III

SUMMARY OF THE PHYSICAL PROPERTIES OF ALTERNARIC ACID, ALTERNARIC ACID MONOHYDRATE, AND METHYL ALTERNATE[a]

Compound	Crystalline form	Molecular formula	Molecular weight	Melting point (°C)	Ultraviolet peaks			Infrared peaks (cm⁻¹)
					Solvent	max	log ε	
Alternaric acid	Colorless needles	$C_{21}H_{30}O_8$	410	138	0.1 N sodium hydroxide	271, 250	4.14, 4.17	1650, 1710, 1732, 3150, 3430
					Water	273, 253, 210	4.04, 3.83, 4.06	
					Ethanol	274, 210	4.04, 4.07	
					Methanol (acidic)	273, 210	3.99, 3.99	
Alternaric acid monohydrate	Colorless thin rectangular plates	$C_{21}H_{32}O_9$	428	135–136	—	—	—	1650, 1710, 1727, 2490, 2610, 3480
Methyl alternate	Colorless prisms	$C_{22}H_{32}O_8$	424	103	0.1 N sodium hydroxide	273, 250	4.10, 4.17	1650, 1710, 1738, 3500, —
					Water	—	—	
					Ethanol	273, 252, 210	4.05, 3.98	
					Methanol (acidic)	273, 210	4.06, 4.08	

[a] After Grove (1952).

of seedlings when applied in nutrient solutions at concentrations of 5-10 μg/ml to young radish, cabbage, mustard, and carrot seedlings, while tomato, pea, and beet were apparently unaffected after 7 to 10 days. Growth was severely retarded when seeds of radish and mustard were grown on agar containing 1-5 μg/ml, while wheat and clover were somewhat more resistant. When introduced into the vascular systems of cut shoots of tomato and potato in concentrations of 2-20 μg/ml, it caused necrotic lesions of the stem, petioles, and leaf blades, very similar in appearance to the lesions associated with some phases of pathogenesis by *A. solani*. It also produced similar lesions on plants outside the host range of *A. solani*, such as *Atropa belladonna* L., *Solanum dulcamara* L., *Urtica dioica* L., and *Chamaenerion angustifolium* (L.) Scop. It increased transpiration in *A. belladonna* cuttings at concentrations as low as 4 μg/ml (Brian *et al.*, 1952). The pathogenicity of various isolates was not correlated with the production of antifungal activity in culture filtrates, so it seems reasonable to consider this metabolite a secondary determinant of pathogenicity. A determination of toxin production in the host by chemical means or a host plant bioassay would add credence to this argument.

Taken as a whole the evidence is quite good that alternaric acid, or an analogous compound, is excreted and participates in pathogenesis during parasitism of tomato or potato plants by *A. solani*. Because it is stable, highly active, and relatively easy to produce, it would be a good choice for biochemical studies to determine its role in pathogenesis.

IV. Seedling Chlorosis of Cotton and Citrus — *Alternaria tenuis*

TENTOXIN (TENUIS TOXIN)

Tentoxin has been isolated from culture filtrates and mycelial mats of *Alternaria tenuis* Auct., the causal organism of an irreversible, variegated seedling chlorosis in cotton, citrus and many other seedlings. The disease was first characterized by Fulton *et al.* (1960) on cotton seedlings during an investigation into the pathogenicity of fungi commonly isolated from diseased cotton seedlings. It was first produced experimentally with sterile culture filtrates and then found on cotton in the field. The disease is characterized by the distinctive chlorosis seen in Fig. 3. The green areas are sharply demarcated from the chlorotic areas, which are yellow rather than white in most of the sensitive species. Affected seedlings remain chlorotic and grow at a reduced rate, depending upon the amount of cotyledonary area affected. Cotton seedlings with more than 35% chlorotic area usually die (Fulton *et al.*, 1965). The fungus has been isolated repeatedly from affected plants in the field, but the etiology has not been com-

FIG. 3. Variegated chlorosis of cotton induced by tentoxin.

pletely worked out. Ryan has theorized that the fungus grows saprophytically, either on the seed coat or upon refuse in the soil, and produces the toxin, which then diffuses into the cotyledons through injuries or upon rupture of the inner seed coat during germination (Ryan *et al.*, 1961). The fungus then infects the roots and hypocotyls of the affected plant. Leaf spots of tobacco and cotton incited by *A. tenuis* usually show only slight chlorotic areas at the margin of lesions. On immature apple leaves that have been inoculated with the fungus, small chlorotic spots are formed that fade away in 3 to 4 months. Mature apple leaves, even though colonized by the fungus, do not show visible symptoms (Taylor, 1965, 1966).

Prior to development of a bioassay for this biological activity, factors affecting the amount and pattern of chlorosis in affected seeds were studied (Templeton *et al.*, 1965, 1967a). Cucumber was selected as the test plant, since it was quite sensitive to the toxin, chlorosis was more uniform than in cotton, and an abundance of uniform good quality, rapidly germinating seeds were available that could be easily handled. Rather than work with the metabolized filtrates that frequently inhibit germination, the active fraction was extracted into diethyl ether to obtain a crude toxin preparation that could be dissolved in water with heating. It was found that the condition of the inner seed coat affected the amount and pattern of chlorotic area. Chlorosis always involved bases of cotyledon when intact seeds were soaked or germinated in the toxin. Likewise, chlorotic bands were produced only at the bases when only the outer seed coat was cut. Seeds that were clipped at the apex to rupture both the inner and outer seed coat developed chlorotic bands at each end of the cotyledon. Seeds with only the outer seed coat removed would not respond to the metabolite unless they were soaked after the natural rupture of the inner seed coat by extension of the radicle during germination. Complete chlorosis resulted when both seed coats were completely removed before the seeds were dipped momentarily into the toxin solution.

The amount and pattern of chlorosis were influenced by the concentration of crude metabolite and the period of exposure. Intact seeds soaked for 1 hour in the crude toxin at concentrations of 0.05 to 1 mg/ml produced chorosis ranging from a trace to 75% of the area of the cotyledons. A similar response was obtained with much lower concentrations (0.005–0.050 mg/ml) when seeds were exposed continuously to the metabolite during the 5-day growth period. With the increase in chlorotic area at the higher concentration, the apical margins of the chlorotic area were more irregular. This was most evident in seedlings grown continuously in the solution, where the chlorotic area radiated toward the apex of the cotyledons along the veins.

Plants were affected by the metabolite only during the first 32 hours of growth. This was determined by exposing seeds or seedlings to a 0.3 mg/ml solution for 1 hour after 1, 16, 24, 32, and 48 hours of germination in the light. Over 95% of the cotyledon was chlorotic when seedlings were grown for up to 32 hours before treatment, and no chlorosis developed if they were treated 48 hours after initiation of germination in this particular lot of seeds. The hypocotyls of seedlings treated after 32 hours were light green and a trace of green was detected at the base of the cotyledons, whereas those treated earlier were yellow. The amount of chlorosis with the treatments at 1, 16, and 24 hours of germination was approximately the same, but extension of the chlorosis along the veins was first evident with the treatment at 24 hours.

The sensitivity of plants to the toxin was increased by holding them in darkness after treatment with the metabolite. This was determined with a 1-hour soak in a 0.3 mg/ml solution followed by dark periods of 0, 24, 48, and 64 hours. The amount of chlorosis increased with increasing periods of darkness, and complete chlorosis of the seedlings resulted when they were held in the dark for 64 hours after treatment. Control seedlings held in the dark for 64 hours and then placed in the light developed green cotyledons.

From these studies, it was concluded that the factors which govern the amount and pattern of chlorosis induced by the toxin are those which contribute to its concentration, its entrance into the cotyledons, its mobility within the cotyledon, and the developmental stage to which the cells in the cotyledons have progressed when the toxin enters them. With this information it was then possible to trace the biological activity through the number of purification steps using the cucumber as a test plant.

Activity was followed in a semiquantitative way by visually rating the amount of chlorosis in ten seedlings allowed to imbibe dilutions of the various preparations for 1 hour, washed with distilled water, and placed on moist filter paper in sterile Petri plates in the light. Readings were made after 5 days at room temperature (Grable *et al.*, 1966; Grable, 1967). After the toxin was crystallized, the assay was quantitated. Chlorophyll content was determined on a fresh weight basis in three replicates of ten seedlings each after 5 days' growth at room temperature in the light. Chlorophyll was extracted with 80% acetone and determined by its adsorption at 663 and 645 mμ according to the procedure of Maclachlan and Zalik (1963). The toxin concentration versus milligrams chlorophyll per gram fresh weight is plotted in Fig. 4.

The culture filtrates from which the toxin was purified were obtained by growing an active strain of *A. tenuis* in 500-ml Erlenmeyer flasks containing 220 ml of medium. The medium contained 50 gm dextrose, 10 gm potassium nitrate, 5 gm potassium monobasic phosphate, 2.5 gm magnesium sulfate, .02 gm ferric chloride, 100 ml V-8 Juice, and 1000 ml distilled water. The medium was inoculated from plugs of mycelium grown on potato dextrose agar and stored at 10°C. After 4 weeks of growth in still culture, the mycelium was separated from the spent medium by filtration through cheesecloth, and the combined filtrates were sterilized

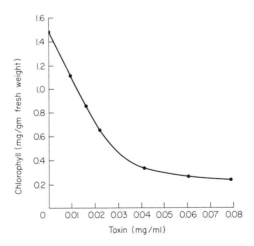

FIG. 4. The effects of tentoxin on chlorophyll of cucumber seedlings.

by autoclaving at 20 psi for 20 minutes. A white flocculant precipitate was removed by filtration before the filtrates were concentrated to 1/10 volume *in vacuo* (35-40°C).

The toxin was purified by extraction of the concentrated filtrates with diethyl ether, chromatography on ion exchange and alumina columns, and then crystallization from benzene. The concentrated filtrate (200 ml), representing 2 liters of original filtrate, was extracted in a Kutscher-Steudel extraction apparatus for 24 hours with diethyl ether. The ether mixture was removed from the apparatus and evaporated in a stream of warm air at room temperature, leaving 200 to 300 mg of brown gummy material per liter of filtrate. This material was taken up in a minimal volume of 80% ethanol and passed through a 0.9 × 15 cm column of cation exchange resin (Dowex 50 × 8) in the hydrogen form prepared according to the procedure of Plaisted (1958) for desalting plant extracts for amino acid chromatography. The toxin was then passed through a column of anion exchange resin, which was converted from the chloride to hydroxide form after the column was bedded in water by passing 1 N sodium hydroxide through it until the chloride ions had been exchanged. The column was repeatedly washed with water followed by washing with 80% ethanol before the ethanolic affluent from the cation exchange resin was passed through it. The active element passed through the column in 1 or 1½ void volumes. The effluent from the column was taken to dryness *in vacuo*, then dissolved in a minimum volume of absolute methanol, and passed through a 0.9 × 10 cm column of basic alumina bedded in absolute methanol. The effluent was collected in 5-ml fractions and the active fractions were combined after having been shown active by bioassay. The band of activity was in the first 1 or 1½ void volumes emerging from this column. The combined active fractions, after removal of the methanol at reduced pressure, were dissolved in a minimum of hot benzene and then refrigerated. Crystals formed in 2 to 3 days and were removed by centrifugation and recrystallization from benzene. The yield averaged 5 mg of crystalline toxin per liter of filtrate. A summary of this procedure is given in Table IV, and the yield and percent recovery for each step is given in Table V (Grable, 1967; Templeton, 1967c). Much of the activity in the residue could be recovered from successive crystallizations. The apparent low recovery in the ether extract was probably due to the difficulty of getting that fraction into solution for the bioassay.

Another purification procedure suitable for large numbers of toxin samples was developed by Saad *et al.* (1971). Toxin prepared by this procedure was spectrophotometerically pure and free of interfering effects, such as stunting and germination inhibition of the bioassay plant. Active filtrates of water extracts of the mycelium were concentrated to 1/10 their

TABLE IV

SUMMARY OF PROCEDURE FOR ISOLATION OF TENTOXIN

1. Filtrate extracted with diethyl ether.
2. Ether extract in 80% ethanol passed through cation exchange resin.
3. Ethanolic effluent passed through anion exchange resin.
4. Ethanol removed, residue taken up in methanol and passed through basic aluminum oxide column.
5. Active fractions combined, taken to dryness, and dissolved in a minimum of hot benzene, then refrigerated.
6. Recrystallized from benzene.

TABLE V

AVERAGE YIELD AND PERCENT RECOVERY OF TENTOXIN AT
VARIOUS STAGES OF PURIFICATION

Fraction	Yield per liter	Recovery (%)
Filtrate	1000 ml	100
Ether extract	215 mg	57
Non-ionic fraction	132 mg	75
Fraction from aluminum oxide	93mg	78
Crystalline toxin	5 mg	17
Residue	43 mg	54

original volume under reduced pressure and then combined with 7 volumes of absolute ethanol. The ethanolic solution was clarified by filtration, and the ethanol was removed by evaporation under reduced pressure. This aqueous solution was adjusted to pH 1.0 with hydrochloric acid and shaken with an equal volume of diethyl ether; the aqueous phase was discarded and the ether phase was shaken with an equal volume of 5% sodium bicarbonate. The aqueous phase was evaporated to dryness under reduced pressure and the residue redissolved in anhydrous diethyl ether (4 ml/liter of original filtrate). The ether solution was passed through a 1.5 × 14 cm column of silic acid (retention volume 21 ml) and the toxin eluted with ethyl acetate:acetone:n-hexane (2:1:1). The bulk of the active material emerged in a peak between 40 and 260 ml, with the maximum activity in retention volumes 7, 8, and 9 (140, 160, and 180 ml). The ultraviolet adsorption spectrum of the most active fraction duplicated that of pure tentoxin.

A gel filtration system has been used to obtain a considerable degree of purification of tentoxin from concentrated culture filtrates. The concentrated filtrate is passed through a 5 × 80 cm column of Sephadex G-10 and the active portion emerges in 3.5 to 4.5 void volumes.

Tentoxin, $C_{22}H_{30}N_4O_4$, molecular weight 414.5, is a cyclic tetrapeptide — cycloleucyl-N-methylalanylglycyl-N-methyldehydrophenylalanyl — as illustrated in Fig. 5. It crystallizes as barrel-shaped or irregular hexagons that melt at 172-175°C. The pure toxin is negative to ninhydrin. The ultraviolet adsorption spectrum is illustrated in Fig. 6, and the molar extinction coefficient is 12, 241. The infrared adsorption spectrum in chloroform has peaks at 1130, 1255, 1510, 1677, 1071, 2960, and 3366 cm^{-1}. The nuclear magnetic resonance and mass spectral data obtained indicate that the sequence of the amino acids is as illustrated, and there is some enzymatic evidence to indicate that leucine is levorotatory. The optical rotatory dispersion spectrum shows optical activity (Templeton et al., 1967b).

Although the complete spectrum of biological activity for tentoxin has not been exhaustively examined, it appears to interfere with chlorophyll formation in specific tissues of certain higher plant species. Only those tissues within the seed (i.e., cotyledon and primary leaves in susceptible species) become chlorotic when seeds or very young seedlings are soaked in the toxin. As these tissues reach a stage of physiological development that roughly coincides with the time they are fully photosynthetically mature, they are no longer sensitive to the toxin. Also, if the development of photosynthetically mature tissue is delayed by holding treated tissue in the dark, toxicity is greatly increased.

Among the higher plants, most dicotyledonous species tested are sensitive, with the exception of tomato and members of the Cruciferae. Among

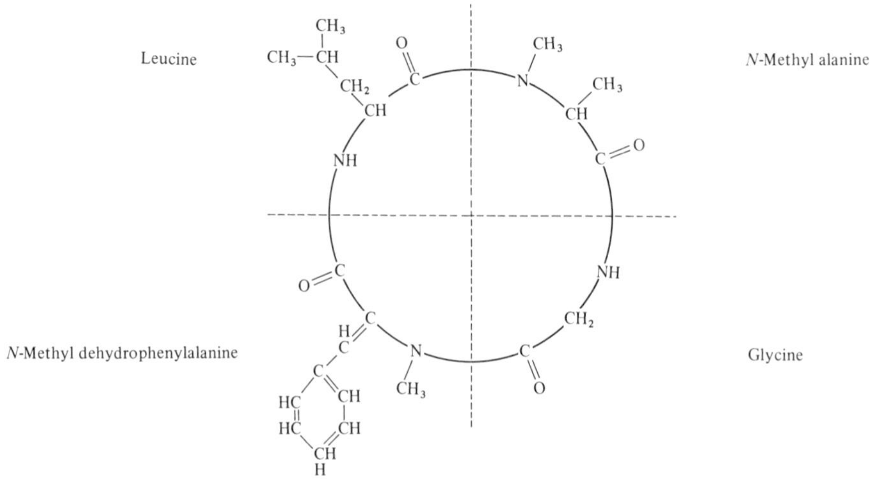

FIG. 5. Tentoxin.

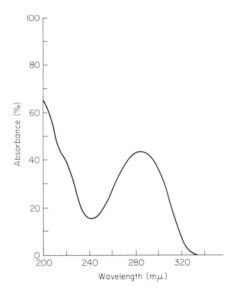

FIG. 6. Ultraviolet absorption spectrum of tentoxin in water.

the monocots tested, only sorghum and crabgrass were sensitive. Loblolly pine, the only Gymnosperm tested, was sensitive even though chlorophyll is present within the seed.

The alga *Euglena gracilis* is not sensitive in broth culture at concentrations up to 100 μg/ml even when bleached by periods of darkness prior to treatment. The growth of bacteria, the yeast *Saccharomyces cerevisiae* (Han) Meyers, and several filamentous fungi are not affected at concentrations that produce 100% chlorosis in cucumber cotyledons.

Halloin and associates (1971) found that tentoxin does not interfere with the conversion of protochlorophyll to chlorophyll when treated cucumber cotyledons were changed from darkness to light. Also they showed that chlorophyll synthesis proceeded in these plants, but at a rate considerably reduced from that of controls. In cabbage seedlings, which do not develop chlorosis when treated with tentoxin, they found it stimulated the rate of chlorophyll synthesis slightly, both before and after exposure to light. Electron microscopy of these tissues revealed considerable starch present in affected plastids of cucumber with little or no development of granalamella or intergranalamellae. No drastic changes were evident in chloroplasts of treated cabbage cotyledons (resistant) held in the light, but, in general, lamellar membranes were not so straight and closely appressed in toxin-treated tissue as in control tissue. With cabbage cotyledons held in the dark, some starch accumulation and formation of prolamellar bodies were noted in toxin-treated tissue along with the disappearance of a substance assumed to be phytoferritin. Mitochondria and nuclei appeared to be normal in both toxin-treated cucumber and cabbage cotyledons.

Halloin *et al.* (1971) suggested that chlorosis induced by tentoxin in cucumber seedlings is due to a reduction of chlorophyll synthesis within each chloroplast, rather than to a nonuniform expression of toxicity among plastids. This apparent specificity for chloroplasts in certain developing seedlings is undoubtedly related to the differences in the physical and metabolic states of these tissues as compared with that of photosynthetically mature tissue in higher green plants. The chloroplast is a semiautonomous self-duplicating organelle with the same origin, form, and function throughout the plant kingdom. Yet, in the germinating seedling, it is in a transitional state developing or at least maturing with sources of energy and metabolites quite distinct from those in the meristem of other photosynthetically active tissue. Thus, the specificity might be due simply to the inhibition of a biochemical pathway that is transient. However, the physical condition of this tissue during this period in terms of density and membrane selectivity is different from species to species and between tissues of young and mature plants of the same species, so that the specificity may be due to factors which govern its mobility to its site of action. In the parasitism of a host plant by the fungus, toxin production by advancing fungal hyphae could overcome to a large degree the mobility restriction.

Studies to determine the relationship between the structure of this cyclic peptide and its mode of action involving the inhibition of chlorophyll formation (chloroplast maturation) would be of considerable interest. Would substitution, rearrangement, or addition of amino acids affect host range or activity? Would removal of a methyl group from one or both of the methylated amino acids affect range or degree of biological activity? What effects would this active peptide or its analogs have on isolated chloroplasts, other organelles, or an isolated membrane system? Does this relatively small peptide alter permeability of specific membranes as has been demonstrated with some of the larger cyclic peptides such as the antibiotic valinomycin?

The specificity for chlorophyll inhibition and the distinctive patterns of chlorosis induced by tentoxin leave little doubt that it is functional in the seedling chlorosis associated with the growth of *A. tenuis* on or near the seed coat of sensitive species. Whether or not it is functional during parasitism by *A. tenuis* is still an open issue. Probably the most direct approach toward an answer to this question lies in understanding its mode of action at the molecular level and then searching for this biochemical lesion or its ramifications in infected tissue.

In any case, tentoxin, because of its peptide nature, its specificity, its stability, and its symptomology, appears to provide an excellent model system from which to generalize about the role of toxins in plant pathogenesis. It is related chemically to the host-specific peptide toxins from *Helminthosporium victoriae, H. turcicum,* and *Periconia circinata,* which are most certainly functioning as primary determinants of pathogenicity. Further, tentoxin causes a symptomatology pattern reminescent of the chlorotic patterns associated with diseases caused by other species of *Alternaria.*

V. Leaf Spot of Zinnia — *Alternaria zinniae*

ZINNIOL

Zinniol is derived from the fungus *Alternaria zinniae* which incites a common leaf spot and seedling blight of zinnia, sunflower, and marigolds (White and Starratt, 1967, Starratt, 1968). Characteristic small brown spots occur on the cotyledons, leaves, stems, and flowers. They begin as tiny necrotic spots, often surrounded by a chlorotic halo, and enlarge to form dark, irregular, reddish brown spots ranging from 2 to 10 mm in width. When zinnia leaves are affected with numerous spots they become brown and dry. Plants with numerous stem lesions often wilt completely, even though the lesions have not entirely girdled the stem (Dimock and Osborn, 1943).

The toxin was detected in undiluted culture filtrates by immersing the tips of the stems of intact or cut zinnia seedlings at the first leaf stage in aliquots of filtrate with sterile unmetabolized medium as a control. Cuttings and seedlings in the filtrate developed shriveled stems and completely wilted leaves with distinct browning in the veins within 24 hours at 27°C in the light. Controls wilted slightly but recovered after 48 hours. Activity was followed through the various purification steps with a similar assay, i.e., diluting preparations with 2 or 4% ethanol solutions to their dilution end points and using corresponding ethanol solutions as controls.

The toxin was produced by incubation of the fungus in still culture for 17-21 days at 26°C on a sucrose-casein hydrolyzate medium at pH 4.9 (after Brian *et al.,* 1949). Standard size Roux bottles containing 150-ml medium were used for batch production. Cultures were maintained on potato-dextrose slants, and inoculum for the Roux bottles was produced in 1 week at 26°C on the sucrose-casein hydrolyzate medium in 200-ml

Erlenmeyer flasks. The week-old cultures were diluted with an equal volume of water (25 ml/flask), blended aseptically for 5 to 10 seconds, and then dispensed in 0.6-ml aliquots into the Roux bottles.

Active filtrates were separated through cheesecloth from mycelial mats that were homogenized in distilled water, then filtered. Both filtrates were extracted separately for 5 minutes with chloroform, which was then siphoned off. The filtrates were washed three times with distilled water, dried with anhydrous sodium sulfate, and then combined. The extracted filtrates had no activity in the bioassay when tested at approximately the original concentration.

Chloroform was removed from the combined extracts *in vacuo* at 30°C leaving a dark viscous oil (3.19g/74 Roux bottles). The oil was taken up in ether (320 ml), extracted first with 5% sodium bicarbonate solution (4 × 25 ml) and then with 2 N sodium hydroxide (4 × 25 ml). These two fractions were acidified, extracted with chloroform, washed, dried, and evaporated, leaving small fractions that were tested by bioassay in 2% ethanol. At 500 ppm the acidic fraction [sodium bicarbonate] caused withering of the stem end in solution and moderate wilting, but no chlorosis or necrosis in the leaves after 12 hours. The phenolic fraction [sodium hydroxide] caused severe wilting and shriveling of leaves, but no marked effect on the stem.

The ether solution (after sodium bicarbonate and sodium hydroxide extraction) was washed with distilled water, dried with sodium sulfate, and evaporated to yield a viscous oil (2.18 gm). This oil was chromatographed over Woelm alumina (Grade V; 136 gm). Elution with benzene gave a homogenous oily fraction (1.09 gm) that was chromatographed on thin-layer Kieselgel (Camag) with 4% methanolic chloroform, detected with 5% ethanolic phosphomolybdic acid solution, and heated at 100°C. This oil was termed zinniol.

Zinniol was characterized by Starratt (1968). It was found to be a penta substituted benzene closely related to qualrilineatin, a fungal inhibitor from the fungus *Aspergillus quadrilineatus* Thom and Raper (Birkinshaw *et al.*, 1957). The structure of zinniol is given in Fig. 7.

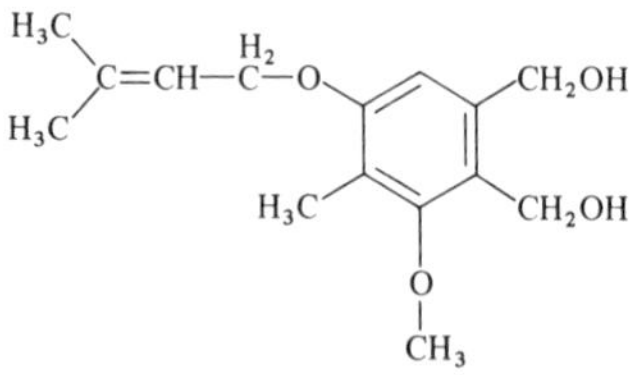

FIG. 7. Zinniol.

The initial preparative steps in characterization involved the formation of crystalline diethyl and dibenzoyl derivatives. The diacetate has the following properties: mp 65-66°C, ν_{max} 1729, 1606, 1588 cm^{-1}; γ_{max} 233, 272, 282 (shoulder) μ (ϵ 9600, 1200, 1100). A comparison of the nuclear magnetic resonance spectra of zinniol and the diacetate indicated the presence of one aromatic proton, a methoxyl group, a methyl group or an aromatic nucleus, and two CH_2OH groups with which the acetyl reacted. The spectra also indicated a 3,3-dimethylallyoxy grouping.

Chromic acid oxidation of zinniol yielded two isomeric phthalides (A and B) that could be reduced with lithium aluminum hydride to zinniol. These phthalides were degraded and resynthesized to confirm the structure of zinniol. A summary of the structural proof is as follows. The dimethylallyoxy substituent of the phthalides was removed by acetylation, and the acetate was hydrolyzed to its respective phenol, which was then methylated and reacted with 1-chloro-3-methyl-2-butene to give the original phthalide.

Phthalide A was found in early fractions during chromatography over alumina (petrol 40-60°C-benzene) of the chloroform extracts of the filtrates. This was considered an artifact, but the phenol of this phthalide is believed to occur in the filtrates, since its presence was detected by acetylation of the sodium hydroxide washing of the ether extract.

Pure zinniol at 500 ppm induced several symptoms in zinnia, including chlorosis, stem withering, and curling of leaf tips when applied to cut stems as in the bioassay. It also inhibited germination of zinnia, tomato, lettuce, watermelon, and carrot seeds. Concentrations of 1000 ppm induced withering of cut seedlings of watermelon, squash, beet, tomato, oat, corn, pea, and bean. The apparent lack of specificity for host plants, the concentrations required for symptom expression, and the nature of the compound suggest that this compound would more likely be considered a secondary determinant of pathogenicity than a primary determinant. Efforts to elucidate the mode of action of zinniol and thereby determine its relation, if any, to pathogenesis by the fungus are greatly augmented by the thorough elucidation of its structure and the synthesis of the various derivatives. The effects of these on the host tissue or a suitable biochemical model system should yield considerable insight into the active portion of the molecule and provide a more direct approach to the biochemical lesion in the host plant.

VI. Summary and Conclusions

Several *Alternaria* toxins have been characterized chemically and their structures completely elucidated (see Table VI). Furthermore, a reason-

TABLE VI
Summary of *Alternaria* Metabolites

Metabolite	Species[a]	Molecular formula	Melting or boiling point (°C)	Description	References
Alternaric acid	1	$C_{21}H_{30}O_8$	135	Colorless needles	Grove, 1952
Phytoalternarin A	2	–	147	Needles	Hiroe and Aoe, 1954
Phytoalternarin B	2	–	–	–	Hiroe and Aoe, 1954
Phytoalternarin C	2	–	235	Needles	Hiroe and Aoe, 1954
Alternariol	3,5,6	$C_{14}H_{10}O_5$	350	Colorless needles	Raistrick *et al.*, 1953
Alternariol methyl ether	3,5,6	$C_{15}H_{12}O_5$	267	Colorless needles	Freeman, 1966
Altertenuol	3	$C_{14}H_{10}O_6$	284	Buff-colored rods	Rosett *et al.*, 1957
Altenusin	3	$C_{15}H_{14}O_6$	202	Colorless prisms	Rosett *et al.*, 1957
Dehydroaltenusin	3	$C_{15}H_{12}O_6$	189	Yellow needles	Rosett *et al.*, 1957
Altenuic acid I	3	$C_{15}H_{14}O_8$	183	Colorless needles	Rosett *et al.*, 1957
Altenuic acid II	3	$C_{15}H_{14}O_8$	240	Colorless plates	Rosett *et al.*, 1957
Altenuic acid III	3	$C_{15}H_{14}O_8$	198–202	Colorless prisms	Rosett *et al.*, 1957
Tenuazonic acid	3	$C_{10}H_{15}NO_8$	117 (bp)	Straw-colored gum	Stickings, 1958
Tentoxin	3	$C_{22}H_{30}N_4O_4$	172–175	Barrel-shaped hexagons	Templeton *et al.*, 1967c
Altenin	1	$C_9H_{14}O_6$	–	Yellow liquid	Sugiyama *et al.*, 1966b
Succinic acid	2	$C_4H_6O_4$	185–187	Monoclinic prisms	Sugiyama *et al.*, 1966b
Mannitol	4	$C_6H_{14}O_6$	165–166	Orthorhombic needles	Starratt, 1968
Zinniol	4	$C_{15}H_{21}O_4$	–	Oil	Starratt, 1968
α,β-Dehydro-curvularin	6	$C_{16}H_{18}O_5$	230–232	Plates	Starratt and White, 1968

[a] Key: (1) *A. solani;* (2) *A. kikuchiana;* (3) *A. tenuis;* (4) *A. zinniae;* (5) *A. cucumerina;* (6) *A. dauci.*

ably large group of untested metabolites have been isolated and identified from this fungal genus, and together they represent several classes of chemical compounds. Collectively, they are chemically related to compounds that have biochemical effects as diverse as membrane disruption, protein synthesis inhibition, hormonal effects, and activity as antimetabolites. Perhaps, on one hand, it is premature to ascribe to any of them a role during parasitism of plants by *Alternaria* species, but on the other hand, one cannot ignore their chemical and symptomological similarities to toxins of other plant pathogens, nor can one ignore the specificity of certain of them for host or organelles. Undoubtedly, in the future, the *Alternaria* toxins will be subjected to wider, more intensive investigations whose results, it is anticipated, will lead to new and important discoveries into the role of toxin action during fungal pathogenesis of plants.

ACKNOWLEDGMENT

The author wishes to express his sincere appreciation to Drs. J. M. Halloin and A. M. Saad for providing manuscripts of their papers on tentoxin and to Dr. R. P. Scheffer for providing translations of several Japanese papers on phytoalternarin.

REFERENCES

Bartles-Keith, J. R., and Grove, J. F. (1959). *Proc. Chem. Soc.* p. 398.
Birkinshaw, J. H., Chaplen, P., and Lahoz-Oliver, R. (1957). *Biochem. J.* **67**, 155.
Brian, P. W., Curtis, P. J., Hemming, H. G., Unwin, C. H., and Wright, J. M. (1949). *Nature* **164**, 534.
Brian, P. W., Curtis, P. J., Hemming, H. G., Jefferys, E. G., Unwin, C. H., and Wright, J. M. (1951). *J. Gen. Microbiol.* **5**, 619.
Brian, P. W., Elson, G. W., Hemming, H. G., and Wright, J. M. (1952). *Ann. Appl. Biol.* **39**, 308.
Dimock, A. W., and Osborn, J. H. (1943). *Phytopathology* **33**, 372.
Freeman, G. G. (1966). *Phytochemistry* **5**, 719.
Fulton, N. D., Bollenbacher, K., and Moore, B. J. (1960). *Phytopathology* **50**, 575 (abstr.).
Fulton, N. D., Bollenbacher, K., and Templeton, G. E. (1965). *Phytopathology* **55**, 49.
Grable, C. I. (1967). M. S. Thesis, University of Arkansas.
Grable, C. I., Templeton, G. E., and Meyer, W. L. (1966). *Phytopathology* **56**, 897 (abstr.).
Grove, J. F. (1952). *J. Chem. Soc.* pp. 4056.
Halloin, J. M., deZoeten, G. A., Gaard, G., and Walker, J. C. (1970). *Plant Physiol.* **45**, 310.
Hiroe, I. (1952). *Ann. Phytopathol. Soc. Japan* **16**, 127.
Hiroe, I., and Aoe, S. (1954). *J. Fac. Agr., Tottori Univ.* **2**, 1.
Hiroe, I., Nishamura, S., and Sato, M. (1958). *Trans. Tottori Soc. Agr. Sci.* **11**, 291 (in Japanese, English summary).
Maclachlan, S., and Zalik, S. (1963). *Can. J. Botany* **41**, 1053.
Miller, M. W. (1961). "The Pfizer Handbook of Microbial Metabolites." McGraw-Hill, New York, New York.
Mohri, R., Kashima, C., Gasha, T., and Sugiyama, N. (1967). *Ann. Phytopathol. Soc. Japan* **33**, 289.
Mori, R. (1962). *Liberal Arts J., Torrori Univ.* **31**, 53 (in Japanese, English summary).
Plaisted, P. H. (1958). *Contrib. Boyce Thompson Inst.* **19**, 231.
Pound, G. S., and Stahmann, M. A. (1951). *Phytopathology* **41**, 1104.

Pringle, R. B., and Scheffer, R. P. (1964). *Ann. Rev. Phytopathol.* **2**, 133.

Raistrick, H., Stickings, C. E., and Thomas, R. (1953). *Biochem. J.* **55**, 421.

Rosett, R., Sankhala, R. H., Stickings, C. E., Taylor, M. E. U., and Thomas, R. (1957). *Biochem. J.* **67**, 390.

Ryan, G. F., Greenblatt, G., and Al-Delaney, K. A. (1961). *Science* **134**, 833.

Saad, S. M., Halloin, J. M., and Hagedorn, D. J. (1969). *Phytopathology* **59**, 1048 (abstr.).

Starratt, A. N. (1968). *Can. J. Chem.* **46**, 767.

Starratt, A. N., and White, G. A. (1968). *Phytochemistry* **7**, 1883.

Stickings, C. E. (1958). *Biochem. J.* **72**, 332.

Sugiyama, N., Kashima, C., Yamamoto, M., Sugaya, T., and Mohri, R. (1966a). *Bull. Chem. Soc. Japan* **39**, 1573.

Sugiyama, N., Kashima, C., Hosoi, Y., Ikeda, T., and Mohri, R. (1966b). *Bull. Chem. Soc. Japan* **39**, 2470.

Sugiyama, N., Kashima, C., Yamamoto, M., and Mohri, R. (1967a). *Bull. Chem. Soc. Japan* **40**, 345.

Sugiyama, N., Kashima, C., Yamamoto, M., and Mohri, R. (1967b). *Bull. Chem. Soc. Japan* **40**, 2591.

Sugiyama, N., Kashima, C., Yamamoto, M., Takana, T., and Mohri, R. (1967c). *Bull. Chem. Soc. Japan* **40**, 2594.

Tanaka, S. (1933). *Mem. Coll. Agr., Kyoto Imp. Univ.* **28**.

Taylor, J. (1965). *Phytopathology* **10**, 1079 (abstr.).

Taylor, J. (1966). *Phytopathology* **56**, 553.

Templeton, G. E., Grable, C. I., Fulton, N. D., and Bollenbacher, K. (1965). *Phytopathology* **55**, 1079 (abstr.).

Templeton, G. E., Grable, C. I., Fulton, N. D., and Bollenbacher, K. (1967a). *Phytopathology* **57**, 516.

Templeton, G. E., Meyer, W. L., Grable, C. I., and Sigel, C. W. (1967b). *Phytopathology* **57**, 833 (abstr.).

Templeton, G. E., Grable, C. I., Fulton, N. D., and Meyer, W. L. (1967c). *Proc. Mycotoxin Res. Seminar, Washington, D.C., 1967* p. 27. U.S. Dept. Agr., Washington, D.C.

Thomas, R. (1961). *Biochem. J.* **80**, 234.

Walker, J. C. (1952). "Diseases of Vegetable Crops." McGraw-Hill, New York.

White, G. A., and Starratt, A. N. (1967). *Can. J. Botany* **45**, 2087.

CHAPTER 8

A Phytotoxin from *Didymella applanata* Cultures

F. SCHURING AND C. A. SALEMINK

I. Introduction

Raspberry disease is caused by the mold *Didymella applanata* (Niessl) sacc. This fungus was described for the first time in 1875 by von Niessl. He named it *Didymosphaeria applanata,* but Saccardo rebaptized the mold in 1883 with its present name (Saccardo, 1882, 1895). In the United States Peck described, in 1894, the mold as *Sphaerella rubina,* which name was changed by Jaczewski (see Stevens, 1925) to *Mycosphaerella rubina.* In 1931 Koch definitely established the identity of *Mycosphaerella rubina* and *Didymella applanata.*

The fungus occurs in several European countries, France, Germany, Great Britain, Norway, and Switzerland and it is known in the United States. It is one of the causes of the so-called stem disease of the raspberry, which manifests itself by decay or defective budding in the axilla of the leaf stalks. An excellent review on the literature of the organism and the disease it causes has been published by Labruyère and Engels (1963).

There exists a certain connection between the raspberry disease caused by *Didymella applanata* and the presence of the raspberry midge *Thomasiniana theobaldi* (Nyveldt, 1963). This insect (gnat) is about 1.5-2 mm long and belongs to the family Diptera. The female insects deposit their eggs under the epidermis of young raspberry stalks. This is only possible at places that are damaged. The damage can occur as tears through growth or can be caused by frost, by snails, or by mechanical means. The larva (which grows to 3.5 mm in length) emerges from the egg in 1 week. Larvae leave the stalks after approximately 2 weeks and burrow into the soil, but during their 2-week stay in the stalks, they excrete a substance that affects the cell walls of the raspberry plant and provides them access to the sap of the plant. When a large number of larva have been present in the stalks, open wounds remain. In this way easy entry by all sorts of molds is possible.

Pitcher and Webb (1949, 1952; Pitcher, 1952) have mentioned the fungus *Didymella applanata* as one of these molds, though *Didymella applanata* can enter the stalks via the leaves and leaf stalks without the presence of a wound (Kerling and Schippers, 1960). Raspberry plants, attacked by this mold, show dark discolorations, often near the buds, but also near growth tears. The discolored spots become somewhat lighter in summer. Through wrinkling and dying of the buds—or in some cases, through poor development—*Didymella applanata* can severely damage the fruits, which come from the raceme, growing in the axilla of 2-year-old stalks.

At the Willie Commelin Scholten Phytopathological Laboratory in Baarn, Holland, Kerling and Schippers established the fact that the fungus *Didymella applanata* forms a phytotoxin upon culture. The mold can be incubated in a liquid medium. The culture filtrate—obtained after the mycelium is filtered—is toxic to young raspberry sprouts placed in this liquid. Small necrotic spots appear between the veins of the leaves, and the sprouts wilt in about 24 hours.

II. The Cultivation of the Mold

For large-scale culture of *Didymella applanata*—obtained from the Centraal Bureau voor Schimmelcultures, Baarn, Holland—a Czapek-Dox nutrient mixture was used—20 gm D-glucose, 2 gm sodium nitrate, 1 gm potassium biphosphate, 0.5 gm potassium chloride, 0.5 gm magnesium sulfate, and 0.01 gm ferrous sulfate dissolved in one liter of water. The pH of the solution was 6. To this culture mixture, 10 gm Difco casein hydrolyzate, 3 gm Difco yeast extract, and 0.1 gm zinc sulfate were added. The culture liquid was sterilized at 120°C for 20 minutes. The mold was incubated as a spore suspension in this liquid, each culture amounting to 250 ml in a 750 ml Erlenmeyer flask. After 24 hours at 20°C the cultures were shaken over a period of 10 days at 20°C. After the culture solution was decanted, the mycelia were washed twice with 100 ml sterilized water for 10 minutes. Then the mycelia were transferred, under sterile conditions, to a 250 ml Czapek-Dox culture medium to which, however, no casein hydrolyzate or yeast extract had been added, and the mixture was shaken again for 10 days at 20°C.

The culture was filtered after the first 10 days to remove any phytotoxin from the culture liquid itself. This phytotoxicity was due to the casein hydrolyzate, which stimulated mold growth and toxin production. Elaborate experiments proved that casein hydrolyzate could not be replaced by another nonphytotoxic fraction with the same effects, i.e., growth stimulation and toxin production. Addition of pyridoxine, thiamine, biotin, folinic acid, L-asparagine, or L-tryptophan to the Czapek-Dox medium did not noticeably influence the growth of the fungus; Mn^{2+} and Co^{3+} diminished and retarded growth, respectively; Cu^{2+} and Cr^{3+} slightly stimulated the growth of *Didymella applanata*. Only zinc sulfate stimulated growth to an appreciable extent. Shaking of the culture is essential. The Czapek-Dox medium was tested for phytotoxicity after shaking the mycelium in the medium for 1 hour at 20°C, with negative results.

Cultures of *Didymella applanata* induced pronounced phytotoxic effects on raspberry sprouts after the second 10-day period of incubation with shaking. When mycelia—shaken for the first 10 days in the medium, containing casein hydrolyzate and yeast extract—was transferred to sterilized water and shaken for another 10 days, the filtrate showed no phytotoxic effect. This seems to indicate that an extracellular factor is formed by *Didymella applanata* during the second period of shaking in the replacement culture (Czapek-Dox medium), and this factor has not been extracted from the mycelium.

III. Test of Phytotoxicity with Raspberry Plants

The toxicity of the culture filtrates, and any fraction derived therefrom, was tested with raspberry sprouts, which regularly developed throughout the year from the roots of raspberry (*Rubus idaeus*, variety Rode Radboud), which were planted in peat soil in a greenhouse situated in such a way that there would be sunshine from sunrise to noon. In autumn and winter, artificial light was supplied from 6 A.M. to 10 P.M. Four rows of Philips 40-W TLS lamps (color no. 33 daylight) were used at a height of 40 cm, each illuminating about 0.7 m² of the bench on which the plants stood. The temperature was 20°C, and the relative humidity in the greenhouse was about 70%.

Each isolated fraction was divided over five tubes and the content of phytotoxin, if present, brought to a concentration of the same order as that of the original filtrate. In each tube containing 25 ml liquid, the base of a raspberry sprout with three leaves was placed. In most experiments we also tested solutions containing higher concentrations of phytotoxin (two or five times the original concentration). Moreover, the original culture liquid and water (blank) were tested for phytotoxicity.

All sprouts were kept under the same environmental conditions for 18-36 hours. During this time the sprouts placed in phytotoxic solutions started to develop typical, small (1-3 mm) necrotic spots on the leaves between the veins. Wilting of the sprouts also occurred. To make the tests comparable, all solutions were brought to pH 6. A humidity of 70% in the greenhouse was an essential factor; higher humidity prevented the uptake of liquid by the sprouts, and so the effects of phytotoxin could not be measured. Bacterial infection of test solutions and sprouts had to be prevented as did air in the vessels of sprouts, which could be avoided by cutting them under water.

Unfortunately, in spite of all the precautions, no quantitative information on phytotoxicity was obtained (compare Gäumann et al., 1950).

IV. Isolation of Phytotoxin

In developing the bioassay and isolation procedures, we used rather arbitrary methods of separation to establish in which fraction the phytotoxicity was present. After several experiments, isolation was carried out as follows. (Each isolation run was done with 30 liters of culture filtrate.) The filtrate, obtained by filtering the mold culture through an asbestos filter, was passed over charcoal (Norite R II) packed in a column (80.3 cm) that was heated to 50°C, at a rate of flow of 1 liter/hour. The phytotoxic factor was adsorbed onto the charcoal and was eluted — after washing the column with 500 ml water — by passing 5 liters of pyridine-water (1:1 v/v) through the column at a rate of 1 liter/hour.

Through repeated concentration in vacuo and dilution with water a pyridine-free residue was obtained (as demonstrated by the odor of the solution). This residue was then filtered in a Sephadex G 25 coarse column (80.5 cm) with water as eluent. Each hour a fraction of 36 ml was collected; fractions 16-23 contained the toxin.

These active fractions were combined, and the solution was concentrated and chromatographed by elution on charcoal (Norite R II, column 80.3 cm, 25°C) with a water-ethanol gradient. The rate of flow was 60 ml/hour. The toxic material was present in the "hour" fractions 23-33 (starting volume of water: 1 liter). Fractions 23-33 were combined and the solution was also concentrated and partition chromatographed on cellulose (columns 40.5 cm) with butanol-pyridine-water (6:4:3 v/v). The rate of flow was 15 ml/hour. Now the active material was isolated from the hour fractions 40-56, which proved to be chromatographically pure. One liter of culture liquid yielded 0.5-2.0 mg phytotoxin. This substance was found to be active even at dilutions of 1 mg/liter.

The described isolation procedure gave reproducible results (in general, no more variation than three fractions). During the period of trial and error, which at last led to the isolation procedure described above, we obtained negative or nonreproducible results with the following methods:

1. Dialysis: this method is not at all suitable for isolation and purification of the toxin.

2. Precipitation of the toxin with the usual reagents.

3. Extraction with organic solvents and salting-out procedures.

4. Ion exchange.

5. Adsorption of the toxin onto charcoal at room temperature; it is probable that an increase in temperature would accelerate the very slow adsorption.

6. Chromatographic procedures following LKB Uvicord I or II systems.

V. Determination of the Purity of the Toxin

The toxin occurred as one spot only in seven different chromatographic systems.

A. PAPER CHROMATOGRAPHY

Paper chromatography was carried out on Whatman No. 1 paper at room temperature, flow over 40 cm, visualization with periodate-benzidine or with diphenylamine. The results are tabulated below.

Solvent systems	R_f value of the toxin
Butanol-acetic acid-water (4:1:5 v/v)	0.40
Butanol-acetic acid-water (4:1:1 v/v)	0.36
Butanol-pyridine-water (6:4:3 v/v)	0.44

B. THIN-LAYER CHROMATOGRAPHY

The chromatography was carried out on Merck Silica at room temperature, flow over 10 cm, visualization with 30% H_2SO_4. The results are tabulated below.

Solvent systems	R_f value of the toxin
Chloroform-methanol-17% ammonia (2:2:1 v/v)	0.45
Methanol	0.75
Ethanol (96%)	0.90
Propanol	0.70

C. Visualization of the Toxin

1. With Diphenylamine

Four grams diphenylamine, 4 ml aniline, and 20 ml orthophosphoric acid (80%) are dissolved in 200 ml acetone. The chromatograms are sprayed with this mixture and then heated for 4 minutes at 80°C. The toxin appears as a grayish spot.

2. With Periodate-Benzidine

The chromatogram is sprayed with a solution of sodium periodate (0.5% in water) and 5 minutes later with a solution of benzidine (1% in alcohol-acetic acid, 4:1 v/v). The toxin gives a white spot on the blue chromatogram.

3. With Sulfuric Acid

Thin-layer chromatograms are heated, after spraying with sulfuric acid (30% in water), for 10 minutes at 150°C. The toxin appears as a carbon spot.

VI. Structure Elucidation of the Toxin

The toxin is a colorless substance, readily soluble in water. When high concentrations of the toxin are dissolved in water, viscous solutions are formed. The behavior of the compound on acid or base ion-exchange columns (Amberlite IR 120 and Dowex 1X8) indicated a neutral, nonionogenic character. The neutral character has been confirmed by the production of a neutral solution in water.

Elemental analysis [qualitative (Baker and Barkenbus, 1937; Feigl, 1954) and quantitative] gave negative results for nitrogen, sulfur, and phosphorus. Other elements—besides carbon, hydrogen, and oxygen—are not present. Methoxyl groups could not be detected.

The isolated substance has a molecular weight lower than 2500 by gel filtration. The usual methods of molecular weight determination yielded no results. The molecular weight of the toxin, which could be determined with a Mechrolab Osmometer, is established as 1682 ± 150.

1. Infrared Spectroscopy

The infrared spectrum of the colorless solid is shown in Fig. 1. Certain data of this spectrum indicate the presence of a carbohydrate. This fact was supported by periodate, anthrone, and orcinol reactions.

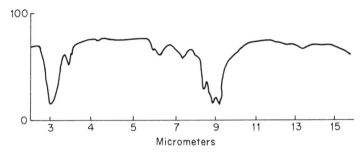

FIG. 1. Infrared spectrum of the isolated toxin, obtained in a Beckman IR-8 instrument equipped with a potassium bromide pellet.

2. MOLECULAR WEIGHT DETERMINATION

Sample: 23.88 mg toxin in 1 ml distilled water.
Apparatus: Mechrolab Osmometer
$D = 0.56$ (0.55-0.55-0.56 balance $0 \rightarrow +\frac{1}{2}$)
Concentration sample: $14.2 \times 10^{-3} M$
$MW = 1682 \pm 150$

3. REACTION WITH ORCINOL

0.1 mg toxin was dissolved in 1 ml water; to this solution was added 1 ml of a solution of 1.5 gm orcinol in 20 ml sulfuric acid (30%), followed by 5 ml sulfuric acid (60%). The mixture was heated for 20 minutes at 80°C. A brown-orange color demonstrated that a carbohydrate was present (Rimington, 1940).

4. REACTION WITH ANTHRONE

0.1 mg toxin was dissolved in 0.5 ml water. To this solution was added 1 ml of a solution of 0.2 gm anthrone in 100 ml sulfuric acid (80%). The mixture was heated for 10 minutes at 100°C. The deep blue color that appeared demonstrated that a carbohydrate was present (Sheelds and Burnett, 1960).

5. OTHER REACTIONS

The naphthoresorcine, benzidine, and ammoniacal silver reactions were negative, proving that the toxin had little or no reduction capacity. After acid hydrolysis of the toxin, three degradation products could be isolated by paper and thin-layer chromatography. Gas chromatographic separation could be realized after trimethylsilylation of the degradation products. One of the products here was D-glucose, but the following monosaccharides or related compounds could not be detected: arabinose, xylose, fructose, mannose, galactose, fucose, sorbitol, mannitol, rhamnose, ascorbic acid, or gluconic acid.

At this stage of research the data made it possible to predict that the toxin could be an oligosaccharide with little or no reduction capacity.

6. ULTRAVIOLET SPECTROSCOPY

The ultraviolet spectrum of the toxin (Fig. 2) proved, however, that the oligosaccharide had to contain groups other than carbohydrate.

7. ACID HYDROLYSIS

Ten milligrams of the toxic material was heated in 10 ml hydrochloric acid (5%) under nitrogen for 3 hours at 100°C. Repeated evaporation *in vacuo* removed most of the hydrochloric acid, and the last traces were removed with silver carbonate.

a. Paper Chromatography of the Hydrolyzate. The chromatography was carried out on Whatman No. 1 paper at room temperature, flow over 40 cm, and visualization with periodate-benzidine or with naphthoresorcine-orthophosphoric acid. The results are tabulated below.

Solvent systems	R_f values in the hydrolyzate	R_f D-glucose
Butanol-acetic acid-water (4:1:5 v/v)	0.22-0.38-0.54	0.22
Butanol-acetic acid-water (3:1:1 v/v)	0.18(w)-0.35-0.85	0.35
Butanol-ethanol-water (5:5:2 v/v)	0.00(?)-0.20.0.28-0.75	0.20
Butanol-pyridine-water (6:4:3 v/v)	0.37-0.47-0.58	0.47
Pyridine-ethylacetate-water (1:2:2 v/v)	0.22-0.46-0.80	0.45

b. Thin-Layer Chromatography. The chromatography was carried out on silica buffered with 0.1 *N* boric acid at room temperature, flow over 10 cm, and visualization with periodate-benzidine or naphthoresorcine-orthophosphoric acid. The results are tabulated below.

Solvent systems	R_f values in the hydrolyzate	R_f D-glucose
Butanol-acetone-water (3:1:1 v/v)	0.45-0.70	0.70
Butanone-acetic acid-water (3:1:1 v/v)	0.35-0.48-0.59	0.59

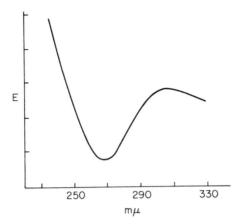

FIG. 2. Ultraviolet spectrum of the isolated toxin, obtained in a Beckman DK-2 instrument; ϵ_{max} (water) = 204 mμ g,1.

c. Silylation of the Toxin Hydrolyzate Products. Five milligrams of the dried acid hydrolyzate were dissolved in 1 ml pyridine to which is added a solution of 0.2 ml trimethylsilyl chloride and 0.4 ml $(CH_3)_3SiNHSi(CH_3)_3$. The mixture is stirred, and after 10 minutes, 5 ml hexane and 5 ml water are added. After complete separation of the layers, the hexane layer is shaken with 5 ml water. The hexane layer is concentrated to 0.5 ml, after drying on anhydrous sodium sulfate.

d. Gas Chromatography of the Silylation Products. Separation of the mixture of the silylated products of the acid hydrolyzate of the phytotoxin was by gas-liquid chromatography (Fig. 3). The retention time of

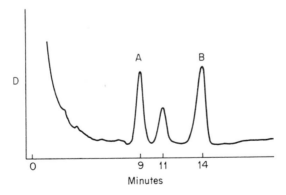

FIG. 3. Gas-liquid chromatogram of the silylated products of the acid hydrolyzates of the phytotoxin. Performed with 5 μl of sample on an F and M model 720 column 6 feet × ⅛ inch, with 5% Se52 on 60-80 mesh W (Chromosorb); thermoconductivity cell detector, injection temperature 210°C, detector temperature 230°C, column temperature 170°C, hydrogen gas flow rate 30 ml/minute.

peak A agrees with that of silylated α-D-glucose, the retention time of peak B with that of silylated D-sorbitol. After the addition of silylated α-D-glucose or silylated D-sorbitol, both peaks appear larger on the chromatogram. These experiments, however, give no decisive proof for the presence of D-glucose or D-sorbitol. The results of the paper and thin-layer chromatograms confirm, however, the presence of D-glucose. For D-sorbitol, the results are contradictory. A comparison of the retention time of a number of silylated sugars and derivatives under similar circumstances is tabulated below.

Sugar	Retention time (minutes)
β-D-Arabinose	3
α-D-Xylose	4
β-D-Fructose	6.8
D-Mannose	6.3 + 9.8 (α or β, β or α)
D-Galactose (most probably)	8
α-D-Glucose	9
D-Sorbitol	14
D-Mannitol	14.6
β-D-Glucose	18

8. Methods and Results of the Methylation and Silylation Experiments

The course of the structure elucidation was determined as follows: the toxin was completely methylated and then hydrolyzed. The fragments in the hydrolyzate were silylated and separated by gas-liquid chromatography. The methylated and silylated monosaccharides thus obtained were examined by mass spectrometry.

The method, mentioned above, was originally only a methylation procedure. An unknown poly- or oligosaccharide was methylated, next hydrolyzed, and chromatographed on paper. The identity of the fractions thus obtained was mostly based on R_f values, as compared to the data of known products. Supplementary data came from molecular weight determinations and partial hydrolysis. Kochetkov (1964; Kochetkov and Cherkov, 1964) improved this method. The fractions of the methylated poly- or oligosaccharide were separated by gas chromatography, and each compound was submitted to mass spectrometry. The method of Kochetkov has the advantage that, in addition to R_f values (in this case retention times), some physical data of the unknown derivatives (from mass spectra) are also obtained. The low validity of the methylated monosaccharides with one or two hydroxyl groups, however, limits the scope of

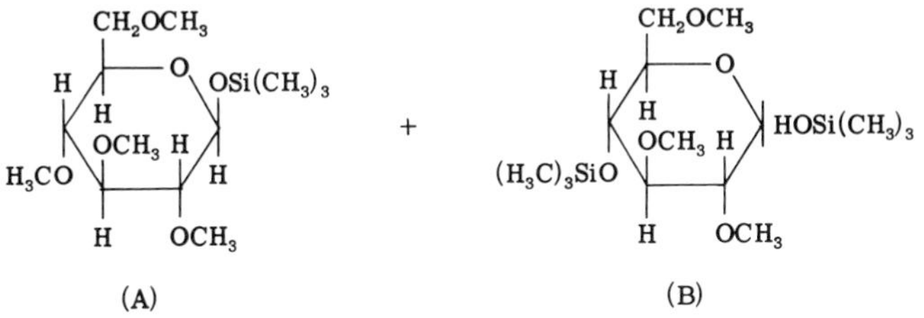

FIG. 4. Cellobiose products of methylation-hydrolysis-silylation; (A) molecular weight 308, mostly β isomer; (B) molecular weight 366, α and β isomer present.

this method. Kochetkov improved his method by deuteromethylating the free hydroxyl groups of the hydrolysis products of the methylated poly- or oligosaccharides.

The method used on the phytotoxin of *Didymella applanata* was that of Sweeley *et al.* (1963) and Wood *et al.* (1965) and is a variant of the method of Kochetkov, with trimethylsilyl chloride used instead of deuteromethyl iodide. Experimental details were controlled with cellobiose and saccharose. The methylation method according to Hakamori (1964) proved to give the best results.

Cellobiose gives the products shown in Fig. 4 after methylation, hydrolysis, and silylation. Compounds A and B were separated by gas-liquid chromatography (SE 30 column). Table I shows some exact masses with the corresponding empirical formulas, derived from the mass spectrum of A.

Two metastables in this spectrum, in good correlation with m/e 192.15 and 176.24, indicate the transition of $223^+ \rightarrow 207^+ + 16$ and $207^+ \rightarrow 191^+ + 16$. According to these data, the peak at m/e 207 represents the fragment $C_6H_{11}O_6Si$. In the spectrum no parent peak was found. The highest mass, 293, represents the parent ($C_{13}H_{28}O_6Si$) minus methyl (308 − 15), as was already reported (Sharkey *et al.*, 1957). Furthermore there is a

TABLE I

SOME EXACT MASSES DERIVED FROM THE MASS SPECTRUM OF COMPOUND A

m/e	Formula	Measured mass	Calculated mass
223	$C_7H_{15}O_6Si$	223.0653	223.0638
193	$C_5H_9O_6Si$	193.0175	193.0169
191	$C_5H_7O_6Si$	191.0018	191.0012
149	$C_5H_9O_5$	149.0453	149.0456
147	$C_6H_{11}O_4$	147.0666	147.0657
133	$C_4H_5O_5$	133.0137	133.0137
104	C_7H_4O	104.0267	104.0262

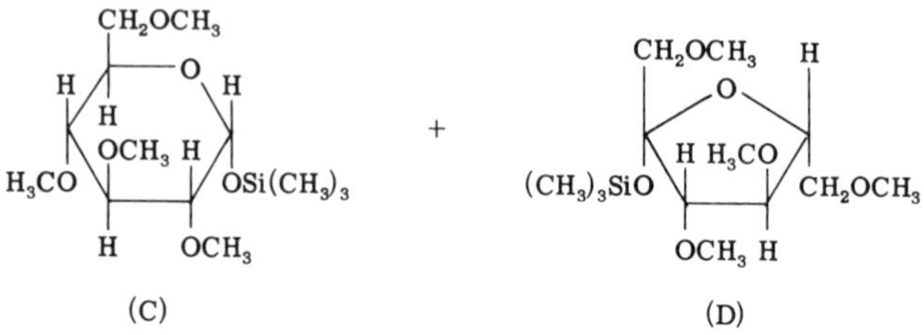

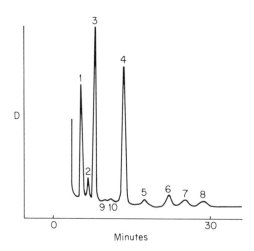

(C)   (D)

FIG. 5. Sucrose products of methylation-hydrolysis-silylation; (A) molecular weight 308, mostly α isomer; (B) molecular weight 358.

repeated loss of methylene (CH_2) and methane (CH_4). The mass spectra of the two forms of B show a fragmentation pattern similar to compound A.

The same experiments were carried out on sucrose and the products shown in Fig. 5 could be obtained. The mass spectrum of C is very similar to that of A: there are only differences in intensity of corresponding peaks. The mass spectrum of D is quite different from that of A, but it also shows a loss of methylene and methane.

The phytotoxin was subjected to the described reactions. The reaction mixture, after methylation, hydrolysis, and silylation, was separated by gas-liquid chromatography (Fig. 6). Fractions 1-8 were collected (fractions 9 and 10 were too small). Some data of the mass spectrum of fraction 3 are shown in Table II.

FIG. 6. Gas-liquid chromatogram of the methylation-hydrolysis-silylation product. Performed with a 100 μl sample on an Aerograph model 1520 column 8 feet × ¼ inch, with 5% SE 30 on 60-80 mesh W (Chromosorb); thermoconductivity cell detector, injection and detector temperature 250°C, column temperature 115°C, hydrogen gas flow rate 55 ml/minute.

TABLE II

SOME DATA OF THE MASS SPECTRUM OF FRACTION 3

m/e	Formula	Measured mass	Calculated mass
295	$C_{10}H_{23}O_6Si_2$	295.1040	295.1033
207	$C_6H_{11}O_6Si$	207.0316	207.0325
191	$C_5H_7O_6Si$	191.0017	191.0012

There are many similarities between this spectrum and that of compound A: peaks 207, 193, 191, 177, 163, 147, 133, and 119. The formula for peaks 207 and 191 are the same for both spectra. This is also true for the metastable 176.24 ($207^+ \rightarrow 191^+ + 16$). All these data make it seem plausible that fraction 3 is a glucose derivative with two trimethylsilyl groups. The position of the two trimethylsilyl groups can be determined as follows.

A metastable at 145.25 indicates the transition $295^+ \rightarrow 207^+ + 88$. The formulas for 295 and 207 are $C_{10}H_{23}O_6Si_2$ and $C_6H_{11}O_6Si$, respectively. The difference, $C_4H_{12}Si$, seems to be $(CH_3)_4Si$, thus indicating that tetramethylsilane has been eliminated. Tetramethylsilane is probably formed from a trimethylsilyl group and a single methyl group, originally from the other silyl group. There are two reasons to support this possibility. In compounds with one trimethylsilyl group no elimination of tetramethylsilane has so far been observed. Moreover, the elimination of tetramethylsilane in this way is energetically favored.

It seems highly probable that such an elimination, not discussed here, can only take place if the trimethylsilyl groups are close together. One of the trimethylsilyl groups has to be attached to carbon atom 1—a glucosidic hydroxyl cannot be present because the compound has no reduction capacity—in the α-position. In the gas chromatogram of the hydrolyzate of phytotoxin only glucose in the α-form was found. The second trimethylsilyl group must, therefore, be attached to the hydroxyl group on carbon atom 2 (Fig. 7).

The data derived from the mass spectrum of fraction 2 are recorded in Table III.

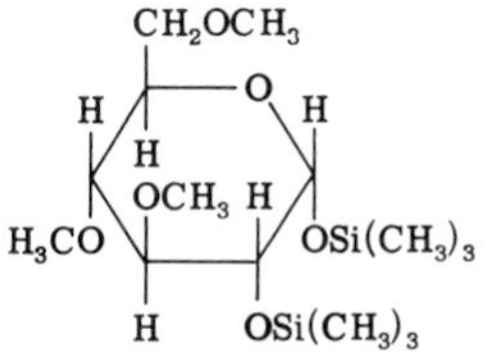

FIG. 7.

TABLE III

DATA DERIVED FROM THE MASS SPECTRUM OF FRACTION 2

m/e	Formula	Measured mass	Calculated mass
295	$C_{10}H_{23}O_6Si_2$	295.1009	295.1034
281	$C_9H_{21}O_6Si_2$	281.0871	281.0876
265	$C_8H_{17}O_6Si_2$	265.0573	265.0564
223	$C_7H_{15}O_6Si$	223.0644	223.0638
207	$C_6H_{11}O_6Si$	207.0342	207.0326
193	$C_5H_9O_6Si$	193.0162	193.0168
133_a	$C_5H_9O_4$	133.0504	133.0501
133_b	$C_4H_9O_3Si$	133.0323	133.0321
133_c	$C_4H_5O_5$	133.0140	133.0137
	$C_3H_9O_2Si_2$	133.0140	133.0141
73	C_3H_9Si	73.0471	73.0473

This spectrum closely resembles that of fraction 3 (peaks 295, 279, 265, 207, 193, 191, 177, 163, 147, 133, and 119, and, respectively, the formulas $C_{10}H_{23}O_6Si_2$, $C_6H_{11}O_6Si$, $C_5H_9O_6Si$, and $C_5H_7O_6Si$). Probably here, too, a glucose derivative with two trimethylsilyl groups is present, and one of these groups is attached to carbon atom 1. The second tri-methylsilyl group cannot be attached to carbon atom 2, for, in that case, the substance would have to be identical with fraction 3. The second group cannot be attached to carbon atom 4 either, because the mass spectrum of fraction 2 is not identical with that of compound B. Attach-ment of the trimethylsilyl group to carbon atom 6 seems to be most im-probable, as this would produce, in the mass spectrum, a peak with mass 103 [$(CH_3)_3SiOCH_2$] (Kim et al., 1967), but this peak is not present. This means that the only remaining possibility is the attachment of the second trimethylsilyl group to carbon atom 3 (Fig. 8). Fractions 7 and 8 must be nearly the same, because the corresponding mass spectra are almost identical.

The masses for fraction 8 were measured (Table IV).

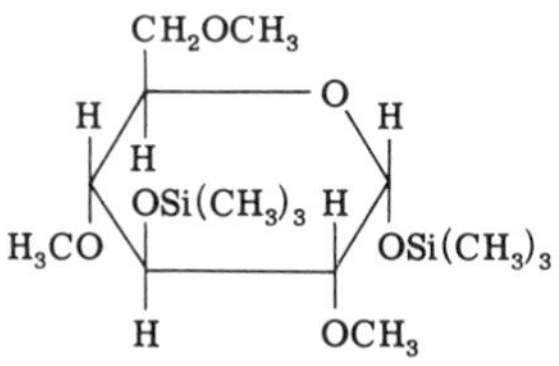

FIG. 8.

TABLE IV
MASSES FOR FRACTION 8

m/e	Formula	Measured mass	Calculated mass
281	$C_8H_{17}O_7Si_2$	281.0528	281.0513
207	$C_6H_{11}O_6Si$	207.0320	207.0325
176	$C_{13}H_{20}$	176.1565	176.1565
133	$C_{10}H_{13}$	133.1019	133.1017
119	C_9H_{11}	119.0861	119.0861
105	C_8H_9	105.0703	105.0704

Peaks 223, 207, 193–191, and the formula $C_6H_{11}O_6Si$, again indicate the presence of a glucose skeleton. Next to this derivative, an extra oxygen atom ($281 = C_8H_{17}O_7Si_2$) and a hydrocarbon part ($176 = C_{13}H_{20}$) are present next to trimethylsilyl groups. Peaks 176, 161, 147, 133, 119, 105 and 91, together with the formulas $C_{13}H_{20}$, $C_{10}H_{13}$, and C_8H_9, show a loss of one methyl and five methylene groups with C_6H_7 remaining. These data can be explained by the presence of the next group at one of the hydroxyl groups of glucose (Fig. 9).

Peak 91 represents the tropylium ion (Formula I), originating from a phenyl group. It is well known (Spiteller, 1966) that in a mass spectrometer tropylium ions can be formed from aromatic compounds along different lines. The presence of a tropylium ion is, moreover, supported by the presence of peak 65 (91 − ethyn). The ultraviolet spectrum of the phytotoxin (Fig. 2) supports the presence of an aromatic ketone.

Tropylium ion ($C_7H_7^+$)

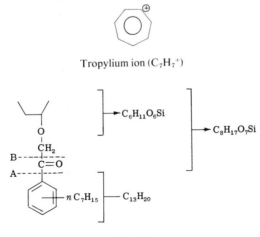

FIG. 9.

CH₂OSi(CH₃)₃

A = CH₃ or X

X = CH₂—C—⟨⟩—C₇H₁₅

Fɪɢ. 10.

CH₂OSi(CH₃)₃

X = CH₂COC₆H₄C₇H₁₅

Fɪɢ. 11.

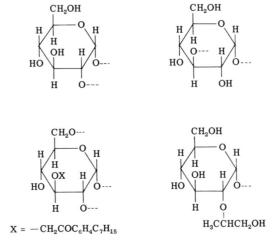

X = —CH₂COC₆H₄C₇H₁₅

Fɪɢ. 12.

Low intensities of peaks with masses larger than 281 made it impossible — until now — to measure masses. Peaks 369 and 355, which are also found in the spectrum of fraction 7, represent formulas $C_{12}H_{29}O_7Si_3$ and $C_{11}H_{27}O_7Si_3$. These data confirm the presence of three trimethylsilyl groups, which can be attached to carbon atoms 1, 2, and 6 (see peak 103 and compare other spectra).

A methoxyl group can be attached either to C-3 or to C-4 of a fragment of the structure in Fig. 10. The presence of a peak at $m/e = 145$, however, gives an indication that $A = CH_3$. The group X must therefore be attached to C-3 (Fig. 11). Fraction 7 and 8 represent the α- or β-form (or β- or α-form) of this structure. The results so far indicate the presence of the structures in the phytotoxin as shown in Fig. 12.

REFERENCES

Baker, J., and Barkenbus, S. (1937). *Ind. Eng. Chem., Anal. Ed.* **9**, 135.
Feigl, F. (1954). "Spot Tests," Vol. II. Amsterdam.
Gäumann, E., Neaf-Roth, S., and Miessher, G. (1950). *Phytopathol. Z.* **16**, 257.
Hakamori, S. (1964). *J. Biochem. (Tokyo)* **55**, 205.
Kerling, L. C. P., and Schippers, B. (1960). Phytopathological Laboratory "Willie Commelin Scholten." Baarn, Holland (unpublished work).
Kim, S. M., Bentley, R., and Sweeley, C. C. (1967). *Carbohydrate Res.* **5**, 373.
Koch, L. W. (1931). *Phytopathology* **21**, 247.
Kochetkov, N. K. (1964). *J. Sci. Ind. Res. (India)* **23**, 468.
Kochetkov, N. K., and Cherkov, O. S. (1964). *Biochim. Biophys. Acta* **83**, 134.
Labruyère, R. E., and Engels, G. M. M. (1963). *Neth. J. Plant Pathol.* **69**, 235.
Nyveldt, W. (1963). *Neth J. Plant Pathol.* **69**, 222.
Peck, C. H. (1894). *Rept. N. Y. State Mus.*, **48**, Part I, 114.
Pitcher, R. S. (1952). *J. Hort. Sci.* **27**, 71.
Pitcher, R. S., and Webb, R. C. R. (1949). *Nature* **163**, 574.
Pitcher, R. S., and Webb, R. C. R. (1952). *J. Hort Sci.* **27**, 95.
Rimington, C. (1940). *Biochem. J.* **34**, 931.
Saccardo, P. A. (1882). *Sylloge Fungorum* **1**, 546.
Saccardo, P. A. (1895). *Sylloge Fungorum* **14**, 527.
Sharkey, A. G., Friedel, R. A., and Lange, S. H. (1957). *Anal. Chem.* **29**, 770.
Sheelds, R., and Burnett, W. (1960). *Anal. Chem.* **32**, 885.
Spiteller, G. (1966). "Massenspektrometrische Strukturanalyse organischer Verbindungen." Weinheim.
Stevens, F. L. (1925). "Plant Disease Fungi," p. 172. New York.
Sweeley, C. C., Bentley, R., Makita, M., and Wells, W. W. (1963). *J. Am. Chem. Soc.* **85**, 2497.
von Niessel, G. (1875). *Oesterr. Botan. Z.* **25**, 129.
Wood, R. D., Rajn, P. K., and Reisen, E. (1965). *J. Am. Oil Chemists' Soc.* **42**, 161.

CHAPTER 9

Compounds Accumulating in Plants after Infection

JOSEPH KUĆ

I. Introduction

This chapter will consider the organic compounds accumulating in infected plant tissues, and their structure, biosynthesis, and influence on the metabolism of plant and infectious agent. Reference will also be made to the relation of such compounds to mechanisms for disease resistance.

Plant tissues have been demonstrated to accumulate a wide array of organic compounds after infection or when placed under a variety of physiological stresses (Kuć, 1961, 1963, 1964, 1966, 1967, 1968; Cruickshank, 1963a,b, 1965, 1966; Cruickshank and Perrin, 1964; Farkas and Kiraly, 1962; Uritani, 1963). A few of the compounds are unique to infection or stress, whereas most are normally present in the plant and markedly accumulate under these conditions. Often the accumulation of compounds can be initiated by extracellular microbial products as well as homogenates and cell-free preparations of microorganisms. The compounds accumulating are generally not microbial in origin, but their production can be initiated by microbial products, and these products can arise because of stress applied to microorganisms. Many terms have been used to describe these compounds, including abnormal metabolites, phytoalexins, and stress metabolites. The use of the term abnormal metabolite is unfortunate, since the accumulation of these compounds is a normal response to a particular set of conditions. The term phytoalexin as defined by Muller and Borger (1940) implies that the compounds have a role in the disease resistance of plants. Indeed, some appear to be a part of the complex mechanism for disease resistance in plants, but others do not appear to have this association. The one condition that is common to their accumulation is metabolic stress. The sources of stress are varied and include infection, temperature, chemical toxicants, hormones, low or high oxygen tension, and nutrient unbalance.

Stress metabolites are synthesized by modifications of either the shikimic acid, acetate-malonate, or acetate-mevalonate pathways, or by joint participation of two of the pathways (Kuć, 1966, 1967). The metabolic alterations necessary for their increased synthesis are profound. The plant is stimulated to mobilize carbon skeletons, energy requirements increase, and enzymes undetected in the healthy plant are synthesized and the activity of others is markedly increased. The potential for the stress response is present in healthy plants but under tight control. The broad nature of the response, affecting many complex metabolic pathways, implies a master control rather than a specific effect on a single enzyme or process. The synthesis of compounds and metabolic alterations require living cells and often are detected in advance of cells penetrated by the microorganism. In the plant-pathogen interaction, the pathogen exerts a profound influence on the metabolism of the plant and, in turn, the plant, by means of the compounds synthesized and the chemical and physical environment that it establishes, influences the metabolism and development of the pathogen. Compounds accumulating in infected plant tissues arise from several possible sources: synthesis by the plant, liberation from glycosides or other bound forms by microbial or plant enzymes, microbial modification of compounds synthesized by the plant, and compounds synthesized by the microorganism. Since infected or otherwise stressed plants and their products are often used as foodstuffs, their effect on animal metabolism should be studied.

II. Irish Potato Tuber

Most of the work concerned with the biosynthesis and accumulation of compounds in potato tubers as a result of infection has been limited to the interaction between *Phytophthora infestans,* the incitant of late blight, and several nonpathogens of potato. Compounds reported to accumulate include chlorogenic and caffeic acid, scopolin, α-solanine, α-chaconine, solanidine, and rishitin (Figs. 1 and 2). All of these compounds, with the

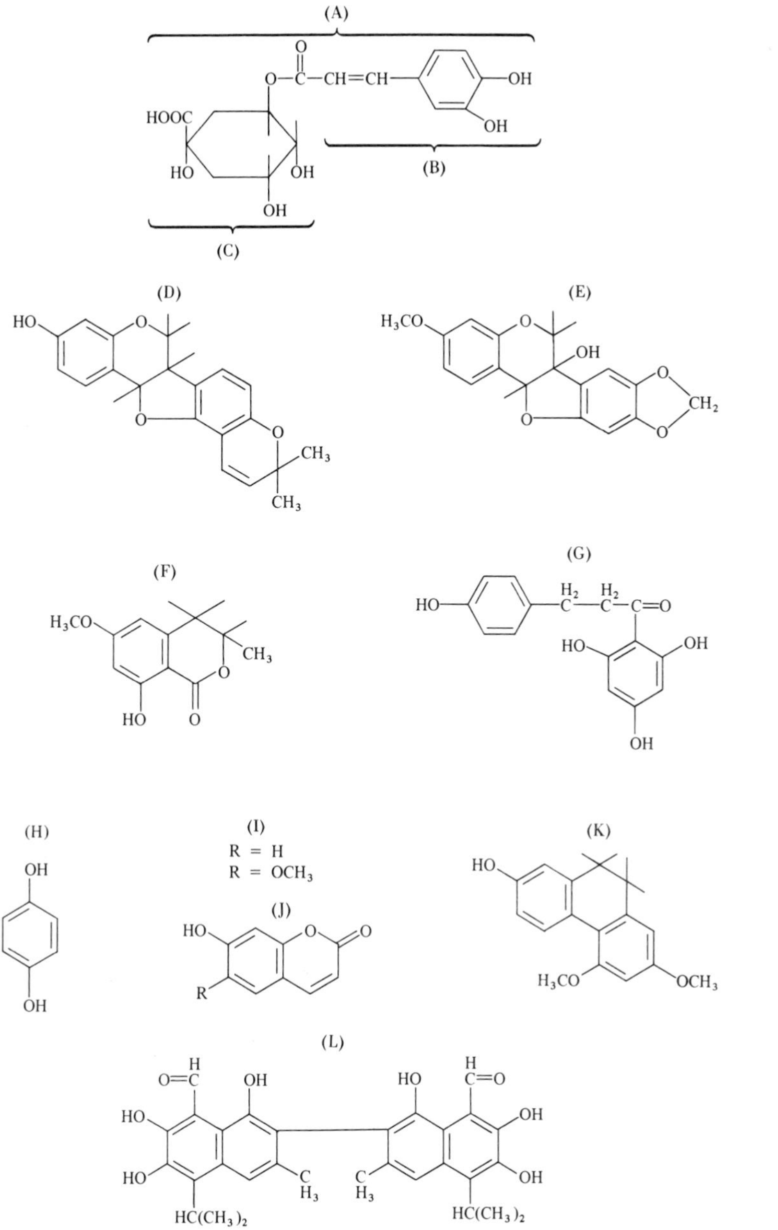

FIG. 1. Phenolic compounds accumulating in plants after infection or injury; (A) chlorogenic acid, (B) caffeic acid, (C) quinic acid, (D) phaseolin, (E) pisatin, (F) 6-methoxymellein, (G) phloretin, (H) hydroquinone, (I) umbelliferone, (J) scopoletin, (K) orchinol, (L) gossypol.

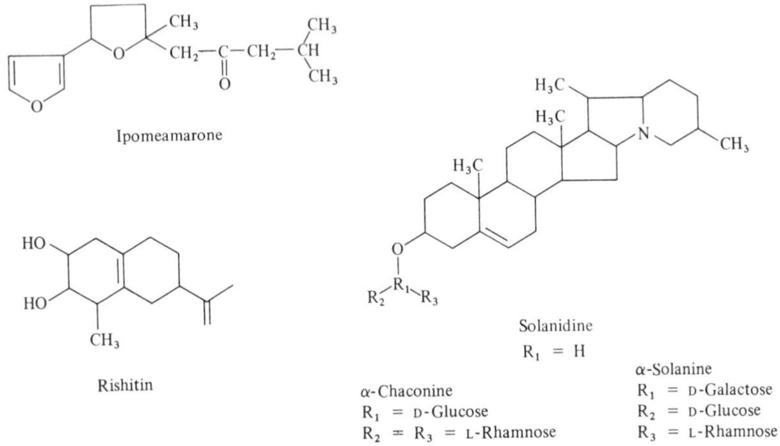

Fig. 2. Nonphenolic compounds accumulating in plants after infection or injury.

exception of rishitin, are normally present in the peel at levels equal to or greater than those produced after infection of peeled tissue.

Slices of fresh fleshy roots and tubers are well suited for studies of the response of plant tissue to stress or infection. They are easily inoculated and held under controlled environmental conditions. After careful sectioning, it is possible to differentiate tissue penetrated by the fungus from that affected but not penetrated. In general, the techniques used for tissue preparation and extraction of roots and tubers are similar. A heavy suspension of asexual spores in water is applied to the surface of slices 1 to 4 cm thick. The concentration of spores and slice thickness are important in determining the extent of response. Dilute spore suspension and slices less than 2- to 5-mm thick limit response. In the former case, this is due to the limited sites of infection, and in the latter the limitation of living cells capable of response. The inoculated tissue is incubated for a period of time, and the infected tissue is separated from the adjacent tissue by slicing. A standard potato peeler is capable of removing slices approximately 1 mm thick and is well suited for sectioning. Fresh tissue and tissue incubated as above but not inoculated serve as controls. Tissue slices are generally cut into 95% ethanol, homogenized, boiled, and the resulting brei is cooled and filtered. The filtrate is used for microbiological assays and qualitative and quantitative studies.

A. Chlorogenic and Caffeic Acids

Kuć et al. (1955, 1956) studied the response obtained by bringing together completely incompatible combinations of fungi and higher plants.

The plant tissues used were potato tubers and carrot and turnip roots, and the fungi studied were *Helminthosporium carbonum, Ceratocystis ulmi,* and *Fusarium oxysporum* f. *lycopersici.* The method of tissue preparation and inoculation were as described above. Chlorogenic and caffeic acid accumulated in the top 2 mm of inoculated potato tuber slices at levels many times that formed in response to slicing without inoculation. The concentration of the acid in the fresh tissue was less than one-tenth of that accumulating after slicing. Extracts of inoculated tissue markedly inhibited the above three nonpathogens of potato, but the chlorogenic and caffeic acid could account for all the inhibitory activity. The acids also accumulated in response to inoculation with two potato pathogens, *Sclerotium rolfsii* and *Fusarium solani* f. *radicola* (Kuc, 1957). Extracts of infected tissue, however, were not appreciably inhibitory to their growth, and the pathogens were less sensitive than the nonpathogens to inhibition by chlorogenic and caffeic acid.

Rubin *et al.* (1947) were among the first to demonstrate a marked increase in the polyphenol content of different varieties of tubers resistant (incompatible) to a single race of the pathogen *P. infestans* as opposed to those susceptible (compatible). The incompatible plants respond rapidly to inoculation, and potato cells penetrated by the fungus collapse within 30 minutes, whereas compatible cells survive for more than 2 days. The principal phenolic accumulating was chlorogenic acid with lesser quantities of caffeic acid. Sokolova *et al.* (1958a) and Tomiyama *et al.* (1955, 1956) infected tubers of a single potato variety with different compatible and incompatible races of *P. infestans.* Only infection with incompatible races resulted in marked accumulation of phenolics. The differences between resistant and susceptible combinations were clear cut, however, only if thick tissue slices were inoculated. Infection of very thin slices with incompatible races did not result in polyphenol accumulation (Tomiyama *et al.,* 1958).

More recent studies by Sakuma and Tomiyama (1967) verified the earlier reports that chlorogenic acid is the principal phenolic accumulating in cut tissue or tissue inoculated with an incompatible race of *P. infestans.* However, they reported that even though a 25-fold increase in chlorogenic acid occurs after inoculation with an incompatible race, it is only one-third that found in cut tissue. Though results with a compatible race were not reported, it appears from earlier reports (Tomiyama *et al.,* 1955, 1956) that the compatible race prevented or more markedly reduced accumulation of phenolics than the incompatible race. Investigations of the distribution of phenolics as a function of distance from the infected cut surface showed that the tissue zone where the metabolism of phenolic compounds is markedly accelerated by infection with an incompatible race was about 10 to 15 cells, or 1.5 to 2 mm in thickness (Sakai *et al.*, 1967). During the first 24 hours after inoculation, the content of total phenols increased more rapidly in this zone than in the corresponding cut tissue. However, after this time interval, the higher the concentration of inoculum the lower the content of total phenols and *o*-diphenols in this zone as compared to the cut tissue. Nevertheless, the rate of incorporation in this zone of D-glucose-U-^{14}C into chlorogenic and caffeic acid was not reduced by inoculation with dense inoculum. The reduced level of chlorogenic and caffeic acid and constant rate of isotope incorporation may be explained by the increase of Klarson lignin in infected tissue as compared to the control. Thus, in infected tissue, phenolic synthesis is accelerated but oxidation and polymerization by phenoloxidases reduces accumulation. The formation of the lignin may be part of the tissue repair mechanism associated with resistance.

Chlorogenic acid and other *o*- or *p*-diphenols are rapidly oxidized to the corresponding *o*- or *p*-quinones by plant and microbial phenoloxidases and peroxidases. Quinones have been suggested to be the inhibitors responsible for restricting development of microorganisms in infected tissue (Farkas and Kiraly, 1962; Okú, 1962; Rubin and Ozeretskovskaya, 1963; Schaal and Johnson, 1955). This hypothesis is supported by the observations that increased synthesis or activation of phenol-oxidizing enzymes in infected tissue is as widespread as the microbially induced synthesis of phenols (Farkas and Kiraly, 1962; Rubin and Aksenova, 1957; Sokolova *et al.*, 1958b). Some evidence has been reported that increased phenol-oxidizing activity arises as a result of the accumulation of appropriate aromatic compounds (Herzmann, 1958). Quinones localized at the site of penetration could inhibit fungi by any or all of a series of mechanisms, and by the same mechanisms serve as localized phytotoxic agents. The phytotoxic effect might explain the rapid collapse of a limited number of plant cells in highly resistant interactions. Quinones can oxidize sulfhydryl groups of enzymes and cofactors and disrupt the delicate oxidation-reduction balance in cells. Rubin and Ozeretskovskaya (1963) report that quinones uncouple phosphorylation from oxidation in necrotized potato tubers. Quinones readily form 1,4 addition products with sulfhydryl compounds, less readily with amino compounds, and they can deaminate amino acids. The protein-precipitating properties of tannins have been used in the tanning process for centuries. A report by Lovrekovich *et al.* (1967) implicates phenolic oxidation products with resistance of tubers to rotting by the bacterium *Erwinia carotovora*. Bacterial pectinase induced polyphenoloxidase formation in the tuber, which resulted in the oxidation and polymerization of phenols and formation of a melanin-like black infection barrier. Resistance was induced by treating healthy tuber slices with pectinase. The virulent bacterium, however, also inhibits phenolic oxidation in rotted tissue by the action of bacterial cell-bound dehydrogenases. In this way, the dehydrogenase activity of *E. carotovora* seems to be a factor in virulence of the bacterium, while the formation of polyphenoloxidase and phenolic oxidation and polymerization products by the host is concerned with restricting infection.

Kuc'(1967) discussed the activation of the pentose and shikimic acid pathways as related to the increased synthesis of phenylpropane derivitives, including chlorogenic and caffeic acid. Increased activity of the pentose pathway (PP) could provide reduced NADP necessary for synthesis via the shikimic and acetate pathways as well as provide erythrose 4P, a precursor in the shikimic acid pathway. Tomiyama *et al.* (1967) treated thin potato tuber discs with a mixture of chlorogenic and ascorbic acids. These slices brown rapidly when not washed after treatment, and the activities of phenoloxidase and glutathione reductase and the content of *o*-diphenols in the tissue increased more rapidly than in untreated controls or unbrowned slices washed immediately after treatment. Activity of PP was increased by the treatment, whereas that of the Emden-Meyerhof-Parnas pathway (EMP) was markedly inhibited. The browning of the discs was accompanied by a reduction of the activity of fructose diphosphate aldolases, while it did not affect 6-phosphogluconate dehydrogenase activity. This suggests that browning shifts the pattern of glucose breakdown to the PP and decreases participation of the EMP and tricarboxylic acid cycle (TCA), resulting in increase phenolic synthesis. This is further substantiated by a decrease in the C_6/C_1 ratio of carbon dioxide evolved from treated discs and increased phenolic content of discs treated with malonate, an inhibitor of the TCA cycle. The investigators suggest that oxidized products of chlorogenic and ascorbic acid are the inhibitors of aldolase and hence shift respiration to PP. Tomiyama (1966) demonstrated that inoculation of potato slices with a compatible race of *P. infestans* 12 to 20 hours before inoculation with an incompatible race prevented browning and hypersensitive collapse of whole cells associated with a high level of resistance.

Using radioactive tracers, it has been established that L-phenylalanine and *trans*-cinnamic acid are precursors of the caffeoyl portion of chlorogenic acid in plant tissues, including the potato tuber (Hanson, 1966; Runeckles, 1963; Reid, 1958; Geissman and Swain, 1957). Levy and Zucker (1960) and Hanson and Zucker (1963) offer evidence to support the following sequence for the synthesis of chlorogenic acid in tuber slices: phenylalanine → shikimic acid → 3-*O*-cinnamoylquinate → 3-*O*-*p*-coumarylquinate → chlorogenic acid. A key enzyme in the biosynthetic pathway is phenylalanine deaminase, which catalyzes the conversion of phenylalanine to cinnamic acid. Zucker (1963, 1965) reported that light-induced synthesis of phenylalanine deaminase and cinnamic acid repressed synthesis. In view of the work by Tomiyama and Kuć and their co-workers discussed earlier, which reported increased phenolic biosynthesis after injury or treatment with various chemicals, it appears that light is merely another of many inducers for phenolic biosynthesis. The regulation of phenolic synthesis has recently been extensively reviewed by Zucker et al. (1967).

B. COUMARINS

Hughes and Swain (1960) demonstrated the accumulation of scopolin in potato tubers infected with *P. infestans*. Blue fluorescent zones were found surrounding areas of infected tissue. The fluorescence in these zones was demonstrated to be due to a 10- to 20-fold increase in scopolin and a 2- to 3-fold increase in chloroginic acid.

C. RISHITIN

Rishitin, a new antifungal terpenoid, has recently been isolated by Tomiyama *et al.* (1968a). Tuber slices 1.5 to 2 mm thick were inoculated with either a compatible or incompatible race of *P. infestans* and incubated for 48 hours. The slices were then homogenized in organic solvents and the rishitin was isolated by column chromatography and thin-layer chromatography using silica gel. Final purification was effected by formation of the crystalline di-3, 5-dinitrobenzoate. Hydrolysis of the benzoate produced a pure crystalline active compound. The compound is a bicyclic norsesquiterpene alcohol with molecular formula of $C_{14}H_{22}O_2$. Its structure was established as one of two isomeric forms by infrared, ultraviolet, nuclear magnetic resonance, and mass spectral data (Katsui *et al.*, 1968). Rishitin could not be detected in fresh tuber tissue or in culture filtrates or extracts of the fungus. A trace of rishitin was found in sliced noninoculated potato tissue and in compatible interactions of host and pathogen, whereas 120 mg/gm fresh weight of tissue was detected in the incompatible or resistant interaction (Fig 3). The ED_{50} of rishitin for growth of germ tubes of *P. infestans* is 2.1×10^{-4} M and spore germination was completely inhibited at 10^{-3} M. The ED_{50} for the growth of germ tubes of *Alternaria kikuchiana* and *Fusarium solani* f. *phaseoli* was 1.3×10^{-4} M and 10^{-4} M, respectively. Two days after infection, the concentration of rishitin in the incompatible interaction is sufficient to markedly retard hyphal growth of *P. infestans* and it may play a role in the disease resistance of tubers to the fungus. Rishitin also profoundly influences the me-

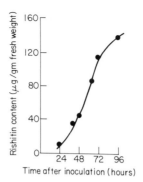

FIG. 3. Accumulation of rishitin in potato tuber slices after inoculation with an incompatible race of *Phytophthora infestans*.

tabolism of plants. The elongation of *Avena* coleoptiles induced by indoleacetic acid (IAA) and wheat leaf tissue by gibberellin (GA$_3$) is completely inhibited by 10^{-3} M rishitin in the test medium. The inhibition could not be completely overcome by higher concentration of IAA or GA$_3$. Rishitin may, therefore, have a role in conditioning a resistant environment in addition to its role as a direct inhibitor of fungal growth.

In further studies, Sato *et al.* (1968) demonstrated that rishitin accumulates in incompatible interactions of potato tubers having R_1, R_2, R_3 and R_4 genes for resistance. In addition, the *r*-gene varieties, susceptible to all the known races of the fungus, accumulated rishitin when dipped into a cell-free homogenate of the fungus. The potential for resistances, therefore, appears to exist in completely susceptible varieties, but this potential is not expressed. This supports the suggestions by Kuć (1968) that resistance often depends on the expression of genetic material rather than the presence of the potential for resistance. It is consistent with disease resistance mechanisms found in animals and is the basis of protection by immunization. Though rishitin does not appear to have a role in the establishment of specificity between host and pathogen (Tomiyama *et al.*, 1968b), it may be the active component restricting growth and development of the pathogen in incompatible interactions. Resistance could be determined by the ability of the host–pathogen interaction to induce rapid rishitin synthesis in tubers, all of which possess the pathways for its synthesis.

Examinations of a series of rishitin derivatives indicated that the hydroxyl group at C-3 is indispensable for antifungal activity (Ishizaka *et al.*, 1969). The activity was intensified by saturating the double bond between the rings of rishitin and/or that of the isopropenyl group at C-7. Activity decreased when oxygenated functional groups were introduced into the side chain. Aromatization of the A ring did not lower biological activity. The antifungal activity of most rishitin derivatives generally paralleled their activity as plant growth retardants. However, some compounds without antifungal activity were active as growth retardants.

D. STEROID GLYCOALKALOIDS

This class of steroids, which contains nitrogen and sugars and includes α-solanine and α-chaconine, has been reported in potato tubers and foliage (Wolf and Duggar, 1946; McKee, 1955; Paseshnichenko, 1957; Prelog and Jeger, 1960; Allen, 1964; Allen and Kuć, 1968; Locci and Kuć, 1967). McKee (1955) reported that α-solanine and α-chaconine were produced by wounded potato tubers and were localized around sites of injury. Paseshnichenko (1957) reported approximately twice as much α-chaconine as α-solanine in whole potato tubers and leaves. Allen and Kuć (1968) demonstrated that the alkaloids were largely restricted to the peel of whole tubers, but injury caused by slicing rapidly increased synthesis and accumulation in peeled tubers. They reported that the principal antifungal components in extracts of potato peels were α-solanine and α-chaconine. They further suggested the importance of these compounds in peel as part of a passive mechanism for disease resistance in potato. Locci and Kuć (1967) repeated the work of Allen and reported a broad spectrum of isoprenoid derivatives, including steroid alkaloids produced in response to infection. The authors demonstrated that the above compounds increase in tubers inoculated with *H. carbonum*, a pathogen of corn. A variety of potato inoculated with two races of *P. infestans* to which it was susceptible did not respond, but inoculation of a variety resistant to the two races of *P. infestans* resulted in a pattern of response similar to that evident in both varieties inoculated with *H. carbonum*. The authors suggested that the accumulation of compounds is due to physiological stress induced by mechanical injury or interaction with microorganisms. A resistant response induces a rapid marked stress, and hence, accumulation of compounds. This pattern is observed in other plant tissues, including sweet potato, bean, and carrot, and it is characterized by the accumulation of compounds synthesized by the activation of the shikimic acid, acetate–malonate, or acetate–mevalonate pathways.

Tomiyama *et al.* (1968b) verified the report of Allen and Kuć (1968) that cutting induces potato tubers to accumulate a high concentration of α-solanine and α-chaconine in tissue close to the cut surface. They reported 538 mg α-solanine plus α-chaconine per kilogram of fresh weight of tissue 48 hours after slicing. Rishitin, α-solanine, and α-chaconine are probably synthesized by the acetate–mevalonate pathway, and it seems reasonable, therefore, that some relationship exists between synthesis of rishitin and the steroid-glycoalkaloids. To test this hypothesis, Tomiyama *et al.* (1968b) sliced potato tubers of a variety resistant to race zero but susceptible to race 1 of *P. infestans*. These were then inoculated with either of the two races, incubated for 48 hours, and then examined for rishitin, α-solanine, and α-chaconine. Rishitin was not detected in fresh tubers or sprouts and occurred in trace amounts in cut slices. Approximately 120 mg of rishitin/kg fresh weight of tissue accumulated in the resistant and 0.44 mg/kg in the susceptible interaction. The level of steroid-glycoalkaloids in the resistant was 3 and in the susceptible 72 mg/kg fresh weight of tissue. Though rishitin accumulated in a very limited area adjacent to the browned lesion, the steroid-glycoalkaloids were distributed in a much wider band of tissue adjacent to the cut or affected surface. It appears that infection with *P. infestans* diverts biosynthesis from α-solanine and α-chaconine to rishitin in the resistant reaction, whereas accumulation of the steroid-glycoalkaloids and rishitin is markedly inhibited in the susceptible reaction. The fungus may have an important role in directing the biosynthesis of stress metabolites repressing or favoring one or the other of the possible pathways inherent in the host.

III. Sweet Potato Root

Infection, injury, or treatment with mercuric chloride or iodoactetate (Uritani, 1963; Akazawa and Wada, 1961) all lead to the accumulation of chlorogenic acid, isochlorogenic acid, caffeic acid, scopoletin, esculetin, umbelliferone, and ipomeamarone in sweet potato root (Figs. 1 and 2). The peel of sweet potato contains all of the above, including ipomeamarone at levels equivalent or greater than that produced by infected peeled tissue. Thus, the stress metabolites are not unique as a response to infection, but reflect increased synthesis of compounds normally localized in the peel of the root (Betaincourt and Kuć, 1970). The work of Uritani and his co-workers (1963) has been directed toward the elucidation of the mechanism of resistance of certain varieties of sweet potato root to *Ceratocytis fimbriata* the incitent of the disease black rot. This has included a study of the compounds accumulating in susceptible and resistant roots after inoculation, the effect of these compounds on the growth and development of the fungus, and the metabolic alterations in the root associated with the synthesis and accumulation of the compounds. The methods of inoculation and extraction are basically as described (see Section II).

A. CHLOROGENIC AND ISOCHLOROGENIC ACIDS

Isochlorogenic acid and chlorogenic acid are the major phenolics accumulating in sweet potato root after inoculation with *C. fimbriata*, and

more isochlorogenic is reported produced than chlorogenic acid (Akazawa and Wada, 1961; Uritani and Miyano, 1955). The layer of tissue
penetrated by the fungus is largely collapsed and low in phenolics due to
their rapid oxidation and polymerization catalyzed by liberated phenoloxidases. The greatest amount of phenolics accumulate in the tissue adjacent to that penetrated by the fungus, and the quantity rapidly decreases
with distance from this tissue. Generally, tissue 5 to 6 mm from the infected zone shows very little accumulation. Fresh tissue contains less
than 1 mg of chlorogenic and isochlorogenic acid per gram fresh weight of
tissue, whereas inoculated tissue contains 10 to 15 mg 72 hours after inoculation (Fig. 4A). Cut slices accumulate chlorogenic and caffeic acids at
less than one-fifth the rate of inoculated slices 72 hours after inoculation.
Though chlorogenic acid and its derivatives generally accumulate to a
higher level in resistant than in susceptible varieties (Akazawa and Wada,
1961), they are not appreciably inhibitory to the growth of *C. fimbriata* at
levels as high as 0.5% *in vitro*. This concentration is considerably greater
than the concentration found in inoculated tissue. Thus, chlorogenic and
isochlorogenic acids do not directly limit development of the fungus in
resistant varieties; however, this does not eliminate the possibility of oxidation or other degradation products as inhibitors of fungal growth *in
vivo*. Another possibility is an indirect role of the acids and their derivatives in disease resistance based on their effect on plant metabolism, including plant hormones (Zucker *et al.*, 1967). The pathway for biosynthesis appears similar to that discussed previously (see Section II,A).
Minamikawa and Uritani (1964) correlated increased phenylalanine and
tyrosine deaminase activity in infected or injured potato slices with increased synthesis of chlorogenic acid, and this is consistent with the observations of Biehn *et al.* (1968a) and Zucker *et al.* (1967).

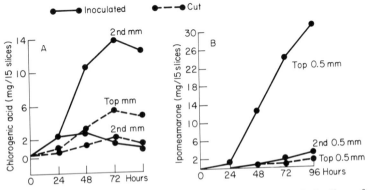

FIG. 4. Accumulation of chlorogenic acid (A) and ipomeamarone (B) in slices of sweet
potato root after inoculation with *Ceratocystis fimbriata*.

B. Coumarins

Umbelliferone, scopoletin, esculetin, and glycosides of umbelliferone and scopoletin also accumulate in sweet potato roots infected with *C. fimbriata* (Minamikawa *et al.*, 1962, 1963; Uritani and Hoshiya, 1953). Unlike chlorogenic acid, umbelliferone accumulates to a greater extent in the penetrated tissue than in the adjacent layer. Umbelliferone is the major coumarin accumulating, but its concentration is only approximately one-thousandth that of the total phenolics. Both umbelliferone and scopoletin accumulate to higher levels in a resistant as compared to a susceptible variety, but umbelliferone accumulation is evident sooner after infection. The compounds markedly inhibit growth of *C. fimbriata in vitro* at a concentration of 0.1%; however, this level may be higher than that found in diseased tissue. The effective concentration of compounds in diseased tissue, i.e., that at the site of fungus development, has not been determined, and the concentrations reported on a fresh weight basis or per slice basis probably reflect a significant dilution. *Ceratocytis fimbriata* degrades umbelliferone and scopoletin in liquid culture to the extent of 20 and 90%, respectively. Neither the transformation of one coumarin to another nor their endogenous formation by the fungus was detected.

C. Ipomeamarone and Other Furanoterpenes

The formation of ipomeamarone in diseased sweet potato root was first reported by Hiura (1941) and the chemical structure was determined by Kubota and Matsuura (1953). Other compounds of similar structure isolated from diseased tissue are ipomeanine and ipomeanic acid. More than ten furanoterpenes were separated by Akazawa (1960) from the oil of diseased plants by thin-layer chromatography using silica gel. Unlike chlorogenic and isochlorogenic acids, ipomeamarone increases markedly in tissue penetrated by the fungus (Fig. 4B). Ipomeamarone appears to be synthesized by the acetate–mevalonate pathway (Akazawa *et al.*, 1962; Imaseki *et al.*, 1964), but the pathway of biosynthesis is not completely elucidated. Tracer studies with acetate-^{14}C reveal that tissue of sweet potato infected with *C. fimbriata* was incapable of synthesizing ipomeamarone from acetate, but it was formed from some intermediate produced from acetate by noninfected tissue in a short time after inoculation (Imaseki and Uritani, 1964). They offer evidence that an intermediate for ipomeamarone synthesis migrates from the noninfected tissue to the infected tissue during the infection process.

Uritani and co-workers suggested that ipomeamarone may be associated with the disease resistance of sweet potato root to *C. fimbriata*. The concentration accumulating is much greater in several resistant as compared to susceptible varieties. The compound accumulates in excess of 1% in diseased tissue and markedly inhibits growth of the fungus at a concentration of 0.1% *in vitro*. Ipomeamarone has at least some selective toxicity toward fungi and is highly toxic to several nonpathogens of sweet potato; but it is only slightly toxic to several pathogens (Nonaka and Kazuomi, 1966). Some doubt is cast on the role of ipomeamarone in disease resistance by the report of Weber and Stahmann (1964). They reported that inoculation of a susceptible variety of sweet potato with a nonpathogenic isolate induced, in a thin layer of tissue around the site of inoculation, an acquired immunity to subsequent inoculation with a pathogenic isolate. Ipomeamarone, however, was not detected in the tissue inoculated with the nonpathogenic isolate. This work does not eliminate a role for ipomeamarone as part of a general disease resistance mechanism against nonpathogens, but it certainly must be admitted that it suggests the presence of other mechanisms for defense. Ipomeamarone and related furanoterpenes also accumulate in sweet potato slices treated with mercuric chloride and iodoacetate (Akazawa and Wada, 1961; Uritani *et al.*, 1960).

IV. Garden Pea and Pisatin

Unlike other plant tissues, only a single compound, pisatin, has been reported to accumulate in the endocarps of detached opened pods inoculated with fungi. The method for detection and assay for pisatin, however, has minimized the possibility of detecting other compounds produced after infection. The presence of an antifungal principle produced by pods of the green bean *Phaseolus vulgaris* L. in response to infection with *Monolinia fructicola* was demonstrated on the basis of biological assays by Muller (1956, 1958). His technique for studying the accumulation of post-infectional compounds was employed by Cruickshank and Perrin (1960) to detect and isolate pisatin from the pea *Pisum sativum* L. The endocarps of detached opened pea pods were inoculated with a fungal spore suspension and the drops of liquid remaining on the tissue containing diffusate were recovered after a suitable period of incubation. The fungitoxic material was extracted from the crude aqueous diffusate by light petroleum ether (bp 55–60°C). The aqueous fraction did not inhibit the growth of fungi. The concentration of aqueous fraction employed and the question as to whether the aqueous fraction was further examined for the accumulation of compounds was not discussed. The structure of pisatin (Fig. 1) was established as 3-hydroxy-7-methoxy-4', 5'-methylenedioxychromanocoumaran (Perrin and Bottomley, 1962). The ED_{50} of pisatin for *M. fructicola* is about $10^{-4} M$, and the concentration for complete inhibition of growth is approximately 2.8×10^{-4} M (Cruickshank and Perrin, 1960). Spore germination was markedly inhibited when spores were treated with 2.8×10^{-4} M pisatin for 48 hours even after they were washed with water. Pisatin has been reported in attached pods inoculated with *M. fructicola* or the pathogen *Ascochyta pisi* at concentrations that prevent germination of the spores of *M. fructicola*.

In further studies, representative groups of microorganisms were assayed for sensitivity to pisatin in agar media (Cruickshank, 1962). Pisatin is a weak antibiotic with a broad biological spectrum. Mycelial growth of *M. fructicola* is three times more sensitive to pisatin than is spore germination. Fungi pathogenic to pea are relatively insensitive to the levels of pisatin accumulating after infection, whereas nonpathogens of pea are generally sensitive. Pisatin formation did not occur as a result of gross mechanical injury (Cruickshank and Perrin, 1963a), although a low concentration of pisatin was reported in uninoculated and incubated open pea pods (Cruickshank and Perrin, 1964). Pisatin formation was stimulated by 19 fungi tested (Fig. 5) and spore-free germination fluids (Cruickshank and Perrin, 1963a). Bacteria did not stimulate the accumulation of pisatin. The major conclusion is that nonpathogens of peas were associated with the formation of pisatin *in vivo* at concentrations in excess of their ED_{50} and pathogens of peas at concentrations below their ED_{50}. The ED_{50} determinations were made *in vitro*. Cruickshank and Perrin further noted that the disease reaction of pea pod endocarp appeared primarily dependent on the sensitivity of the infecting fungus to the concentration of

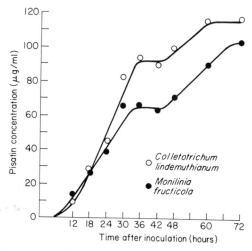

FIG. 5. Accumulation of pisatin in diffusates after inoculation of pea pod endocarp with two nonpathogens of pea.

pisatin formed during the first few days of incubation. In addition to spore-free germination fluids (Cruickshank and Perrin, 1963a), a broad spectrum of metabolic inhibitors also stimulate pisatin formation. They include silver, mercury, and copper salts; p-chloromercuribenzoate; sodium thioacetate; sodium fluoride; sodium azide; sodium cyanide; thioglycolic acid; sodium selenate; and sodium senenite (Perrin and Cruickshank, 1965).

Pisatin formation by pods following treatment at 45°C or anerobic storage was dependent on the duration of the exposure period (Cruickshank and Perrin, 1965a). Pods treated at−20° or 50°C for 3 hours or 100°C for 10 minutes permanently lost their capacity to form pisatin. The endocarp tissues following these treatments were susceptible to *M. fructicola* and various saprophytic bacteria. Inoculation of the endocarp of pods immediately after treatment at 45°C for 2 hours was followed by a susceptible host reaction, and pisatin was not formed. Storage of host tissue prior to inoculation for 3 days, following treatment for 2 hours at 45°C, resulted in marked pisatin production and a level of resistance to *M. fructicola* almost equivalent to that of untreated pea pod endocarps. Exposure of pods to anaerobic storage of 6, 9, or 12 days prior to inoculation also resulted in reduced pisatin production and susceptibility to *M. fructicola* (Cruckshank and Perrin, 1967). Pisatin is localized in infected tissue and does not diffuse to neighboring healthy tissue (Cruickshank and Perrin, 1965a). Cruickshank and Perrin report that pisatin is stable in infected pea pods and in the mycelium of *A. pisi*. Wit-Elshove (1968), however, reported that pisatin is degraded by two pea pathogens but not by three nonpathogens of pea. He suggests that the presence of a mechanism for the detoxication of pisatin may determine whether an organism can parasitize pea and should be considered in addition to the quantitative response of the host. The formation of pisatin by fifty-eight named cultivars, nine numbered lines of *P. sativum* and *P. arvense,* and three other species of *Pisum* following inoculation by *M. fructicola* is further evidence that pisatin has a role in the disease resistance mechanism of pea pods (Cruickshank and Perrin, 1965b).

Biosynthesis of pisatin appears to require participation of the acetate-malonate pathway and the shikimic acid pathway. L-Phenylalanine-U-^{14}C, cinnamic acid-1-^{14}C, and acetate-1-^{14}C were incorporated into pisatin by excised pea pods treated with 3×10^{-3} M cupric chloride (Hadwiger, 1966, 1967). The activity of phenylalanine deaminase, a key enzyme in the biosynthesis of pisatin, increased tenfold when spore suspensions of *M. fructicola* were applied to detached pea pods (Hadwiger 1968a). Actidione inhibited the increase in enzyme activity, the conversion of L-phenylalanine-U-^{14}C to pisatin, and protein synthesis of pea pods treated with inducer. Actinomycin D, however, at concentrations that actively inhibit protein synthesis in other plants, stimulated pisatin synthesis and the above processes, even in the absence of inducers. In further work, Hadwiger (1968b) demonstrated that pisatin production was stimulated by many microbial metabolites, including chloramphenicol, puromycin, cyclohexamide, gliotoxin, phytoactin B, mitomycin C, actinomycin D, ribonuclease, and culture filtrates of *Cochliobolus victoriae* and *Cochliobolus carbonum*. Most of the inducers are microbial metabolites capable of inhibiting protein synthesis and blocking pisatin production at higher concentrations. Since infected pea pods in the field were reported to contain pisatin (Cruickshank and Perrin, 1960), its presence in harvested peas and its metabolic effect on animals deserves further attention.

V. Green Bean and Phaseolin

Using techniques established for working with pisatin and pea, Cruickshank and Perrin (1963b) and Perrin (1964), isolated phaseollin from detached bean pods *Phaseolus vulgaris* L. inoculated with *M. fructicola*. They established the structure as 7-hydroxy-3',4'-dimethylchromanocoumarin (Fig. 1). Though work with phaseolin has not been as extensive as with pisatin, the evidence supports its role in the disease resistance mechanism of bean pods (Cruickshank, 1963b).

Phaseolin and an unidentified compound were reported produced by green bean seedlings in response to infection by the pathogen *Rhizoctonia solani* (Pierre, 1966; Pierre and Bateman, 1967). Both compounds were toxic to *R. solani in vitro*, and the authors suggested that the compounds are associated with a mechanism responsible for restricting the size of lesions in susceptible plants. Similar substances are found in bean seedlings inoculated with the pathogens *Thielaviopois basicola* and *Fusarium solani* sp. *Phaseoli* (Pierre, 1966). The first report associating phaseollin in disease resistance mechanisms with varietal resistance was by Rahe *et al.* (1969a). They reported that etiolated bean seedlings produced phaseolin following inoculation with a varietal nonpathogenic (resistant interaction) or varietal pathogenic (susceptible interaction) race of *Colletotrichum lindemuthianum*, the incitant of bean anthracnose. The resistant or fleck response 60 to 72 hours after inoculation was accompanied by the production of much higher levels of phaseolin than was apparent with the appearance of susceptible symptoms or lesions 72 to 96 hours after inoculation. Phaseolin production appears associated with the appearance of symptoms, but the reason for fungus containment in the resistant interaction has not been established. Inoculation with two nonpathogens of bean,

Helminthosporium carbonum and *Alternaria* sp., induced hypersensitive symptoms 24 hours after inoculation, but phaseolin was not detected even 120 hours after inoculation with *H. carbonum*. A number of unidentified phenolics were shown to accumulate in infected tissue in addition to phaseolin, and some of these were unique for the response to inoculation with *H. carbonum* or *Alternaria* sp. Further reports by Rahe *et al.* (1969b) and Rahe and Kuć (1969) suggest that phaseolin and other phenolics associated with visible cell collapse are not the primary sources of protection in a disease resistance mechanism.

Plants inoculated with the two nonpathogens or a varietal nonpathogenic race were protected against subsequent infection by a varietal pathogenic race of *C. lindemuthianum* (Rahe and Kuć, 1969). They also demonstrated protection against *C. lindemuthianum* by infecting seedlings with a varietal pathogenic race of the fungus that is heat inactivated in the plant after a period of incubation. These observations suggest a similarity between basic mechanisms for disease and immunization in plants and animals.

The mechanism by which a fungus triggers increased phaseolin biosynthesis is not completely elucidated. Recently, Cruickshank and Perrin (1968) isolated a polypeptide, designated as monilicolin A, from the mycelium of *M. fructicola* grown in shake culture in the dark for 96 hours. This polypeptide added to pea pods at concentrations as low as 2.5×10^{-9} M stimulated phaseolin biosynthesis. The median effective dose in relation to the response of the tissue to a conidial suspension of *M fructicola* was $8 \times 10^{-9} M$. Under similar assay conditions using about $10^{-6} M$ monilicolin, half pods of broad bean and pea did not accumulate viciatin or pisatin, respectively. Monilicolin does not appear to be fungitoxic or phytotoxic at levels that stimulate phaseolin production. Structural similarities between pisatin and phaseolin suggest that the latter is synthesized via joint participation of the shikimic acid and acetate-malonate pathways, and a recent report supports this suggestion (Hess and Schwochau, 1969).

VI. Soybean

At least one compound has been reported to accumulate in soybean pod cavities and foliage following inoculation with fungi. The compound has not been completely characterized. Using techniques of inoculation and extraction similar to those used in work with pisatin and phaseolin (see Sections IV and V), Uehara (1958) and Nonaka *et al.* (1966) reported the accumulation of fungitoxic compounds in open soybean pods *Glycine max* L. inoculated with four nonpathogens of soybean. Nonaka and co-workers did not characterize the compounds other than to report solubility properties and ultraviolet absorption spectra. The compounds were slightly soluble in water and the spectra of the compounds produced in response to the fungi were similar, exhibiting an absorption maximum at 285 mμ.

Many problems arise using the pod technique in an attempt to establish a role in disease resistance for compounds accumulating after infection. The interior of a pod is certainly not a normal site through which a microorganism gains entry into the pod, and the use of detached pods makes the conditions even more artificial. Not all tissues of a plant are equally susceptible to disease, and the portal of entry for a microorganism is often very specific. Reports using attached pods, leaves, roots, and tubers inoculated on the normal outer surface are rare. A less artificial set of conditions was reported by Klarman and Gerdemann (1963a,b). Extracts were leached from soybean plants by passing sterilized strings through uninoculated wounds in the hypocotyl or wounds inoculated with either the pathogen *Phytophthora sojae* or two *Phytophthora* nonpathogenic to soybean. One end of each string was placed in a 50-ml beaker of distilled water to a point just above the soil surface. Water passed freely through the wounds and evaporated from the strings. Strings were removed 3 days after inoculation by clipping off the free ends adjacent to the hypocotyls and fungitoxic components were eluted from the strings with water. A soybean variety resistant to *P. sojae* became susceptible when post-infectional compounds were leached from the wound via the string. Nonpathogens of soybean or a varietal nonpathogenic race of *P. sojae* induced the production of a fungitoxic compound, but a varietal pathogenic race of *P. sojae* did not. Nonpathogens of soybeans induced the production of fungitoxic substances regardless of the soybean's susceptibility or resistance to the pathogen *P. sojae*. Plants susceptible to *P. sojae* were protected against the pathogen by a fungitoxic compound produced in a resistant inoculated variety. The fungitoxic compound was transported from the resistant to the susceptible plant by the string technique in sufficient quantity for protection (Chamberlain and Paxton, 1968). Heating soybean plants just before inoculation caused them to temporarily lose their ability to produce fungitoxic compounds and they became susceptible to several nonpathogens (Chamberlain and Gerdemann, 1966).

Paxton and Chamberlain (1969) do not believe that the production of fungitoxic substances after infection is the sole mechanism for disease resistance in soybean. They present evidence for at least two types of disease resistance — resistance in plant tissue 0 to 2 weeks old in which the production of fungitoxic compounds is important and resistance in older tissue recognized by woody stems and greatly reduced production of fungitoxic compounds following infection.

Biehn et al. (1968a) reported phenolics accumulated in etiolated seedlings of soybean, green bean, and lima bean inoculated with the pathogen of corn H. carbonum. Numerous pinpoint pits characteristic of hypersensitive resistance were evident on the seedlings 24 to 30 hours after inoculation. Phenolics accumulating after inoculation of lima bean and green bean differed from those produced by soybean after inoculation with H. carbonum. The same phenolics accumulated when soybean seedlings were inoculated with H. carbonum, M. fructicola, Trichoderma viride, Cercospora sojiiana race 1, or Alternaria sp., but the quantity depended upon the fungus used for inoculation. Twenty-four to twenty-nine hours after inoculation with H. carbonum, total phenolics increased to levels to four to five times above those in uninoculated seedlings. The initial increase in phenolics accompanied an increase in phenylalanine deaminase. Phenylalanine deaminase activity in uninoculated seedlings increased to only 20% of that in inoculated seedlings. Increased synthesis of the enzyme is suggested by the observation that treatment of seedlings with solutions of 0.1 M ethionine or 0.01 M benzimidazole prior to inoculation with H. carbonum reduced phenylalanine deaminase activity 55 to 65%. One of the major phenolic compounds accumulating in soybean after inoculation with H. carbonum or Alternaria sp. inhibited the growth of these fungi 70 to 90%, as well as inhibiting C. sojiina, M. fructiola, and T. viride (Biehn et al., 1968b). The inhibition was obtained with concentrations of compound equal to or less than those accumulating in inoculated tissue. On the basis of spectrophotometric and solubility data, it appears that the inhibitor accumulating in soybean may be closely related to phaseolin and pisatin (Klarman and Sanford, 1968). Both have a chromanocoumaran skeleton and both are produced by legume pods in response to infection. The inhibitor appears to be the major component contributing to the fungitoxicity of extracts and a second inhibitor was not detected in bioassays even when a tenfold concentration of crude extract was examined.

VII. Carrot

The metabolic response of carrot root to infection or injury does not result in the accumulation of a single compound. Chlorogenic acid, 6-methoxymellein (Figs. 1 and 6), and other unidentified compounds, some of which fluoresce under ultraviolet radiation, have been shown to accumulate (Condon and Kuć, 1960, 1962; Condon et al., 1963; Sandstedt, 1967). Histological studies indicate that 6-methoxymellein and chlorogenic acid reach fungitoxic levels around infected sites within 24 hours after inoculation with several fungi nonpathogenic to carrot. The contain-

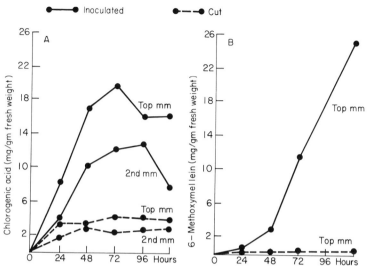

FIG. 6. Accumulation of chlorogenic acid (A) and 6-methoxymellein (B) in slices of carrot root inoculated with *Ceratocystis fimbriata*.

ment of the fungi in the infected tissue coincides with the accumulation of the inhibitors. The 6-methoxymellein completely inhibits the growth of *C. fimbriata* and *T. basicola* at 1×10^{-3} *M*, and at 5×10^{-4} *M* inhibits *C. fimbriata* and *T. basicola* 80 and 60%, respectively. Chlorogenic acid is inhibitory to *T. basicola* (Hampton, 1962), but it does not inhibit *C. fimbriata* at levels accumulating in the carrot. The 6-methoxymellein does not arise as the result of the hydrolysis of a glycoside or other conjugate by host or microbial enzymes. Freezing of carrot slices or autoclaving followed by inoculation does not give rise to 6-methoxymellein or chlorogenic acid. The resistance of carrot root slices to some but not all fungi nonpathogenic to carrots is lost by holding slices at 43° to 45°C for 3 hours immediately prior to inoculation, and production of chlorogenic acid and 6-methoxymellein is markedly reduced to nonfungitoxic levels. Allowing heat-treated carrots to remain at room temperature for 2 to 3 days prior to inoculation restored resistance and the production of chlorogenic acid and 6-methoxymellein to fungitoxic levels. Heat treatment did not affect the resistance of carrot root to *Alternaria solani*, *Botryosphaeria ribis*, and *H. carbonum*, but it did reverse resistance to *Cladosporium cucumerinum*, *Fusarium oxysporum* f. *lycopersici*, and *Glomerella cingulata* (Sandstedt, 1967). Carrot slices inoculated with *C. fimbriata* incubated for 3 days, autoclaved, and reinoculated with *C. fimbriata* did not support growth of the fungus. *Ceratocytes fimbriata* overgrew freshly cut slices that were autoclaved or frozen and then inoculated as well as those held at room temperature for 3 days prior to autoclaving and inoculation. Inoculation of intact carrot roots with *C. fimbriata* via the peel resulted in the accumulation of 1×10^{-3} to 1×10^{-2} molal 6-methoxymellein in the peel 72 hours after inoculation. Not all mechanisms of disease resistance in carrot root, however, depend upon isocoumarin or chlorogenic acid accumulation. Spores of *Ceratocystis fagacearum, C. pilifera,* and *C. coerulescens* did not germinate on the surface of carrot slices, and these fungi did not stimulate the accumulation of chlorogenic acid or 6-methoxymellein (Herndon *et al.,* 1966). The validity of associating the accumulation of 6-methoxymellein with the disease resistance of carrot roots was strengthened by the use of seven isolates of *C. fimbriata* and two of *T. basicola* as well as five different varieties of carrot (Herndon *et al.,* 1966).

The accumulation of 6-methoxymellein in carrot can also be induced by chemicals (Condon *et al.,* 1963), cold treatment (Dodson *et al.,* 1965; Sondheimer, 1957), and ethylene (Chalutz *et al.,* 1969). The compound is responsible for the condition of bitter carrot in storage. Accumulation of 6-methoxymellein, therefore, is not a specific response to infection but rather a response to stress and in this respect its accumulation is consistent with data obtained from other host–fungus interactions.

Chlorogenic acid in carrot root probably is synthesized by pathways discussed earlier (see Sections II and III). Biosynthesis of 6-methoxymellein is not completely established, but it appears to proceed via the acetate–malonate pathway. Malonate-2-^{14}C, acetate-2-^{14}C, and D-glucose-U-^{14}C are readily incorporated into the compound by thin carrot discs. L-Phenylalanine-U-^{14}C, though taken up by discs, was not incorporated into the isocoumarin 40 hours after treatment. Thus, the shikimic acid pathway does not appear involved in the biosynthesis. Structural considerations also suggest the acetate–malonate pathway as the most likely route for synthesis (Gatenbeck and Mosbach, 1959; Mosbach, 1960). Orsellinic acid and 6-methylsalicylate (Birch *et al.,* 1955) are synthesized from acetate and malonate, and these compounds have the same arrangement of substituent groups about the aromatic ring as 6-methoxymellein and both appear synthesized from three molecules of malonyl-CoA and one molecule of acetyl-CoA (Bentley and Keil, 1961; Bu'Lock and Smalley, 1961).

VIII. Apple

Most of the work with apple fruit or foliage has been concerned with the biochemical changes of the host after inoculation with *Venturia inaequalis,* the incitant of apple scab. Phenolics found in apple leaves, fruit, bark, and roots include phloridzin, leucoanthocyanins, epicatechins, catechins, quercetin, cyanidin, 3-hydroxyphloridzin, *p*-coumarylglucose, kaempferol, *p*-coumarylquinic acid, isochlorogenic acid, and chlorogenic acid (Bradfield *et al.*, 1952; Hutchinson *et al.*, 1959; Richmond and Martin, 1959; Williams, 1955, 1956, 1960a,b, 1961; Fisher, 1966). Aside from metabolic changes occurring to phloridzin, little is known concerning the quantitative and qualitative changes of other phenolic compounds following inoculation with *V. inaequalis*. This is further complicated by the fact that the location of the phenolic compounds in the leaf has not been established, and little or nothing is known concerning the metabolic pathways operative in the pathogen. Kirkham (1954) demonstrated that water-soluble constituents extracted from apple leaves with ethyl acetate strongly inhibited the growth and sporulation of *V. inaequalis*. An inorganic source of nitrogen in the medium increased and an organic source decreased inhibition. The water-soluble constituents from a susceptible or a resistant apple variety injected into leaves of a susceptible variety markedly reduced the area of leaf tissue covered by lesions. The effect of the fraction was to reduce susceptibility of apple leaves, but it did not change the disease reaction of the leaf tissue, i.e., it did not make a susceptible leaf resistant. Protection was reduced or lost by injecting urea with the fraction. The investigators suggested that organic sources of nitrogen aided in the degradation of phenols in the fraction, either by the plant or fungus, into nonfungitoxic products. In subsequent studies, Kirkham and Flood (1956) demonstrated that cinnamic acid was highly inhibitory to spore germination and growth of *V. inaequalis* and reduced susceptibility of the host when injected into apple shoots. The presence of cinnamic acid in untreated host tissues was not reported. In general, hydroxylation of cinnamic acid reduced its toxicity. *o*-Hydroxycinnamic acid was the most toxic of the hydroxy derivatives tested. The quantity of hydroxy derivatives of cinnamic acid in apple leaves other than chlorogenic acid, isochlorogenic acid, and *p*-coumarylquinic has not been reported. Further studies of the major components of the phenolic fractions of resistant and less resistant apple and pear varieties (Kirkham, 1957a) suggested no qualitative differences related to the resistance or susceptibility of the varieties tested. There were suggestions, however, of a possible correlation between quantitative differences and resistance. The existence of an optimal balance between phenolic and nitrogen content for development of the pathogen in a host-parasite interaction was presented (Kirkham, 1957b). It was suggested by Flood and Kirkham (1960) that phenolics in the host are important in the physiology of the host-pathogen interaction, but resistance to *V. inaequalis* was not necessarily due to compounds or a mixture of compounds unique for resistant plants.

A fluorescent phenolic compound was reported to increase in the peel of fruits and leaves of several apple selections infected with either *V. inae-qualis* or *Podosphaera leucotricha* (Barnes and Williams, 1960). Diseased leaves of Jonathan, Gallia Beauty, Golden Delicious, Geneva, and selection 384-1 contained a higher concentration of the fluorescent compound than healthy leaves. In the race differential selections Geneva and 384-1, the concentration was higher in the susceptible host-parasite interaction than in the resistant interaction. The increase appeared to be localized in and around infected areas and was detected from 40 hours to 23 days after inoculation. In uninoculated tissues, the compound occurred in trace quantities with some increase with age, but concentrations were always higher in infected tissues of corresponding age. Mechanical injury failed to increase the concentration of the fluorescent compound, and it was not detected on chromatograms of culture filtrates or extracts of *V. inaequalis*. Since whole leaves were extracted and resistant lesions are much smaller than susceptible ones, the concentration of fluorescent compound per unit infected area would be higher in resistant inoculated tissue. The increase in concentration as early as 40 hours after inoculation, prior to any symptom expression on susceptible or resistant hosts, indicates that the compound arises as a result of a physiological interaction between host and pathogen and not merely as a result of gross injury to the host. Recently, Hunter and Kirkham (1968) investigated the fluorescent compounds found in the leaves of a susceptible apple variety and a variety that is resistant under British orchard conditions. Extracts were made from the leaves of the two varieties and compared chromatographically and electrophoretically. A larger number of compounds which fluoresce bright blue in the presence of ammonia vapor were apparent in extracts of the resistant as compared to the susceptible variety. Two of these fluorescent compounds were separated from the bulk of the leaf extract and were shown to inhibit the growth of *V. inaequalis in vitro* and decrease susceptibility of trees when injected *in vivo*. The same compounds increased in the leaves of the resistant variety within a short time after inoculation.

Since phloridzin is the most abundant glycoside found in the leaves of apple, most of the studies of the chemical changes following infection have centered around changes occurring to this compound and its various degradation products.

Noveroske (1962) conducted a lengthy study of the nature and role of host constituents associated with the host-pathogen interaction. Leaf washings from heavily inoculated leaves of the hypersensitive selection *M. atrosanquinea* 333-9 inhibited spore germination of *V. inaequalis*, whereas washings from uninoculated leaves of the same selection and leaves of selection 384-1 exhibiting sporulating lesions did not. The use of spore germination as an assay for fungitoxicity is subject to criticism. *Venturia inaequalis* germinates on the cuticle of susceptible and resistant hosts and, with the exception of hosts with hypersensitive resistance, shows considerable growth beneath the cuticle. Growth rather than spore germination would be a far better assay for the presence of fungitoxic materials in apple leaf; however, the slow growth of the fungus makes it difficult to study the effect of unstable inhibitors for other than a fungicidal effect. Extracts of unoxidized leaf tissue of eight different apple selections were found to be more inhibitory to spore germination of the fungus when assayed in the light than in the dark (Noveroske *et al.*, 1964b). Solutions of phloridzin and extracts of unoxidized tissue held in the light were toxic when assayed in the dark, demonstrating that the effect of light was on components of the solutions rather than the fungus. It was suggested that toxicity in extracts of unoxidized tissue resulted from the autoxidation of phloridzin. Toxicity in extracts from oxidized plant tissue did not change when assayed in the light or dark.

In further studies, Noveroske *et al.* (1964b) found that the toxicity to spore germination of extracts of leaves varied with the length of time tissue was autolyzed prior to extraction. Toxicity increased and then decreased, phloridzin decreased, and phloretin (Fig. 1) first increased and then decreased, paralleling changes in toxicity. Autolyzing leaves in the presence of the reducing agent sodium metabisulfite reduced toxicity, phloretin accumulated, and phoridzin hydrolysis was unaffected. Autolysis in the presence of the β-glycosidase inhibitor glucono-1,5-lactone prevented phloretin accumulation, reduced toxicity levels about 50%, and reduced the rate of phloridzin hydrolysis. It was concluded that both phloridzin and phloretin are oxidized by enzymes in apple leaves. This was substantiated by experiments with enzyme preparations from apple leaves incubated with phloridzin in the presence or absence of the above inhibitors (Noveroske *et al.*, 1964a). Intermediates, suggested to be the dihydroxy analogs of phloridzin and phloretin, were detected when crude enzyme preparations of apple leaves were incubated with phloridzin. No correlation could be found between phloridzin content, activity of β-glycosidase or phenoloxidase, and resistance with eight selections of apple. Since a potential to restrict *V. inaequalis* appeared to exist in resistant as well as susceptible hosts, the mechanism of resistance was suggested to reside in the sensitivity of host tissue to metabolites produced by the fungus. Noveroske (1962) and Noveroske *et al.* (1964b) obtained a diffusate from germinating spores of the fungus which caused collapse of leaf tissue; however, collapse of both susceptible and resistant tissue was observed.

Raa (1968) and Raa and Overeem (1968) repeated the work of Nover-

oske (1962) and Noveroske *et al.* (1964a,b) and confirmed (1) that there
was no relation between phloridzin content of the host and resistance, (2)
the presence of phenoloxidase and β-glycosidase in resistant and susceptible varieties, (3) the conversion of phloridzin to phloretin by leaf homogenates, and (4) the oxidation of both phloridzin and phloretin by leaf homogenates. They further confirmed the work of Holowczak *et al.* (1962),
indicating that phloretin and not phloridzin, inhibited the growth of *V.
inaequalis*. They disagree with Noveroske and co-workers that oxidation
products of phloretin are the principal inhibitors of the fungus, and
present evidence that oxidation products of phloridzin rather than phloretin are of prime importance in the conversions of phloridzin catalyzed by
phenoloxidase and β-glycosidase. In incubation mixtures of phloridzin
with acetone powders of apple leaves, phloridzin was shown to be oxidized to 3-hydroxyphloridzin and further to the corresponding *o*-quinone.
The optimum pH for both reactions is between 4 and 5 with a sharp maximum. The 3-hydroxyphloridzin does not accumulate in the incubation
mixture unless reducing substances are present during oxidation. The *o*-
quinone of phloridzin is extremely reactive and undergoes spontaneous
oxidative coupling reactions with the phlorogluinol nucleus of phloretin.
Thus, the phloretin level decreases as phloridzin is oxidized. The *o*-qui-
none can also react with amino groups of amino acids. Simultaneous with
oxidation, enzymatic hydrolysis of phloridzin takes place with formation
of glucose and phloretin. Phloretin is oxidized, however, at only 7 to 14%
of the rate of phloridzin. The oxidation product of phloretin is 3-hydroxy-
phloretin. The report by Noveroske *et al.* (1964a) that phloridzin was
quantitatively transformed into phloretin before appreciable oxidation
took place may be explained by a difference in the preparation of the
crude enzymes. As suggested by Raa (1968), it appears that the relative
proportion of β-glycosidase to phenoloxidase in the enzyme preparations
employed by Noveroske and associates was higher than that used by Raa.
The reason for the discrepancy may be that Noveroske's group studied
the transformation reactions of phloridzin in dilute aqueous extracts of
homogenized leaves, whereas Raa used undiluted preparations. Since the
β-glycosidase from various sources dissolves readily in water, and the
extraction of the polyphenoloxidase occurs more slowly, it is possible that
the ratio of phenoloxidase to β-glycosidase is higher in undiluted leaf
preparations than in the dilute extracts used by Noveroske and co-
workers. A scheme for the transformation reactions of phloridzin in the
presence of apple leaf enzymes is presented by Raa (1968) and Raa and
Overeem (1968).

The importance of phloridzin and its transformations as reported by
both Raa and Noveroske and co-workers have bearing on the disease re-

sistance problem only if the toxic oxidation products can be shown to occur in resistant but not susceptible hosts. Since it appears that host cell collapse is associated with these transformations, a metabolite produced by the fungus capable of causing the collapse of cells in resistant but not susceptible varieties is suggested. It is uncertain, however, whether the collapse of host cells is due to such a metabolite or rather to the formation of oxidation products of phloridzin and phloretin leading to cell collapse. An extracellular metabolite produced by the fungus may cause a loss of compartmentalization of enzymes and substrate in the epidermal cells, initiating the mixing of phloridzin, phenoloxidase, and β-glycosidase. This leads to the formation of fungitoxic and phytotoxic oxidation products and host cell collapse. It is also uncertain whether the fungus is restricted prior to cell collapse and phloridzin transformations.

IX. Pear

The accumulation of fungitoxic compounds in pear following infection has largely been limited to the interaction of pear foliage and the bacterial pathogen *Erwina amylovora,* the incitant of fire blight. Hildebrand and Schroth (1963, 1964a,b) reported that the liberation of hydroquinone (Fig. 1) from arbutin at levels inhibitory to the growth of *E. amylovora* was part of the disease resistance mechanism of pear leaves, stems, and flowers. The variation in inhibitory activity was primarily related to β-glycosidase activity rather than arbutin concentration. β-Glycosidase activity was low in tissues most susceptible to fire blight including the nectary and inner parts of blossoms, the midrib and petioles of leaves, and the bark of stems. Arbutin hydrolysis in these tissues was slow, and inhibitory levels of hydroquinone were not reached. Rapid oxidation and polymerization of hydroquinone by phenoloxidases or peroxidases coupled with slow release from arbutin prevented its accumulation. High β-glycosidase activity was demonstrated in exterior parts of blossoms, blades of leaves, and the wood of stems. These tissues are generally less susceptible to rapid development of the bacterium. The results reported by Smale and Keil (1966) are in apparent disagreement with those of Hildebrand and Schroth. Smale and Keil worked with leaf homogenates of the resistant varieties Kieffer, Magness, and Moonglow and susceptible varieties Dawn, Bartlett, and DeVoe. The relationship between antibacterial activity and hydroquinone concentration in homogenates of Kieffer indicates a resistance mechanism based upon the hydrolysis of a major portion of arbutin immediately after injury and the transitory accumulation of hydroquinone at a time during the oxidation by phenoloxidases and peroxidases that permits maximum expression of antibacterial activity. With

18 minutes of oxidation, a repression of bioactivity of the residual or unmetabolized hydroquinone occurred. Magness homogenates reflect the presence of an inhibitor or antagonist of hydroquinone that is slowly and not completely metabolized during the 18-minute oxidation interval. Homogenates, oxidized for 18 minutes, contain sufficient antagonist to repress the inhibitory activity of hydroquinone by 50%. The hydroquinone content and antibacterial activity of Moonglow homogenates represents still another type of resistance mechanism. The effect of an antagonist of hydroquinone is marked in homogenates oxidized 6 minutes but is essentially absent in homogenates oxidized for 18 minutes. Varieties Dawn, Bartlett, and DeVoe appear susceptible to fire blight, primarily because antibacterial levels of hydroquinone are not attained. In Dawn, this appears to result from the rapid oxidation of hydroquinone preventing accumulation. It appears that resistance cannot be attributed to a single factor but is associated with no less than the following: (1) the presence of large amounts of arbutin and free hydroquinone in unoxidized leaf homogenates, (2) accumulation and persistence of antibacterial levels of hydroquinone released by the action of β-glycosidase during oxidation, and (3) disappearance, following injury, of an unidentified antagonist and/or the appearance of a synergist of hydroquinone.

X. Orchid

Gaümann (1963, 1963–1964), Gaümann and Kern (1959a,b), Gaümann and Hohl (1960), and Gaümann et al. (1960) reported that *Orchis militaris* and *Loroglossum hircinum* synthesize two fungitoxic compounds, orchinol and hircinol, following injury or inoculation with various microorganisms.

Orchis militaris produces 50 to 100 times more orchinol than hircinol, and *L. hircinum* produces approximately 100 times more hircinol than orchinol. Both substances are produced by the tissues of the bulbs, roots, and stalks under aerobic but not anaerobic conditions. Orchinol has been characterized as 9,10 dihydro-2-4-dimethoxy-6-hydroxyphenanthrene (Fig. 1), whereas hircinol has the same carbon skeleton but with one methoxyl and two hydroxyl groups for which positions have not been established. Both substances have not been detected in intact and sterile bulb tissues but are reported to arise only after infection or injury. Microorganisms vary in their ability to induce synthesis of orchinol and hircinol and differ in their sensitivity to the compounds. *Rhizoctonia versicolor*, *Nectriopsis solani*, *Aureobasidium pullulans*, and *Endomyces albicans* are killed by 10^{-5} M hircinol. Production of the compounds in response to injury may have an important role in protecting orchid bulbs, which are covered by only a tender thin skin. Readily injured in the soil, the bulb response to injury could prevent organisms from penetrating wounds. *Ophrys araneifera* and *Ophrys arachnitiformis* synthesize fungitoxic compounds other than orchinol or hircinol after infection or injury.

XI. Cotton

Introduction of conidia of *Verticillium albo-atrum* into boll cavities or xylem vessels of excised stems of *Gossypium hirsutum* or *Gossypium barbadense* induced a marked accumulation of ether-soluble phenolic compounds after 24 to 72 hours, and the predominant phenolic was identified as gossypol (Fig. 1), 8,8'-dicarboxyaldehyde-1,1',6,6',7,7'-hexahydroxy-5,5'-diisopropyl-3,3'-dimethyl-2,2'-binaphthalene (Bell, 1967). Gossypol is normally found in glands distributed throughout leaves, stems, and root cortices of most cotton varieties (Adams *et al.*, 1960). From 1 to 3 days following inoculation of stem sections and from 1 to 4 days in intact cotton plants, the rate of accumulation of gossypol and related compounds was directly related to host resistance and inversely related to virulence of the pathogen (Bell, 1969).

Resistance of varieties of *Gossypium* spp. to a defoliating strain of *V. albo-atrum* increased as mean incubation temperatures were increased from 22 to 32°C (Bell and Presley, 1969a). At 22°C, all varieties were susceptible, at 32°C all were resistant, and at 25° to 29°C susceptible and resistant varieties gave expected reactions. The rate of accumulation of gossypol and related compounds was maximum at 27.5° to 35°C. Below 27.5°C the rate decreased rapidly, becoming negligible at 15°C. The rate of conidial formation by the fungus was greatest at 20° to 25°C and decreased at 15° and 30°C. The ratio, rate of gossypol accumulation to rate of fungal growth or conidial production, increases markedly as temperature is increased from 20° to 30°C. The speed of gossypol accumulation relative to the speed of secondary colonization by *V. albo-atrum* appears to be an important factor in determining wilt resistance in cotton. Further work by Bell and Presley (1969b) demonstrated that heat-killed conidia or conidia inhibited by heat in stem sections stimulated gossypol synthesis and increased resistance. Inoculation with avirulent strains of the fungus also protected against subsequent inoculation with virulent strains. During a 72-hour incubation at 27.5°C, 10^8 heat-killed conidia/ml induced a 20-fold greater concentration of gossypol in a resistant as compared to various susceptible or tolerant varieties. Enhanced gossypol accumulation in a resistant selection as compared to a susceptible selection was also observed in response to infiltration of stem sections with cupric chloride. The induction of host resistance to the fungus appears to require activation of genetic information which results in the synthesis of enzymes necessary for increased synthesis and accumulation of gossypol. The data strongly imply that genes for resistance are product genes which enhance speed of production or the number of compounds produced after infection or stress. This is consistent with the hypothesis advanced by Kuć (1968) and Hadwiger and Schwochau (1969) and the report by Yarwood et al. (1969). Contrary to the hypothesis of Hadwiger and Schwochau (1969), avirulent and virulent strains of V. albo-atrum did not selectively activate gossypol synthesis. All strains of the fungus induced gossypol synthesis in all varieties of cotton studied. Like the many stress metabolites discussed in preceding sections, gossypol synthesis is activated by low concentrations of toxic chemicals, wounding, or chilling. If gene activation occurs, it appears due to minor cellular injury. Specific derepressors, if they occur, would probably be released by injured host tissue. Partial cellular integrity appears to be necessary for the synthesis of stress metabolites, since excessive doses of compounds which cause severe and essentially irreversible cell injury either prevent or markedly reduce synthesis. Irreversible damage to membranes by high levels of fungal toxin or sensitivity to toxin may prevent the accumulation of stress metabolites resulting in susceptibility, whereas low levels of toxin or insensitivity to toxin may induce synthesis of the compounds, resulting in resistance.

XII. Other Plant Tissues and Stress Metabolites

Alfalfa plants were shown by pathogenicity tests and histological studies to be highly resistant to *Helminthosporium turcicum* and *Colletotrichum phomoides*, although conidia of these fungi germinated and produced appressoria on alfalfa leaves (Higgins and Millar, 1968). *Stemphylium botryosum* and *Stemphylium loti* were, respectively, pathogenic and weakly pathogenic on alfalfa. An antifungal compound was present in diffusates obtained by incubating spore suspensions of *H. turcicum* and *C. phomoides* on excised alfalfa leaves. The compound was extractable from aqueous solution with carbon tetrachloride, reacted positively with tetrazatized benzidine reagent, and had absorption maxima in the ultraviolet at 280 and 285 mμ. The growth of the above pathogens and nonpathogens was inhibited to the same degree by the compound; *S. botryosum* did not stimulate and *S. loti* stimulated accumulation of only a low level of compound as compared to that formed in response to the nonpathogens. A fungitoxic substance was isolated from safflower leaves (Aldwinckle, 1969) and hypocotyls inoculated with *Phytophthora drechsleri* (Thomas and Allen, 1969). The compound was extractable from aqueous solution with diethyl ether and the compound and crude extracts containing the compound were highly inhibitory to the growth of the fungus. Thomas and Allen (1969) identified the compound as 3,11-tridecadiene-5,7,9-triyne-1,2-diol.

Matta *et al.* (1969) demonstrated increased total phenol and orthodihydric phenols in tomato leaves and stems inoculated with two strains of *F. oxysporum* nonpathogenic on tomato as compared to the pathogen *F. oxysporum* f. *lycopersici.* Inoculation with the nonpathogenic *Fusaria* protected plants against subsequent infection with the pathogen. The protective effect increased to a maximum 2 to 3 days after inoculation and then decreased and disappeared after 10 days. The variation with time after inoculation of the protective effect and accumulation of phenolics was similar. The authors suggest that the acquired resistance of tomato plants is reflected by, and perhaps depends upon, the increase of phenolic substances. The phenolics might act by inhibiting growth of the pathogen or by inducing more complex metabolic changes, such as accumulation of growth substances, which lead to the formation of barriers to infection in the xylem vessels. The relationship between acquired immunity and accumulation of fungitoxic compounds was also demonstrated by Siradhana *et al.* (1969) using *Peperomia* spp. and virulent and avirulent isolates of *Phytophthora nicotianae* var. *parasitica.* The structure of the fungitoxic compounds was not reported.

Fluorescent compounds accumulate in tissues adjoining infected xylem vessels of tobacco plants infected by the bacterium *Pseudomonas solonacearum* (Sequeira, 1969). Scopolin and scopoletin are primarily responsible for the increased fluorescence, and hydrolysis of scopolin accounts for only part of the increase in scopoletin. The major portion of the increase appears due to increased synthesis of scopolin and scopoletin. An increase in peroxidase activity and the induced formation of new isozymic peroxidases was reported in tobacco following injection of heat-killed cells or cell-free extracts of the pathogen *Pseudomonas tabaci* into leaves (Lovrekovich *et al.*, 1968). Injection of leaves with dead cells, extracts, or commercial peroxidase increased resistance to the pathogen. A positive correlation between the level of peroxidase activity in tobacco leaves and resistance to *P. tabaci* was reported, and this was associated with the age of leaves. The role of peroxidase in conditioning resistance is unclear and no evidence for an antibody–antigen reaction is presented. The effect of the enzyme is probably on plant metabolism which in turn results in the establishment of a resistant environment. Nevertheless, the possible direct effect of enzyme on bacterium, perhaps resulting in the oxidation of sites on the microbial cell wall or membrane, cannot be excluded.

XIII. Summary

The accumulation of compounds in plant tissues following stress or infection is a general and widespread observation. Some of the compounds accumulate in a broad spectrum of plants (e.g., chlorogenic acid, caffeic acid, and coumarins), whereas others are specific for a given plant (e.g., pisatin, phaseolin, 6-methoxymellein). Profound alterations in metabolism are associated with their synthesis and accumulation. Many enzymes and pathways are activated and new enzymes arise which were not detected in the unstressed plant. The inclusive nature of the response eliminates the involvement of only a single operon and suggests regulation at a "master switch" or control site affecting many operons. The ability of plants to respond similarly to different forms of stress, including high and low temperature, suggests the initial inducer of change or control is present in the unstressed plant. This factor may either be in an inactive form or compartmentalized away from the "master switch." Extracellular metabolites produced by microorganisms can initiate physiological stress in plants, and the rapidity and magnitude with which stress is induced can be a factor in determining the ultimate outcome of the host-pathogen interaction. Rapid induction of stress or incompatibility between microorganism and plant with the maintenance of some cell integrity results in rapid accumulation of fungitoxic stress metabolites and may result in the containment of the microorganism. Delayed plant response, resistance of the microorganism to the stress metabolites or ability to detoxicate them, and rapid death of cells because of hypersensitivity to microbial toxins or enzymes may result in disease. The key to stress response can be alteration in membrane permeability. Temperature stress, chemical toxicants, and microbial metabolites capable of inducing the stress response also affect leakage of metabolites from plant tissues.

Of equal importance to the induction of the stress response is the ultimate return of metabolism to normal. Enzyme activation or increased synthesis and the accumulation of metabolites is often transitory. Many investigators cited in this chapter report a return to normal as early as 48 hours after infection or stress. The ability to produce a constant and fresh supply of inducer appears essential for continued response. It appears the "switch" for the stress response rapidly turns off unless stress is maintained. This is consistent with the observation that wounding of plant tissue results in much less of a response than that observed following infection.

The above conjecture is consistent with the observation that susceptibility is the exception in nature and resistance the rule. It is the rare combination of plant and microorganism that does not cause physiological stress and hence resistance. The adaptation of the metabolic systems in plant and microbe, at least until the latter has developed or reproduced, may be the key and this adaptation is rare.

An important conclusion of the studies of stress metabolites and acquired immunity in plants is that structural genes for resistance are often found even in susceptible plants, and expression of genetic information and not its presence is the key to resistance. This is consistent with our knowledge of acquired immunity in animals. Indeed, this author sees emerging from the literature a thread of similarity in evolutionary development between mechanisms for disease resistance in plants and animals. The application of techniques of "immunization" may help in the practical control of disease in plants.

REFERENCES

Adams, R., Geissman, T. A., and Edwards, J. D. (1960). *Chem. Rev.* **60**, 555.
Akazawa, T. (1960). *Arch. Biochem. Biophys.* **90**, 82.
Akazawa, T., and Wada, K. (1961). *Plant Physiol.* **36**, 139.
Akazawa, T., Uritani, I., and Akazawa, Y. (1962). *Arch. Biochem. Biophys.* **99**, 52.
Aldwinckle, H. S. (1969). *Phytopathology* **59**, 1015 (abstr.).
Allen, E. H. (1964). *Phytopathology* **54**, 886.
Allen, E. H., and Kuć, J. (1968). *Phytopathology* **58**, 776.
Barnes, E. H., and Williams, E. B. (1960). *Phytopathology* **50**, 844.
Bell, A. A. (1967). *Phytopathology* **57**, 759.
Bell, A. A. (1969). *Phytopathology* **59**, 119.
Bell, A. A., and Presley, J. T. (1969a). *Phytopathology* **59**, 1141.
Bell, A. A., and Presley, J. T. (1969b). *Phytopathology* **59**, 1147.
Bentley, R., and Keil, J. G. (1961). *Proc. Chem. Soc.* p. 111.
Betaincourt, S., and Kuć, J. (1970). Manuscript prepared as a note for *Phytopathology.*
Biehn, W. L., Kuć, J., and Williams, E. B. (1968a). *Phytopathology* **58**, 1255.
Biehn, W. L., Williams, E. B., and Kuć, J. (1968b). *Phytopathology* **58**, 1261.
Birch, A. J., Massy-Westropp, R. A., and Moye, C. J. (1955). *Australian J. Chem.* **8**, 539.
Bradfield, A. E., Flood, A. E., Hulme, A. C., and Williams, A. H. (1952). *Nature* **170**, 168.
Bu'Lock, J. D., and Smalley, H. M. (1961). *Proc. Chem. Soc.* p. 209.
Chalutz, E., DeVay, J. E., and Maxie, E. C. (1969). *Plant Physiol.* **44**, 235.
Chamberlain, D. W., and Gerdemann, J. W. (1966). *Phytopathology* **56**, 70.
Chamberlain, D. W., and Paxton, J. D. (1968). *Phytopathology* **58**, 1349.
Condon, P., and Kuć, J., (1960). *Phytopathology* **50**, 267.
Condon, P., and Kuć, J. (1962). *Phytopathology* **52**, 182.
Condon, P., Kuć, J., and Draudt, H. N. (1963). *Phytopathology* **53**, 1244.
Cruickshank, I. A. M. (1962). *Australian J. Biol. Sci.* **15**, 147.
Cruickshank, I. A. M. (1963a). *J. Australian Inst. Agr. Sci.* **29**, 23.
Cruickshank, I. A. M. (1963b). *Ann. Rev. Phytopathol.* **1**, 351.
Cruickshank, I. A. M. (1965). *Proc. Symp. Phytopathol. Ger. Acad. Agr. Berlin* **74**, 313.
Cruickshank, I. A. M. (1966). *World Rev. Pest Control* **5**, 161.
Cruickshank, I. A. M., and Perrin, D. R. (1960). *Nature* **187**, 799.

Cruickshank, I. A. M., and Perrin, D. R. (1963a). *Australian J. Biol. Sci.* **16**, 111.
Cruickshank, I. A. M., and Perrin, D. R. (1963b). *Life Sci.* **9**, 680.
Cruickshank, I. A. M., and Perrin, D. R. (1964). *In* "Biochemistry of Phenolic Compounds" (J. B. Harborne, ed.), pp. 511-544. Academic Press, New York.
Cruickshank, I. A. M., and Perrin, D. R. (1965a). *Australian J. Biol. Sci.* **18**, 817.
Cruickshank, I. A. M., and Perrin, D. R. (1965b). *Australian J. Biol. Sci.* **18**, 829.
Cruickshank, I. A. M., and Perrin, D. R. (1967). *Phytopathol. z.* **60**, 335.
Cruickshank, I. A. M., and Perrin, D. R. (1968). *Life Sci.* **7**, 449.
Dodson, A. R., Fukui, N., Ball, C. D., Carolus, R. L., and Sell, H. M. (1965). *Science* **124**, 984.
Farkas, G. L., and Kiraly, Z. (1962). *Phytopathol. z.* **44**, 105.
Fisher, D. J. (1966). *Long Ashton Agr. Hort. Res. Sta., Univ. Bristol, Ann. Rept.* p. 255.
Flood, A. E., and Kirkham, D. S. (1960). *In* "Phenolics in Plants in Health and Disease" (J. B. Pridham, ed.), pp. 81-85. Pergamon Press, Oxford.
Gatenbeck, S., and Mosbach, K. (1959). *Acta Chem. Scand.* **13**, 1561.
Gaümann, E. (1963). *Compt. Rend.* **257**, 2372.
Gaümann, E. (1963-1964). *Phytopathol. Z.* **49**, 211.
Gaümann, E., and Hohl, H. R. (1960). *Phytopathol. Z.* **38**, 93.
Gaümann, E., and Kern, H. (1959a). *Phytopathol. Z.* **36**, 1.
Gaümann, E., and Kern, H. (1959b). *Phytopathol. Z.* **36**, 347.
Gaümann, E., Neusch, J., and Rimparr, R. H. (1960). *Phytopathol. Z.* **38**, 274.
Geissman, T. A., and Swain, T. (1957). *Chem. & Ind. (London)* p. 984.
Hadwiger, L. A. (1966). *Phytochemistry* **5**, 523.
Hadwiger, L. A. (1967). *Phytopathology* **57**, 1258.
Hadwiger, L. A. (1968a). *Neth. J. Plant Pathol.* **74**, 163.
Hadwiger, L. A. (1968b). *Arch. Biochem. Biophys.* **126**, 731.
Hadwiger, L. A., and Schwochau, M. E. (1969). *Phytopathology* **59**, 223.
Hampton, R. E. (1962). *Phytopathology* **52**, 413.
Hanson, K. R. (1966). *Phytochemistry* **5**, 491.
Hanson, K. R., and Zucker, M. (1963). *J. Biol. Chem.* **238**, 1105.
Herndon, B. A., Kuć, J., and Williams, E. B. (1966). *Phytopathology* **56**, 187.
Herzmann, H. (1958). *Phytopathol. z.* **34**, 109.
Hess, S. L., and Schwochau, M. E. (1969). *Phytopathology* **59**, 1030 (abstr.).
Higgins, V. J., and Millar, R. L. (1968). *Phytopathology* **58**, 1377.
Hildebrand, D. C., and Schroth, M. N. (1963). *Nature* **197**, 513.
Hildebrand, D. C., and Schroth, M. N. (1964a). *Phytopathology* **54**, 59.
Hildebrand, D. C., and Schroth, M. N. (1964b). *Phytopathology* **54**, 640.
Hiura, M. (1941). *Gifu Norin Semmon Gakko Gakujutsu Hokoku* **50**, 1.
Holowczak, J., Kuć, J., and Williams, E. B. (1962). *Phytopathology* **52**, 1019.
Hughes, J. C., and Swain, T. (1960). *Phytopathology* **50**, 398.
Hunter, L. D., and Kirkham, D. S. (1968). *1st Intern. Congr. Plant Pathol., 1968* Abstr., p. 93.
Hutchinson, A., Taper, O. D., and Towers, G. H. N. (1959). *Can. J. Biochem. Biophys.* **37**, 901.
Imaseki, H., and Uritani, I. (1964). *Plant Cell Physiol. (Tokyo)* **5**, 133.
Imaseki, H., Takei, S., and Uritani, I. (1964). *Plant Cell Physiol. (Tokyo)* **5**, 119.
Ishizaka, N., Tomiyama, K., Katsui, N., Murai, A., and Masamune, T. (1969). *Plant Cell Physiol. (Tokyo)* **10**, 183.
Katsui, N., Murai, A., Takasugi, M., Imaizumi, K., and Masamune, T. (1968). *Chem. Commun.* p. 43.

Kirkham, D. S. (1954). *Nature* 173, 690.
Kirkham, D. S. (1957a). *J. Gen. Microbiol.* 17, 120.
Kirkham, D. S. (1957b). *J. Gen. Microbiol.* 17, 491.
Kirkham, D. S., and Flood, A. E. (1956). *Nature* 179, 422.
Klarman, W. L., and Gerdemann, J. W. (1963a). *Phytopathology* 53, 863.
Klarman, W. L., and Gerdermann, J. W. (1963b). *Phytopathology* 53, 1317.
Klarman, W. L., and Sanford, J. B. (1968). *Life Sci.* 7, 1095.
Kubota, T., and Matsuura, T. (1953). *J. Chem. Soc. Japan, Pure Chem. Sect.* 74, 101, 197, 248, and 668.
Kuć, J. (1957). *Phytopathology* 47, 676.
Kuć, J. (1961). *Mededel. Landbouwhogeschool Opzoekingssta. Staat Gent* 26, 997.
Kuć, J. (1963). *Conn. Agr. Expt. Sta., New Haven, Bull.* 663, 20.
Kuć, J. (1964). *In* "Phenolics in Normal and Diseased Fruits and Vegetables" (V. C. Runeckles, ed.), pp. 63-81. Plant Phenolics Group North American, Norwood, Massachusetts.
Kuć, J. (1966). *Ann. Rev. Microbiol.* 20, 337.
Kuć, J. (1967). *In* "The Dynamic Role of Molecular Constituents in Plant-Parasite Interaction" (C. J. Mirocha and I. Uritani, eds.), pp. 183-202. Bruce Publ., St. Paul, Minnesota.
Kuć, J. (1968). *World Rev. Pest Control* 7, 42.
Kuć, J., Ullstrup, A. J., and Quackenbush, F. W. (1955). *Science* 122, 1186.
Kuć, J., Ullstrup, A. J., Henze, R. E., and Quackenbush, F. W. (1956). *J. Am. Chem. Soc.* 78, 3123.
Levy, C. C., and Zucker, M. (1960). *J. Biol. Chem.* 235, 2418.
Locci, R., and Kuć, J. (1967). *Phytopathology* 57, 1272.
Lovrekovich, L., Lovrekovich, H., and Stahmann, M. A. (1967). *Phytopathology* 57, 737.
Lovrekovich, L., Lovrekovich, H., and Stahmann, M. A. (1968). *Phytopathology* 58, 193.
McKee, R. K. (1955). *Ann. Appl. Biol.* 43, 147.
Matta, A., Gentile, I., and Giai, I. (1969). *Phytopathology* 59, 512.
Minamikama, T., and Uritani, I. (1964). *Arch. Biochem. Biophys.* 108, 573.
Minamikawa, T., Akazawa, T., and Uritani, I. (1962). *Nature* 195, 726.
Minamikawa, T., Akazawa, T., and Uritani, I. (1963). *Plant Physiol.* 38, 493.
Mosbach, K. (1960). *Acta Chem. Scand.* 14, 457.
Muller, K. O. (1956). *Phytopathol. Z.* 27, 237.
Muller, K. O. (1958). *Australian J. Biol. Sci.* 11, 275.
Muller, K. O., and Borger, H. (1940). *Arb. Biol. Abt. (Anst.-Reichsanst.), Berl.* 23, 189.
Nonaka, F., and Kazuomi, Y. (1966). *Agr. Bull. Saga Univ.* 22, 39.
Nonaka, F., Isayama, S., and Furukawa, H. (1966). *Agr. Bull. Saga Univ.* 22, 51.
Noveroske, R. L. (1962). Doctoral Thesis, Purdue University, Lafayette, Indiana.
Noveroske, R. L., Kuć, J., and Williams, E. B. (1964a). *Phytopathology* 54, 92.
Noveroske, R. L., Williams, E. B., and Kuć, J. (1964b). *Phytopathology* 54, 98.
Okú, H. (1962). *Phytopathol. Z.* 44, 39.
Paseshnichenko, V. A. (1957). *Biochemistry (USSR) (English Transl.)* 22, 929.
Paxton, J. D., and Chamberlain, D. W. (1969). *Phytopathology* 59, 775.
Perrin, D. R. (1964). *Tetrahedron Letters* No. 1, 29.
Perrin, D. R., and Bottomley, W. (1962). *J. Am. Chem. Soc.* 84, 1919.
Perrin, D. R., and Cruickshank, I. A. M. (1965). *Australian J. Biol. Sci.* 18, 803.
Pierre, R. E. (1966). Ph.D. Thesis, Cornell University, Ithaca, New York.
Pierre, R. E., and Bateman, D. F. (1967). *Phytopathology* 57, 1154.
Prelog, V., and Jeger, O. (1960). *Alkaloids* 7, 343-361.

Raa, J. (1968). Doctoral Thesis, University of Utrecht, Utrecht, The Netherlands.
Raa, J., and Overeem, J. C. (1968). *Phytochemistry* 7, 721.
Rahe, J. E., and Kuć, J. (1969). *Phytopathology* 59, 1045 (abstr.).
Rahe, J. E., Kuć, J., Chuang, C. M., and Williams, E. B. (1969a). *Neth. J. Plant Pathol.* 75, 58.
Rahe, J. E., Kuć, J., Chuang, C. M., and Williams, E. B. (1969b). *Phytopathology* 59, 1641.
Reid, W. W. (1958). *Chem. & Ind.* (*London*) p. 1439.
Richmond, D., and Martin, J. T. (1959). *Ann. Appl. Biol.* 47, 583.
Rubin, B. A., and Aksenova, V. A. (1957). *Biochemistry* (*USSR*) (*English Transl.*) 22, 191.
Rubin, B. A., and Ozeretskovskaya, O. L. (1963). *Biochemistry* (*USSR*) (*English Transl.*) 22, 61.
Rubin, B. A., Arzichowskaja, V. A., and Proskurnikova, T. A. (1947). *Biokhimiya* 12, 141.
Runeckles, V. C. (1963). *Can. J. Biochem.* 41, 2249.
Sakai, R., Tomiyama, K., Ishizaka, N., and Sato, N. (1967). *Ann. Phytopathol. Soc. Japan* 33, 216.
Sakuma, T., and Tomiyama, K. (1967). *Ann. Phytopathol. Soc. Japan* 33, 48.
Sandstedt, K. J. (1967). M.S. Thesis, Purdue University, Lafayette, Indiana.
Sato, N., Tomiyama, K., Katsui, N., and Masamune, T. (1968). *Ann. Phytopathol. Soc. Japan* 34, 140.
Schaal, L., and Johnson, G. (1955). *Phytopathology* 45, 626.
Sequeira, L. (1969). *Phytopathology* 59, 473.
Siradhana, B. S., Schmitthenner, A. F., and Ellett, C. W. (1969). *Phytopathology* 59, 405.
Smale, B. C., and Keil, H. L. (1966). *Phytochemistry* 5, 113.
Sokolova, V. E., Saveleva, O., and Rubin, B. A. (1958a). *Dokl. Akad. Nauk SSSR* 123, 335.
Sokolova, V. E., Saveleva, O., and Rubin, B. A. (1958b). *Dokl. Akad. Nauk SSSR* 123, 251.
Sondheimer, E. (1957). *J. Am. Chem. Soc.* 79, 5036.
Thomas, C. A., and Allen, E. H. (1969). *Phytopathology* 59, 1053 (abstr.).
Tomiyama, K. (1966). *Ann. Phytopathol. Soc. Japan* 32, 181.
Tomiyama, K., Takase, N., Sakai, R., and Takakuwa, M. (1955). *Ann. Phytopathol. Soc. Japan* 20, 59.
Tomiyama, K., Takare, N., Sakai, R., and Takakuwa, M. (1956). *Res. Bull. Hokkaido Agr. Expt. Sta.* 71, 32.
Tomiyama, K., Takakuwa, M., and Takase, N. (1958). *Phytopathol. Z.* 31, 237.
Tomiyama, K., Sakai, R., Otani, Y., and Takemori, T. (1967). *Plant Cell Physiol.* (*Tokyo*) 8, 1.
Tomiyama, K., Sakuma, T., Ishizaka, N., Sato, N., Katsui, N., Takasugi, M., and Masamune, T. (1968a). *Phytopathology* 58, 115.
Tomiyama, K., Ishizaka, N., Sato, N., Masamune, T., and Katsui, N. (1968b). *In* "Biochemical Regulation in Diseased Plants or Injury," pp. 287-292. Phytopathol. Soc. Japan, Tokyo.
Uehara, K. (1958). *Annals Phytopathol. Soc. Japan* 23, 225.
Uritani, I. (1963). *Conn. Agr. Expt. Sta., New Haven, Bull.* 663, 4-19.
Uritani, I., and Hoshiya, I. (1953). *J. Agr. Chem. Soc. Japan* 27, 668.
Uritani, I., and Miyano, M. (1955). *Nature* 175, 812.
Uritani, I., Uritani, M., and Yamada, H. (1960). *Phytopathology* 50, 30.
Weber, D. J., and Stahmann, M. A. (1964). *Science* 146, 929.
Williams, A. H. (1955). *Nature* 175, 213.

Williams, A. H. (1956). *Chem. & Ind.* (*London*) p. 1306.

Williams, A. H. (1960a). *Chem. & Ind.* (*London*) p. 934.

Williams, A. H. (1960b). *In* "Phenolics in Plants in Health and Disease" (J. B. Pridham, ed.), pp. 3-7. Pergamon Press, Oxford.

Williams, A. H. (1961). *J. Chem. Soc.* 4, 33.

Wit-Elshove, A. (1968). *Neth. J. Plant Pathol.* 74, 44.

Wolf, M. J., and Duggar, B. M. (1946). *J. Agr. Res.* 73, 1.

Yarwood, C. E., Hatiro, I., and Batra, K. K. (1969). *Phytopathology* 59, 596.

Zucker, M. (1963). *Plant Physiol.* 38, 575.

Zucker, M. (1965). *Plant Physiol.* 40, 779.

Zucker, M., Hanson, K. R., and Sondheimer, E. (1967). *In* "Phenolic Compounds and Metabolic Regulation" (B. J. Finkle and V. C. Runeckles, eds.), pp. 68-93. Appleton, New York.

CHAPTER 10

The Toxic Peptides of *Amanita* Species

THEODOR WIELAND AND OTTO WIELAND

I. Occurrence, Analysis, and Characterization of the Toxins

The deadly poisonous green mushroom *Amanita phalloides* (Vaill. ex Fr.) Secr., the "Grüne Knollenblätterpilz," is sometimes called the "death cap" or "deadly agaric"; it grows in Central Europe from July until the end of October and is associated with deciduous trees, particularly with beeches in open forests that are not too dry. It develops from an egg-shaped white ball within 1 to 2 days to a mushroom 10-15 cm tall. The slightly vaulted cap reaches diameters of up to 12 cm; it is smooth, more or less deep olive green, and often patterned with darker, radially extending filamentous streaks. The lamellae are white, the stem, which sometimes shows pale greenish cross stripes, bears at its uppermost part a large white cuff that ends in a larger tuber surrounded by a leafed sheath. The toxic mushroom has no specific smell or taste, in contrast to the related nontoxic *Amanita citrina* (Schaeff.) Gray. This yellow mushroom is easily discernible by its odor of raw potato. The white toxic variety *Amanita verna* (Bull. ex Fr.) Pers. ex Vitt. appears during the summer, in deciduous forests; its occurrence in Europe is not common. It is the "destroying angel" or "deadly agaric" of the North American continent and contains the same toxic peptides that *A. phalloides* does. The likewise white toxic *Amanita virosa* must be considered as an additional individual (see C. A. Bauer, 1971). The other varieties in North America *Amanita tenuifolia* and *Amanita bisporigera*, which are extremely rich in toxins (Block *et al.*, 1955b; Tyler *et al.*, 1966), have not yet been observed in Europe to our knowledge. Most interestingly, *Amanita* toxins have also been found in some species of *Galerina* by Tyler *et al.* (1963). See also Benedict and Brady (1967) and the article following.

249

The white or green *Amanita* causes more than 95% of fatal mushroom poisoning. These mushrooms are mistakenly confused with the delicious champignon *Tricholoma equestre* (the white forms with *Agaricus campestris*), which appears in similar places, but which bears from the beginning faint pink, and later darker red, lamellae; it has a typical anise-like smell. The high percentage of fatalities occurs from *A. phalloides* and related mushrooms because the first symptoms of intoxication — vomiting and diarrhea, which are not caused by the lethal toxins — do not become apparent until several hours after ingestion, and during this time the deadly toxins have already reached the liver and kidneys. We know with certainty that the slow-acting amatoxins cause death by specifically destroying the liver cells. Their fatal action, however, starts shortly after the toxins reach the organ and is irreversible.

The investigation of the poisonous substances of *A. phalloides* started at the beginning of the last century. A survey on history, chemistry, and pharmacology was given by T. Wieland and O. Wieland (1959) and, more recently, T. Wieland (1967). Lynen and U. Wieland (1937) in the Munich Chemical Laboratories of the Bavarian State succeeded in isolating a toxic material in crystalline form which they called "phalloidin." Four years later, H. Wieland and Hallermayer (1941) obtained an even more toxic component and named it "amanitin." Both these toxins are representatives of two families of poisonous peptides. Later amanitin was resolved into a neutral (α-amanitin) and an acidic (β-amanitin) toxin, and, to the poisonous substances already discovered, γ-amanitin, ϵ-amanitin, amanin, and the nontoxic amanullin were added. This group (including amanullin) is called the family of amatoxins. The phallotoxins include phalloidin, phalloin, phallacidin, phallisin, and phallin B. The schematic chromatogram of a variety of substances, including the nontoxic amanullin, is represented in Fig. 1.

Paper chromatography has proved very useful for the analysis of crude mushroom extracts and of the various isolated fractions. After numerous experiments, a mixture of butanone, acetone, and water (30:3:5 v/v) was found to allow separation of nearly all of the phytotoxins (Fig. 1). The solvent mixture has been modified by addition of some n-butanol by Block *et al.* (1955a). In thin-layer chromatographic analysis on silica gel G, mixtures of butanone and methanol (1:1) or of n-butanol, acetic acid, and water (4:1:1) were successful (Tyler *et al.*, 1966).

The recent literature on solvents used in chromatographic analysis, and a rapid procedure for the recognition of α-amanitin in mushroom extracts by thin-layer chromatography, has been reviewed by Palyza and Kulhanek (1970).

The amatoxins (except amanin) give an immediate violet color with

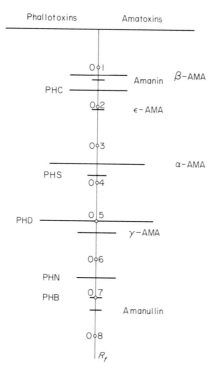

FIG. 1. Diagrammatic representation of a descending paper chromatogram in butanone-acetone-water (30:3:5) of the identified components of *A. phalloides*. The length of the bars roughly indicates relative amounts. Abbreviations as in Table I.

cinnamic aldehyde in an atmosphere of hydrochloric acid gas; the phallotoxins give a delayed weaker blue color, and amanin a brown one that becomes blue after 1-2 hours. With diazotized sulfanilic acid (Pauly reagent), the first group forms an intense red compound, whereas the phallotoxins give only a weak yellow color. The much greater sensitivity of the amanitins to color reagents may be the reason why the spot of phalloidin has not been observed in mushroom analyses of other laboratories. Here, a column chromatographic separation combined with ultraviolet spectroscopy of the eluent as applied for analysis of one *A. verna* mushroom by T. Wieland *et al.* (1966) seems indispensable. In the poisonous species there are about 10 mg phalloidin, 8 mg α-amanitin, 5 mg β-amanitin, and 1.5 mg γ-amanitin per 100 gm of fresh tissue (corresponding to about 5 gm dry weight). Similar values for α- and β-amanitin have been reported by Tyler *et al.* (1966) for *A. phalloides* and *A. verna,* but a higher content (up to 5 mg/gm dry weight) for *A. bisporigera.* Since the lethal dose of the amanitins is lower than 0.1 mg/kg body weight for human beings, it is conceivable that the toxin content of one mushroom weighing 50 gm (about 7 mg of amanitins) may be sufficient to kill a man.

In cases of poisoning, a prominent feature is always the long period of latency between ingestion and the appearance of the first symptoms. Usually 10 to 24 hours have elapsed before abrupt violent emesis and diarrhea begin, and the illness sometimes culminates in rapid death, with cholera-like manifestations. These gastrointestinal signs have nothing to do with the toxic peptides. If this phase is survived, a transient and false remission takes place, and gradually more pronounced signs of injury to parenchymatous organs appear, chiefly in the liver. A hemolytic agent, which is precipitated by organic solvents (methanol) and is labile on heating, was obtained from *A. phalloides* as early as 1891 by Kobert and was called "phallin." It has recently been reinvestigated by Fiume (1967), who purified it to a considerable extent, found it nondialyzable, and stated that, in addition to a hemolytic action, a cytotoxic effect on cultures of cells of KB line and of human amnion cells occurred. This substance, whose chemical nature has not yet been established, is presumably destroyed in the gastrointestinal tract and, therefore, cannot be responsible for lethality of the mushroom.

II. Chemistry of the Toxic Peptides

The *Amanita* toxins are colorless substances; some of them are crystalline; soluble in methanol, pyridin, dimethylsulfoxide, liquid ammonia; more or less soluble in water, ethanol, and water-containing isopropanol or butanol; and insoluble in weakly polar organic solvents.

They can be divided into two main groups: The phallotoxins — mainly comprising phalloidin, phallacidin, phalloin, phallisin (and phallin B) — and the amatoxins — consisting of α-, β-, γ-, and ε-amanitin and amanin (see Fig. 1). The basic differences between these toxin groups lie in some chemical features and in the rapidity of their action. The members of the phalloidin group act quickly, and, at higher dose levels, death of mice or rats occurs within 1 or 2 hours. In contrast, the action of the amanitin group is delayed, so that even with very high doses it is not possible to reduce the lethal interval to less than 15 hours. The toxicity is, however, just the opposite: the amatoxins are 10 to 20 times more toxic than the phallotoxins and are therefore the major poisonous constituents of *A. phalloides*. It may be added that there are differences in sensitivity of different animal species to the toxins, e.g., rats are able to survive fivefold amanitin doses as compared with mice, but are more sensitive to phallotoxins ($LD_{50} \approx 0.75$ mg/kg). For guinea pigs, the LD_{50} is as high as 3.5 mg phalloidin/kg.

TABLE I

DIFFERENCES BETWEEN PHALLOTOXINS AND AMATOXINS

Characteristic	Phallotoxins					Amatoxins					
	PHC	PHS	PHD	PHN	PHB	α-	β-	γ-	$\sqrt{\epsilon}$-AMA	AMN	AMU
Toxixity, LD_{50} (mouse, mg/kg)	2.5	2.5	2	1.8	15	0.3	0.4	0.15	0.5	0.5	∞
Time to death (hours)	1-4					15-100					
Number of amino acids	7					8					
γ-Hydroxy amino acid	Leucine					Isoleucine					
Indole derivate	Tryptophan					6-Hydroxytryptophan			Trp No 6-OH		
Transannular bridge	Thioether					Sulfoxide					
Color reaction cinnamic aldehyde HCl	Blue					Violet				Blue-violet	
λ_{max} (nm) [log ϵ]	292 [4.15]					302 [4.1]				285 [4.1]302	

Abbrev.: PHC = phallacidin, PHS = phallisin, PHD = phalloidin, PHN = phalloin, PHB = phallin B, AMA = amanitin, AMN = amanin, AMU = amanullin.

Both of the toxins are cyclopeptides which have in common a sulfur-containing bridge formed by the coupling of cysteine sulfur with the indole nucleus of a tryptophan, and a γ-hydroxylated amino acid. They differ as summarized in Table I, which also contains some data for the nontoxic amanullin.

A. PHALLOTOXINS

The group of phallotoxins consists of at least five members (Fig. 1) with a common cyclic heptapeptide skeleton as shown in Formula I. Upon hydrolysis with 6 N hydrochloric acid, six different amino acids are liberated, and in the process the thioether bridge is split into L-β-oxindolyl-3-alanine (II) and L-cysteine (III).

The characteristic ultraviolet spectrum of phallotoxins (Fig. 2) is due to the 2-thioether of tryptophan. Compounds of this type can be synthesized by coupling β-substituted indoles with sulfenyl chlorides. The 2-thioether obtained from β-indolylacetic acid and methylsulfenyl chloride (CH_3SCl) is identical to phalloidin (curve 2) in its spectrum. Curve 1 represents amanin, which is also a derivative of tryptophan, but — as a sulfoxide — belongs to the family of amatoxins (Section B, below).

a. Dethiophallotoxins (IV). When treated with Raney nickel in boiling methanol, the thioether bridge of the phallotoxins is opened hydrogenolytically to yield nontoxic dethio compounds. Their hydrolysis yields tryptophan (V) instead of II, and alanine instead of cysteine, thus demonstrating the former thioether structure chemically.

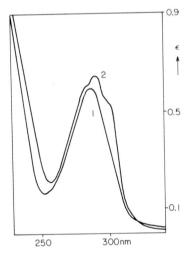

FIG. 2. Ultraviolet spectra in methanol of (1) amanin (XIXe) and (2) phalloidin (1a).

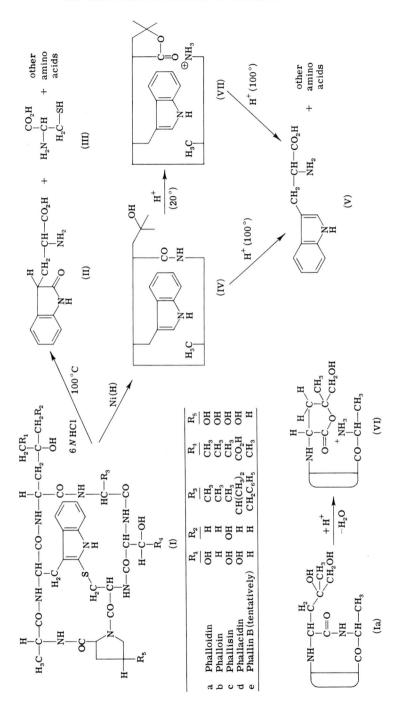

b. Secophallotoxins (VI). In all toxic *Amanita* peptides, including the amatoxins, a γ-hydroxylated amino acid occurs. The γ-hydroxy group in the side chain causes a selective and mild splitting of the peptide bond (50-80% trifluoroacetic acid at 20°C, 2 hours) by its neighboring group effected through γ-lactone formation. The seco compounds (VI) obtained by such a partial hydrolysis are totally nontoxic as are the dethio compounds (IV).

From a hydrolyzate of phalloidin (Ia) the γ-lactone hydrochloride of one diastereoisomer of γ, δ-dihydroxyleucine has been isolated; to this we ascribed the erythro structure (VIII). This structure has been confirmed by nuclear magnetic resonance spectra, in which the signals of methyleneprotons of the CH₂OH group showed equivalence (singlet). Conversely, in the diasteromeric aminolactone obtained by synthesis, these methylene protons cause a doublet; they are no longer equivalent, because of a strong hydrogen bonding of —CH₂OH to H_3N^+— (T. Wieland et al., 1968a). In the same paper it has been shown that the (2S,4R)-4,5-dihydroxyleucine structure (VIII) also applies to the native amino acid bound in the cyclopeptide. Phalloin (Ib) yields the lactone of γ-hydroxyleucine (IX). This compound, among other aminolactones, was also discovered recently in hydrolyzates of gelatin by T. Wieland and Dölling (1966). In phallisin (Ic) a trihydroxylated amino acid, γ,δ,δ′-trihydroxyleucinelactone (X), was shown to be present by Gebert et al. (1967). As a result of periodate oxidation aspartic acid was formed from the open acid of compound X by an excessive glycol-splitting reaction. The structural formula of (X) has been confirmed by chemical synthesis starting from L-aspartic acid (Weygand and Mayer, 1968). Thus it appears that all phallotoxins contain γ-hydroxyamino acids derived from L-leucine.

(VIII)   (IX)   (X)

(XI)   (XII)   (XIII)

c. Secodethiophallotoxins (VII). The selective opening of the peptide ring at the carboxylic site of γ-hydroxylated amino acids also takes place with the dethiophallotoxins. Here, an open heptapeptide is formed which in all cases was submitted to the Edman degradation procedure. This yielded, in the case of phalloidin, successively the phenylthiohydantion derivatives of L-alanine, D-threonine, L-alanine (from cysteine), L-allohydroxyproline, L-alanine, and L-tryptophan, thus demonstrating the presence of structure Ia. This sequence was confirmed by Das (1968) in Lederer's laboratory by mass spectrometric analysis of permethylated secodethiophalloidin, in which the N-terminal had been protected by acetylation. A molecule ion peak m/e = 940, and the peak of several fragments, were definite proof of the structure as demonstrated.

The only amino acids common to all five phallotoxins are L-alanine and the coupling product, called tryptathionine, of L-cysteine and L-tryptophan. Other amino acids present in the different toxins are L-allohydroxyproline (XI) in nearly all of them (Ia–d) except in phallin B, which contains proline instead; D-threonine (XII) in Ia–e; L-valine and D-erythrohydroxyaspartic acid (XIII) in phallacidin (Id); and phenylalanine in phallin B (Ie).

1. STRUCTURE AND TOXICITY

The bicyclic shape gives to the molecules of the phallotoxins a relatively stable conformation in which the side chains are susceptible to several well-defined chemical changes without alteration of the geometry of the peptide ring. Destruction of the ring, however, can be achieved by heating the toxic molecules with Raney nickel (to form the dethio compounds IV) or by treating them with acid under not too energetic conditions. The seco compounds (VI) thus formed are as nontoxic as the dethio compounds; alteration of the shape of the molecules apparently causes loss of toxicity.

On the other hand, chemical changes at the periphery of the intact molecules do not destroy toxicity. Thus, by treating phalloidin with acetic anhydride, a mixture of acetyl derivatives was obtained by Faulstich (1969) of which the monoacetyl compound showed full, and the diacetyl compounds showed partial, toxicity. Here the alcoholic hydroxyl groups of allohydroxyproline, threonine, and particularly, the δ-hydroxy of γ,δ-dihydroxyleucine can be esterified. The possibility exists, however, that *in vivo* the acetyl residues are split off enzymatically giving rise to the formation of the true toxin. The same holds for the toxic effect of ketophalloidin (XIV), the oxidation product of phalloidin by periodate (T. Wieland and Schöpf, 1959), which shows a toxicity in the same range.

Ketophalloidin can be hydrogenated to the toxic demethylphalloin (XV) by means of sodium borohydride (T. Wieland and Rehbinder, 1963), and it cannot be shown that an analogous mechanism is not operating in the liver of the treated animal. By using sodium borohydride-³H a rather strongly labeled phalloidin-like poison (XV-³H) was prepared by Puchinger and Wieland (1969a); it was useful in following the fate of the toxin in rats (Puchinger and Wieland, 1969b).

A similarly toxic dithiolane (XVI) was prepared by reacting XIV with ethylenedithiol-³⁵S and the also toxic δ-tosylphalloidin (XVII) from phalloidin (Ia) with tosylchoride (T. Wieland and Rehbinder, 1963). Recently, the oxime of XIV and a tosylhydrazone have been prepared and found to be toxic. Finally, the sulfur atoms of the dithiolane ring of compound XVI were replaced by hydrogen, keeping the thioether bridge intact. The resulting norphalloin (XVIII) proved even more toxic than phalloidin (T. Wieland and Jeck, 1968). Therefore, a γ-hydroxy group in a side chain of phallotoxins is not a prerequisite for toxicity. The total synthesis of norphalloin carried out by Fahrenholz et al. (1971) in the authors' laboratory confirmed the structural formula XVIII on the lines of classic chemistry. On oxidation of phalloidin by hydrogen peroxide in glacial acetic acid, two sulfoxides were obtained, which probably differed, the one from the other, by the configuration of their sulfoxide groups. Only one of them proved to be toxic, whereas the sulfone obtained by further oxidation of both of the sulfoxides was toxic (Faulstich *et al.*, 1968).

B. AMATOXINS

The amatoxin group has at least six members, for which the common cyclic octapeptide skeleton as elaborated by T. Wieland and Gebert (1966) is shown in formula XIX. The ultraviolet spectra (Fig. 3) have a maximum at 302 nm, which is shifted to about 320 nm when the phenolate is formed by adding a little alkali. To this group also belong the "outsiders" amanin (XIXe) and amanullin (XIXf). Amanin differs from β-amanitin (XIXb) only by its lack of the phenolic hydroxyl group in position 6 of the indole nucleus. Therefore, the ultraviolet spectrum of amanin (Fig. 2) differs from the amatoxins (Fig. 3), but the toxicological (slow) action is the same.

With amanin it was recognized by Faulstich *et al.* (1968) that the amatoxins differ from phallotoxins also in the nature of the sulfur-containing transannular bridge. The differences in the ultraviolet absorption curves (Fig. 2) pointed to a difference of the chromophoric systems: amanin and the amatoxins turned out to be sulfoxides. A sulfoxide prepared by Faulstich *et al.* (1968) from phalloidin by oxidation with hydrogen peroxide in

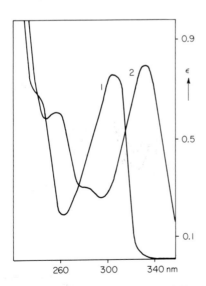

FIG. 3. Ultraviolet spectra in methanol of α-amanitin (XIXa) (1) without and (2) with added sodium hydroxide.

acetic acid exhibited an ultraviolet spectrum nearly identical with that of amanin (Fig. 4). On the other hand, O-methyl-α-amanitin ($R_4 = OCH_3$ in XIXa), obtained from α-amanitin with diazomethane, could be deoxygenated at its sulfoxide group by treatment with Raney nickel (Pfaender, 1966) to give a still toxic thioether with a spectrum very similar to that of a model compound, 2-thioethyl-3-methyl-6-methoxyindole, which had been prepared by T. Wieland and Grimm (1965).

α-Amanitin (XIXa) is the amide of the carboxylic acid β-amanitin (XIXb), and γ-amanitin (XIXc) is the amide of ϵ-amanitin (XIXd) (T. Wieland and Buku, 1968). During acid hydrolysis, the amino acids that do not form the sulfoxide bridge are liberated in the original state: glycine (2 moles), L-aspartic acid, L-hydroxy-proline, L-isoleucine, and, varying among the members, L-γ-hydroxyisoleucine as lactone (XX) from γ-amanitin or ϵ-amanitin and γ,δ-dihydroxyisoleucine (XXI) from α- or β-amanitin and from amanin. Thus, the hydroxylated amino acids of amatoxins are derived from isoleucine.

β-Amanitin, as a carboxylic acid, could be reacted with bovine serum albumin by Boehringer (1959) to give a conjugate that was intended as an antigen when it was injected into rabbits. It proved, however, too toxic in spite of prolonged dialysis to remove the low molecular weight toxin. A quantitative investigation by Cessi and Fiume (1969) of a conjugate containing 2,3 moles of β-amanitin per mole rabbit serum albumin revealed an approximately tenfold increase in toxicity in mice as compared with β-amanitin.

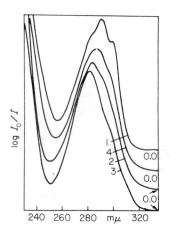

FIG. 4. Ultraviolet spectra in methanol of (1) phalloidin (Ia), (2) phalloidin sulfoxide, (3) skatyl-(2)-ethylsulfoxide (XXV), and (4) amanin (XIXe).

It has been assumed that the high molecular weight derivative passes through the liver much more slowly than the low molecular weight toxin, thus being effective at lower concentrations. In mice, but not in rats, α- and β-amanitin act primarily on kidney tubuli (Fiume *et al.*, 1969, see p. 271).

Amanullin (XIXf) is totally nontoxic. It contains no γ-hydroxylated side chain, but instead a second molecule of isoleucine (T. Wieland and Buku, 1968). In contrast to the phallotoxins, the presence of a γ-hydroxy group is apparently a prerequisite for toxic activity. This view received support from recent oxidation experiments on O-methyl-α-amanitin with periodate (T. Wieland and Fahrmeir, 1970). Here a degradation of the side chain of γ,δ-dihydroxyisoleucine gave an aldehydic derivative [—CH(CH$_3$)—CHO], which was nontoxic. By reduction with sodium borohydride a toxic γ-hydroxylated derivative [—CH(CH$_3$)—CH$_2$OH] was recovered.

The identification of the natural γ-hydroxyisoleucine of XIXc with one of the eight possible diastereoisomeric antipodes as (2*S*,4*R*)-4-hydroxyisoleucine (XX) has been made chiefly by nuclear magnetic resonance spectroscopy (T. Wieland *et al.*, 1968a); the corresponding γ,δ-dihydroxyamino acid of α-amanitin (XIXa) has an analogous structure (XXI).

As in the phallotoxins (I), there are also nontoxic seco compounds (XXII) formed by treatment of amatoxins (XIX) with trifluoroacetic acid at room temperature. Similarly, the dethio compounds (XXIII) obtained with Raney nickel in the amanitin series are nontoxic.

The process of removing sulfur from the molecule led T. Wieland and Gebert (1966) to several products of hydrogenation. Among other reactions, a partial hydrogenation of the indole part took place. To obtain a homogeneous derivative to perform degradation studies, a thorough hydrogenation was carried out with hydrogen and platinum catalyst. It was found that perhydrodethio-α-amanitin contains a saturated ring system, octahydrotryptophan, in which the phenolic hydroxyl group also has been substituted by hydrogen.

The seco compound XXIV obtained from it by treatment with 80% trifluoroacetic acid at 20°C was subjected to mass spectrometry after *N*-acetylation and permethylation by Das (1969); the accepted sequence of amino acids by the demonstration of several typical fragment masses was observed.

The chromophoric part of the amatoxins is the 1-sulfoxide of 6-hydroxyindole (Fig. 3). 6-O-Methylamatoxins give principally the same ultraviolet curves as the hydroxy compounds, but with a little minimum at

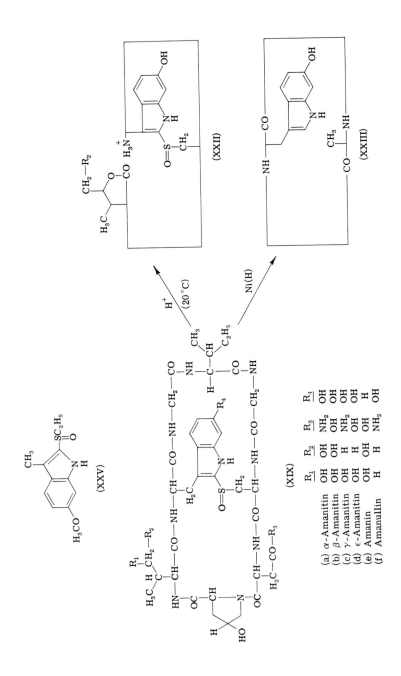

	R_1	R_2	R_3	R_1
(a) α-Amanitin	OH	OH	NH_2	OH
(b) β-Amanitin	OH	OH	OH	OH
(c) γ-Amanitin	OH	H	NH_2	OH
(d) ϵ-Amanitin	OH	OH	OH	OH
(e) Amanin	OH	OH	OH	H
(f) Amanullin	H	H	NH_2	OH

(XXIV)

(XXI)

(XX)

(XXVI)

(XVIII)

$\xrightarrow[100°C]{6 N \, HCl}$

+

(XXVII)

about 308 nm, thus showing two neighboring maxima. This curve agrees exactly with that of the sulfoxide XXV prepared by the oxidation of T. Wieland and Grimm's (1965) thioether mentioned on p. 260 (Faulstich *et al.*, 1968).

Upon acid hydrolysis, the sulfoxide bridge of the amatoxins is broken in a way different from that in phallotoxins. Hydrolysis is observed, predominantly, which results in 6-hydroxytryptophan (XXVI) and cysteic acid (XXVII), a product of disproportionation of the sulfenic acid originally formed. The tryptophan derivative XXVI is acid labile and can be detected only in traces.

C. ANTITOXIC ANTAMANIDE

Antamanide is an antitoxin discovered by T. Wieland *et al.* (1968b, 1969c) in extracts of *A. phalloides*. The antitoxin, when administered at least simultaneously with phalloidin to the white mouse, counteracts a 100% lethal dose of 5 mg/kg of this poison; the protective dose is 0.5 mg/kg. Antamanide is a cyclic decapeptide consisting of only four different L-amino acids—alanine, valine, phenylalanine, and proline—in a molar ration of 1:1:4:4. The sequence of amino acids leading to formula XXVIII has been determined by Prox *et al.* (1969), mainly by mass spectrometry of peptide methylesters of a partial methanolysis, which were separated by gas chromatography after trifluoroacetylation. The total synthesis of antamanide has been performed, by several routes of peptide synthesis, by T. Wieland *et al.* (1969a,b) and by König and Geiger (1969).

(XXVIII)

From infrared spectra, vapor pressure determinations, optical rotatory properties, and electrical conductivity and potential measurements in the presence and absence of Na$^+$ ions, it was concluded that antamanide forms a distinct complex with Na$^+$, which is more stable than the complex with K$^+$ (T. Wieland *et al.*, 1970).

Since the first syntheses of antamanide, several variants with exchanged amino acids in different positions have been prepared. Of these it is striking that when the valine residue of position 1 is replaced by alanine, a marked decrease in antitoxic activity is observed (T. Wieland *et al.*, 1971).

Conformational formulas for antamanide (Tonelli *et al.*, 1971) and its sodium complex (Ivanov *et al.*, 1971) have recently been proposed on the basis of nuclear magnetic resonance and optical rotatory data and minimal energy calculations.

III. Toxicology of *Amanita* Toxins

A. CLINICAL SYMPTOMATOLOGY AND PATHOLOGICAL ANATOMY

After a characteristic latency period of 10–20 hours following ingestion of the mushroom (an equivalent of about 50 gm fresh weight of *A. phalloides* may be lethal for an adult person), intoxication is manifested by violent emesis and diarrhea. If this acute phase, which may result in rapid death, is survived, a transient remission takes place. This is followed by increasing signs of liver injury with swelling and tenderness of the liver and is often accompanied by icterus. The serum glutamate-oxaloacetate and glutamate-pyruvate transaminase values may be extremely elevated. Lesion of the kidneys is manifested by the appearance of protein, casts, and erythrocytes in the urine. Death occurs during hepatic coma, 2–4 days after ingestion of the fatal meal.

Among the postmortem findings, the alterations in the large parenchymatous organs, especially the fatty degeneration and destruction of the liver, appear most important. The lesions are similar to those occurring in other hepatopathies, especially in degeneration due to phosphorus poisoning and in acute yellow liver atrophy — severe fatty degeneration with partial or complete necrosis of the liver cells. Some cases are characterized by hemorrhagic alterations with capillary stasis and hematoma-like extravasates.

In addition to the liver, the kidney also presents pathological alterations. These include all the signs of nephrosis, with cloudy swelling, fatty degeneration, and necrosis and nuclear atrophy, especially of the convoluted tubules and the ascending limb of the loop of Henle. As discussed later, the kidney lesions after mushroom poisoning can be traced back to the selective action of amatoxins on this organ. The adrenals as well as the heart and skeletal muscles may also show degenerative changes after amanita poisoning.

B. EXPERIMENTAL INVESTIGATIONS WITH PURE TOXINS

1. TOXICOLOGICAL FEATURES

The symptomatology and the pathological findings in animals after experimental poisoning with the chemically pure peptides differ in some respects from those observed in men. Thus, gastrointestinal symptoms are completely lacking, whether the toxins are administered by mouth or parenterally. The sensitivity of mice and rats to both phalloidin and amanitin varies with the strain. Therefore, only approximate average toxicological data can be given. In mice the LD_{50} for phalloidin is in the range of 2-3 mg/kg and the LD_{100} between 4 and 7 mg/kg. It is of little effect upon the toxicity whether the toxins are administered parenterally, intravenously, or subcutaneously, but the toxins are only slightly resorbed when applied perorally in rats and mice. Amatoxins are several times more toxic than phallotoxins (cf. Table I). α-Amanitin invariably causes death in a dose of 0.5-1 mg/kg, in mice. Moreover, the two classes differ greatly in the rapidity of their action. Phallotoxins act quickly, and, with multiples of the LD_{100}, death occurs within 1-2 hours. Conversely, the action of amatoxins is delayed, so that with even very high doses the lethal interval cannot be reduced to less than 15 hours.

The important finding of Fiume (1965) that newborn rats are much less sensitive to phalloidin has been thoroughly investigated. It was found that in fifteen 2-week-old rats all animals survived a dose of 10 mg/kg phalloidin, whereas all of a group of 5-week-old animals died after 5 mg phalloidin/kg (O. Wieland and Szabados, 1968). In these studies, up to 360 mg phalloidin/kg were tolerated by 2-week-old rats that had already survived a 20-80 mg/kg dose in the first week of their life. The apparent phalloidin tolerance lasts until the eighteenth day after birth. Then sensitivity develops gradually until 100% mortality is reached at an age of 35-38 days (Fig. 5) (Szabados, 1970). It seems noteworthy that there is no tolerance of the newborn rat to amanitin. If 1-week-old rats are repeatedly injected every week with 10 mg phalloidin/kg they remain resistant to the toxin beyond the critical eighteenth day. Animals thus treated over 10 weeks developed normally and showed no signs of injury. On autopsy, however, the livers presented severe cirrhosis-like changes, which were also confirmed on histological examination.

2. PATHOLOGICAL FINDINGS

The deleterious action of phalloidin upon the liver is already obvious from the gross appearance of the organ after intoxication. The liver is greatly enlarged, dark red in color, and very brittle. These changes are due to an excessive accumulation of blood in the liver. On histological

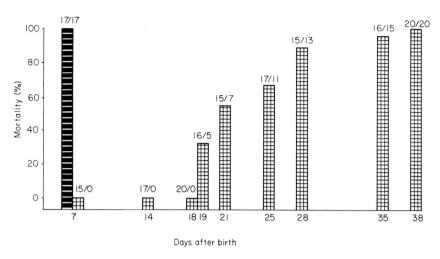

FIG. 5. Disappearance of phalloidin tolerance of newborn rats with increasing age. The figures on top of each bar represent the ratio of animals injected to animals killed. Phalloidin, 5 mg/kg, IP; α-amanitin, 1 mg/kg, IP (From O. Wieland and Szabados, 1968.)

examination, this final stage of acute hemorrhagic necrosis is preceded by the formation of numerous nonfatty vacuoles, which begin at the periphery of the lobule and then extend to the central zone. Electron microscopically, the first structural alterations were observed 10–20 minutes after phalloidin administration to mice or, in the isolated rat liver, after addition of phalloidin to the perfusion medium (Fiume and Laschi, 1965; Miller and O. Wieland, 1967) (Fig. 6).

Surprisingly, very similar structural alterations were observed in the livers of newborn rats during the phase of phalloidin tolerance by Siess *et al.* (1970). From the electron micrographs, it is difficult to compare these lesions quantitatively with those occurring in the livers of adult rats. However, by biochemical analysis, a less pronounced drop in glycogen and also a smaller increase in blood content has been observed in the livers of newborn rats after phalloidin poisoning (O. Wieland and Szabados, 1968). This would indicate that the hepatotoxic effect of phalloidin, although present, is comparably less in the young animals.

As already suggested from other investigations, phalloidin seems to act specifically on the liver. Other tissues, such as kidney, heart muscle, diaphragm muscle, and pancreas, showed no morphological alterations upon electron microscopic examination after phalloidin poisoning (Siess *et al.*, 1970).

In contrast to phallotoxins the action of the amatoxins generally is directed against both the liver and the kidney, the rat being an exception

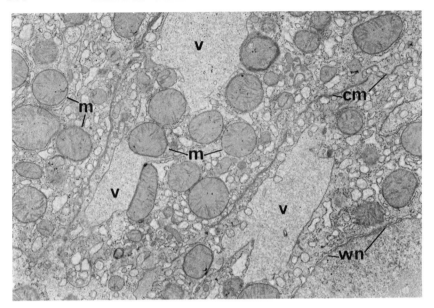

FIG. 6. Electron micrograph of mouse liver 30 minutes after phalloidin poisoning (14500×). CM = cell membrane. WN = wall of cell nucleus, V = typical giant vacuoles within the endoplasmic reticulum, M = mitochondria. Note the huge vacuoles clearly arising from the tubular membrane of the endoplasmic reticulum. The ribosomes are no longer attached to the membrane surface, the liver mitochondria do not show any morphological lesions. (From O. Wieland and Szabados, 1968.)

(see below). The first ultrastructural lesions in these organs are observed in the nuclei 15 minutes after administration. These lesions are character-ized by fragmentation of nucleoli (Fiume and Laschi, 1965) including internucleolar chromation and condensation of euchromation (Fig. 7) (Fiume et al., 1969; Marinozzi and Fiume, 1971). As described later in more detail, these nuclear alterations result from an inhibition of RNA synthesis produced by amanitin. Changes in the cytoplasm appear late (48 hours) after amanitin poisoning and evolve rapidly toward necrosis. In mouse kidney, only the proximal tubules are affected indicating that re-absorption of the toxin filtered through the glomeruli is an important factor in the development of nephrosis (Fiume et al., 1969). This view is confirmed by the observation that β-amanitin covalently bound to albu-min (which does not filter through the mouse glomeruli) does not affect the kidney tubules. Concomitantly, the toxicity of the albumin-bound peptide for the liver increases ten times (Cessi and Fiume, 1969). No lesions were found in rat kidney after injection of α-amanitin. This has been explained by Fiume et al. (1969) as being caused by an incapacity of the epithelial cells in rat kidney tubules to reabsorb the toxin from

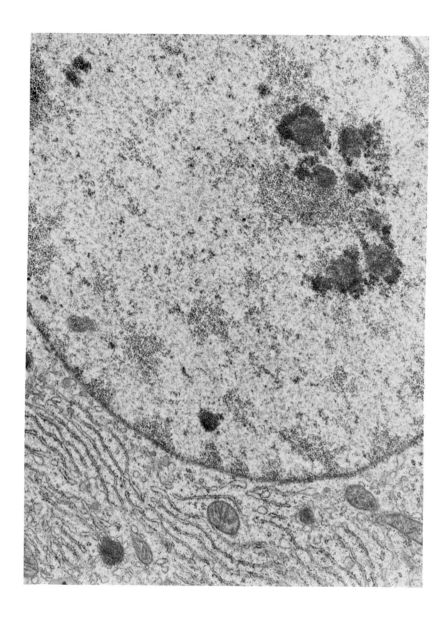

FIG. 7. Ultrastructural lesions of mouse liver cell nuclei after amanitin poisoning. (From Fiume and Laschi, 1965.)

preurine. The kidney lesions observed in man after *Amanita* poisoning seem to indicate that tubular reabsorption of amanitin occurs in the human kidney as it does in the mouse kidney.

3. SPECIES SENSITIVITY TO THE *Amanita* TOXINS

The following animals were found to be highly sensitive to the toxins: guinea pig, rabbit, rat, mouse, horse, goat, sheep, monkey, and pigeon (Dessy and Francioli, 1938). Poikilothermic animals, having a more sluggish metabolism, are much less sensitive to these poisons. In invertebrates (*Helix pomatia, Limax arborum*), toxic effects were observed only with α-amanitin, 100 mg/kg of which was lethal, whereas phalloidin was well tolerated (T. Wieland, 1957). In contrast to earlier studies, a cytopathogenic action of α-amanitin added to *in vitro* cultures of various types of cells such as human amnion cells, carcinomatous cells of the KB-Eagle line, and Lp 3 fibroblasts has been observed by Fiume *et al.* (1966). α-Amanitin did not inhibit proliferation or replication of the following bacteria and viruses: *Escherichia coli, Staphylococcus aureus, Bacillus subtilis,* poliovirus type 2, parainfluenza virus type 3, vaccinia virus, and the virus causing bovine infective rhinotracheitis (Fiume *et al.,* 1966). Interestingly, influenza virus (FDP strain "Rostock") replication has been found to be inhibited by α-amanitin (Rott and Scholtissek, 1970).

4. ORGAN DISTRIBUTION AND BINDING STUDIES WITH LABELED PHALLOIDIN

The fate of phalloidin was studied after injection of either mercaptophalloin-[35]S (Rehbinder *et al.,* 1963) or demethylphalloin-[3]H (Puchinger and T. Wieland, 1969b), both toxic derivatives of phalloidin, into rats or mice. In the rat, the distribution of mercaptophalloin-[35]S 2 hours after subcutaneous injection was as follows: liver, 57.2%; kidney, 2.7%; spleen, 0.2%; blood, 6.0%; lung, 0.9%; heart, 0.2%; skeletal muscle, 9.4%; brain, 0.1%. About 10% of the injected dose is excreted in the urine (Rehbinder *et al.,* 1963). The preferential accumulation of the toxin in the liver was also demonstrated by autoradiography with demethylphalloin-[3]H. The rate of uptake of demethylphalloin-^{3}H was studied in perfusion experiments of the isolated rat liver. The kinetics display the features of a saturation curve with a half saturation of the liver within the first 5 minutes. In the presence of antamanide (see p. 264), which protects animals against intoxication, phalloidin uptake by the isolated rat liver is markedly reduced (Puchinger, 1968) (Fig. 8).

For an understanding of the primary toxic action, the intracellular distribution of phalloidin seems of special interest. After fractional centrifugation and repeated washings of different liver cell fractions, the radio-

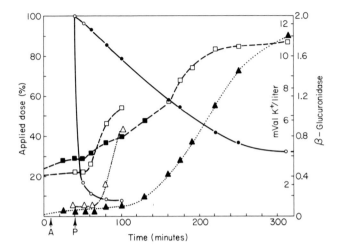

FIG. 8. Absorption of 0.25 mg/100 ml demethylphalloin-^{3}H by a perfused rat liver without (O – O) and with (● – ●) 5 mg/100 ml antamanide given 30 minutes before toxin (P). K$^+$ release without (□ – □) and with (■ – ■) antamanide, β-glucuronidase without (△ – △) and with (▲ – ▲) antamanide (A).

active label from mercaptophalloin-^{35}S was preferentially retained in the microsomal sediment (Rehbinder *et al.,* 1963). There, the peptide seemed to be bound to a high molecular, presumably ribosomal, fraction from which it could be quantitatively recovered in the original form by methanol extraction (Puchinger and T. Wieland, 1969b). As yet there is no experimental evidence for any metabolic degradation of phalloidin. Due to the lack of availability of radioactive labeled amanitin there is little known about the fate of the amatoxin group in the animal body.

5. MECHANISM OF ACTION OF THE AMATOXINS

A considerable number of pathological-biochemical findings in the course of experimental *Amanita* poisoning has accumulated in the literature and has been reviewed elsewhere (T. Wieland and O. Wieland, 1959; O. Wieland, 1965). Since then some further progress has been made in elucidating the mechanism of action of the toxic polypeptides. It now seems to be clearly established that the two toxins phalloidin and amanitin differ principally in the primary mechanism by which they produce their cytopathogenic effects. This fact could already have been derived from the distinct difference in the morphological alterations following the administration of the toxins. In the case of phalloidin it is the structure of the endoplasmic reticulum of the liver that is affected. The primary action of α-amanitin is directed toward the cell nucleus (Fiume and Laschi,

1965). In the course of intoxication, the cytoplasm will be affected, also, and it is cellular necrosis that within 2-4 days leads to death.

The discovery of euchromatin condensation in liver and kidney nuclei led to a study of the action of α-amanitin on the loops of lampbrush chromosomes. α-Amanitin produced a retraction of normal loops in oocytes of *Triturus cristatus carnifex* (Mancino *et al.*, 1970). It also shrunk puffs of giant chromosomes in salivary glands of *Chironomus* larvae (Beermann, 1971). Euchromatin of liver and kidney cell nuclei as well as lampbrush chromosome loops and giant chromosome puffs are the metabolically active parts of chromosomes; the observed changes have been attributed to the inhibition of RNA synthesis by α-amanitin (see below). Fragmentation of nucleoli in liver and kidney cells is thought to be due to condensation of chromosomes containing nucleolar organizers (Marinozzi and Fiume, 1971). This hypothesis is supported by the finding that in oocyte nuclei of *Triturus*, where nucleoli are free in the nucleoplasm and not morphologically connected to the chromosomes, α-amanitin does not change nucleolar morphology (Bucci *et al.*, 1971).

Changes in liver and kidney cell cytoplasm appear late (48 hours) after amanitin poisoning and evolve rapidly to necrosis.

More detailed information on the biochemical action of amanitin came from the finding of Fiume and Stirpe (1966) that, in mouse liver nuclei, the RNA content decreases progressively during the first 24 hours of intoxication of the animals with α-amanitin. The same authors showed that RNA synthesis in isolated liver nuclei of mice was seriously impaired after both *in vivo* and *in vitro* administration of the toxin, suggesting that amanitin inhibits the enzyme RNA polymerase (Stirpe and Fiume, 1967; Novello and Stirpe, 1969). Definite proof has been obtained from experiments with RNA polymerase solubilized from rat liver nuclei. In these studies, α-amanitin in a concentration of 10 ng/ml added *in vitro* inhibited RNA polymerase activity by 60-70% (Fig. 9) (Novello *et al.*, 1970). There is also good evidence that this inhibition results from interaction with the enzyme itself and not with DNA or other components. Thus, the mode of action of the amanitins differs from those inhibitors of RNA polymerase that act by binding to DNA—such as actinomycin D, aflatoxin, chromomycin, etc.—but resembles that of rifamycins on bacterial RNA polymerase. However, in contrast to rifamycin, α-amanitin inhibits RNA polymerase even after the reaction had begun, i.e., after the initiation complex has formed. These results have been fully confirmed by Jacob *et al.* (1970a) and Seifart and Sekeris (1969). Moreover, it has been shown by these workers that the effect of α-amanitin is specific for liver RNA polymerase, whereas the enzyme

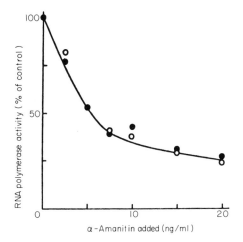

FIG. 9. *In vitro* inhibition by α-amanitin of RNA polymerase solubilized from rat liver nuclei; ● = native DNA, O = heat-denatured DNA. (From Novello *et al.*, 1970.)

from *E. coli* is not inhibited at even higher concentrations of the toxin. Recently the RNA polymerase activity of rat liver nuclei could be separated into two forms, a polymerase I from nucleoli and a polymerase II from nucleoplasm (Roeder and Rutter, 1970). Only the latter enzyme, and likewise a sea urchin RNA polymerase II, are strongly inhibited by α-amanitin (Lindell *et al.*, 1970; Jacob *et al.*, 1970b). Another polymerase of animal origin, namely the RNA polymerase activity B (corresponding to II) from calf thymus, has been found to be inhibited by α-amanitin by the same mechanism as the liver enzyme (Kedinger *et al.*, 1970; Meihlac *et al.*, 1970). This selective inhibition allowed the simultaneous measurement of RNA polymerase I and II in nuclei isolated from rat liver (Novello and Stirpe, 1970). In comparison to actinomycin D, α-amanitin is some five hundred times more effective in inhibiting RNA polymerase. An RNA polymerase of yeast is also sensitive, although the amount of toxic peptide required for complete inhibition is about one hundred times higher than for the mammalian enzymes (Dezelee *et al.*, 1970). Inhibition of the synthesis of an AMP-rich RNA was also observed with plant cells (*Petroselinum sativum*) in the presence of 17 μg/ml of α-amanitin (U. Seitz and U. Seitz, 1971). Similarly one of the three RNA polymerases from the aquatic fungus *Blastocladiella emersonii* is sensitive to the drug (Horgen and Griffin, 1971). In contrast to nuclei of mammalian cells the RNA synthesizing system in mitochondria is not affected by α-amanitin (Menon, 1971). This is also true for the mitochondrial RNA polymerase from *Neurospora crassa* (Küntzel and

Schäfer, 1971). This subject, to March 1970, has been reviewed by Fiume and T. Wieland (1970).

In contrast to the selective inhibitory effect of α-amanitin on RNA polymerases *in vitro*, administration of the drug to intact rats severely reduces the synthesis of both ribosomal and DNA-like RNA (Niessing *et al.*, 1970; Jacob *et al.*, 1970). This difference between the lack of effect on RNA synthesis by isolated nucleoli and the effectiveness of the toxin *in vivo* suggests that ribosomal RNA formation in the intact cell is under an extranucleolar control mechanism sensitive to amanitins.

Because of its specific action α-amanitin can be used as a tool in studies of hormonal effects on enzyme syntheses. Thus the inhibition of induction of tyrosine transaminase in rat liver by cortisol can, at least in the early phases, be related to *de novo* synthesis of mRNA (Sekeris *et al.*, 1970). The induction of δ-aminolevulinate synthetase in chick embryo livers by, e.g., etiocholanone is also inhibited by the drug (Incefy and Kappas, 1971). Dibutyryl cyclic AMP stimulates the incorporation of uridine-^{3}H into nuclear RNA (rat liver) in two consecutive waves. The first consists mainly of 45 S, 32 S, and 18 S ribosomal RNA, and synthesis is completely inhibited by 0.2 μg/ml of actinomycin D. Conversely, the second wave consists mainly of heterogeneous RNA ranging from 7 S to 18 S, and synthesis is completely inhibited by 0.04 μg/ml of α-amanitin (Jost and Sahib, 1971).

The inhibitory action of amanitin on liver RNA polymerase seems to offer an explanation for some of the biochemical lesions observed earlier in the liver following amanitin poisoning. Thus the decrease in protein content of the blood of poisoned animals (T. Wieland and Dose, 1954) and the striking decrease in the activity of glycogen synthetase in liver and, to a lesser extent in skeletal muscle, of mice 24-48 hours after amanitin poisoning (Bonelli *et al.*, 1967) could be explained by the defective synthesis of proteins due to an impairment of RNA synthesis by amanitin. Some of the metabolic disturbances in amanitin poisoning, such as the drop in liver glycogen, ATP, DPN, the changes in serum protein pattern, and the leakage of potassium from hepatocytes (T. Wieland and O. Wieland, 1959), may also originate from defective protein synthesis, although no special enzymes within these metabolic areas have so far been shown to be involved.

6. ACTION OF PHALLOTOXINS

The primary action of phalloidin is much less understood than that of α-amanitin. So far, no special enzyme system affected by phalloidin has been clearly defined. As already pointed out, the primary attack of this peptide seems to be directed against certain structures of the endoplasmic reticulum of the liver. This may be deduced from the morphological alterations characterized by the early occurrence of small vacuoles which increase, due to confluation, in the course of intoxication and finally result in huge lacunae filling the space of the endoplasmic reticulum (Fig. 6). Other cell structures, such as the nucleus, the mitochondria, or the plasma membranes, do not show notable alterations upon electron microscopic investigation (Miller and O. Wieland, 1967; Siess *et al.*, 1970). It must be emphasized that the morphological picture is not unique for phalloidin poisoning but may also develop following the administration of a variety of hepatotoxic agents. In liver perfusion experiments, a release of potassium into the medium about 10 minutes after the addition of phalloidin has been observed by Frimmer *et al.* (1967). Tetraethylammonium, a rather specific inhibitor of the so-called potassium canniculi, prevents the loss of potassium when added with the phalloidin (Frimmer and Weil, 1969a). *In vitro* studies on isolated plasma membranes from rat liver showed no inhibition of K^+ or Na^+-stimulated ATPase activity by phalloidin (Hegner *et al.*, 1970). From these and other observations it has been suggested that phalloidin, like valinomycin and other "ionophorous antibiotics," increases the rate of potassium permeation through the potassium canniculi without interfering with the active ion transport system of the cell membrane, but not even a minimal complexing affinity of either component can be observed in *in vitro* experiments (Burgermeister, 1969). No interaction of phalloidin with "black membranes" has been found (Tosteson *et al.*, 1968).

Another indication of a membrane-directed action of phalloidin came from the finding of an early cessation of bile flow and bromsulphthalein excretion in the perfused rat liver after addition of phalloidin to the medium (O. Wieland, 1965; Matschinsky and O. Wieland, 1960). This functional lesion has its morphological correlate in the form of ultrastructural changes of the bile capillaries as observed later in the course of phalloidin poisoning (Siess *et al.*, 1970). In this context it seems of interest that the increased potassium release just discussed can be prevented by the administration of choleretic drugs (Frimmer and Weil, 1969b). An immediate increase of about 10% in oxygen consumption after phalloidin administration was observed in an erythrocyte-free, perfused rat liver preparation by Jahn (1970). The correlation of this effect, the very first indication of intoxication, with the succeeding events mentioned above is not yet clear.

Another early symptom of phalloidin poisoning is the inhibition of glycogen synthesis from glucose which has also been demonstrated in liver perfusion experiments (O. Wieland, 1965; Meyer, 1966). Studies on the activity of glycogen synthetase could not conclusively establish that the defective synthesis was due to a specific interaction of phalloidin with the enzyme (Matschinsky and O. Wieland, 1961). Hepatic glycogen synthesis

is known to depend on certain structural interrelationships between glycogen, glycogen synthetase, and the smooth surface vesicles of the endoplasmic reticulum (Leloir and Cardini, 1957; Leloir and Goldenberg,
1960; Porter and Bruni, 1959; Hizukuri and Larner, 1964). Thus, the de-
fect in glycogen biosynthesis may be just one of several metabolic disturbances that originate from a common primary attack of phalloidin on the
structure of the endoplasmic reticulum. Inhibition of protein synthesis as
observed with microsomes isolated from the livers of phalloidin-poisoned
animals by von der Decken *et al.* (1960) seems to be of special interest in
this connection.

a. Role of Lysosomes in Amanita Poisoning. It has been suggested that
phalloidin might act primarily on the membranes of liver lysosomes,
thereby leading to an activation and release of lysosomal enzymes. This
view has been suggested by results from histological examinations (Rita
et al., 1967) and also from biochemical findings that seemed to indicate an
acceleration of autolytic reactions relative to the metabolism of nucleotides and proteins within the liver after phalloidin administration (Matschinsky and O. Wieland, 1960). Furthermore, lysosomal enzymes such
as cathepsin, β-glucuronidase, and N-acetyl-glucosaminidase have been
shown by Frimmer *et al.* (1967) to be rapidly released into the medium
when phalloidin was added to perfused isolated rat liver. On the other
hand, attempts to demonstrate an increase in the release of these enzymes
by incubation of lysosomal preparations with phalloidin *in vitro* have
failed (Frimmer *et al.,* 1967; O. Wieland and Hasslinger, 1969). This
would make it appear unlikely that the lysosomal membrane represents a
primary target of phalloidin action. Accordingly, from electron microscopic studies, no further support for an involvement of the lysosomes in
phalloidin intoxication could be obtained by Miller and O. Wieland (1967)
and Siess *et al.* (1970). Finally, a strong argument against the lysosomal
concept has been obtained from histochemical studies of acid phosphatase, a typical lysosomal marker enzyme. If increased autolytic activity is
responsible for the vacuolar degeneration, one would expect to be able to
localize some activity of acid phosphatase within the content of the vacuoles. This however, has, never been demonstrated (Siess *et al.,* 1970).

At present it seems that the lysosomal enzymes are not of primary importance in the pathogenesis of phalloidin poisoning. Lysosomes may,
rather, be damaged secondarily in the course of the intoxication, and this
may contribute to the progressive destruction of the liver.

b. Toxification Hypothesis of Phalloidin Poisoning. As already mentioned (see p. 266) newborn rats are resistant to phalloidin (but not to
amanitin) until the eighteenth day after birth. It has been suggested by
Fiume (1965) that phalloidin, by way of drug metabolizing enzymes, has
first to be metabolized in order to become toxic, and that the respective

enzyme(s) is (are) not yet present in the liver during the postnatal period. This view seemed further to be supported by the findings that treatment of rats and mice with liver-damaging chemicals affecting microsomal function, such as carbon tetrachloride, sodium cinchophen, thioacetamide, and alloxan, resulted in an increased tolerance to phalloidin (Verne *et al.*, 1950; Matschinsky, 1959; Floersheim, 1966a,b; Szabados, 1970). Induction of phalloidin tolerance by repeated toxin injections (see p. 266) may also be mentioned in this context. There are, however, several reasons that render the toxification hypothesis very unlikely. First, no increase in phalloidin sensitivity was seen after pretreatment of rats with inducers of microsomal enzymes, such as barbiturates, 3,4-benzpyrene, 2-O-methylcholantrene, and others (Szabados, 1970). Second, there is no principal difference in the ultrastructural appearance of the liver between the newborn (tolerant) rats and the older rats that succumb to the action of phalloidin (Siess *et al.*, 1970). Third, all attempts to demonstrate a phalloidin metabolite in liver have thus far been unsuccessful. Thus, in experiments with radioactive-labeled toxin (demethylphalloin-³H) nearly 100% of the compound was recovered from the liver 1-2 hours after injection into the mice in the original form (Puchinger and T. Wieland, 1969b); moreover, demethylphalloin-³H was not metabolized on incubation with oxygen and rat liver ribosomes in a NADPH-regenerating system in these studies. From all these observations it can be concluded that phalloidin exerts its toxic action upon liver cells without any metabolic transformation. The phenomenon of phalloidin tolerance in the newborn rat undoubtedly bears on the understanding of phalloidin action. In this connection it seems of special interest that the appearance of a liver-specific antigen has been reported by Frank (1968) that follows the time course of phalloidin sensitivity of newborn rats closely. It seems an attractive hypothesis that this liver-specific antigen, which is part of the endoplasmic reticulum, involves in some way the specific receptor site for phalloidin action. This would explain not only the liver specificity of the toxin, but also its suggested primary action on the endoplasmic reticulum of the liver cell. Moreover, the apparent phalloidin tolerance of the newborn animals could be explained on a quantitative basis by assuming that the number of specific receptor sites has to reach a critical level in order to bind that amount of toxin needed for fatal liver damage.

C. THERAPEUTIC ASPECTS OF *Amanita* POISONING

As already reported (see p. 264) *A. phalloides* contains another peptide called antamanide, which exerts a specific antitoxic action on phalloidin. The therapeutic use of this cyclic decapeptide, which can also be synthesized, seems rather restricted since it must be given before or at least at the time of toxin administration. Nevertheless, for theoretical reasons, the protective action of antamanide is of great interest. Studies on intact mice and on the perfused rat liver have shown that antamanide prevents the uptake of phalloidin by the liver (Fig. 8) probably by interfering with the binding of the toxin at the respective structural sites. At the same time, antamanide reduced the phalloidin-induced release of potassium. Several sulfhydryl compounds, especially α-lipoate and coenzyme A, as described by Frimmer and Weil (1969a), counteract the potassium output due to phalloidin. Also, silymarin, the antihepatotoxic principle of *Silybum marianum* (L.) Gaertn., has been reported to be an antagonist of phalloidin and, less effectively, of α-amanitin by Vogel (1968). The therapeutic value of these compounds in *Amanita* poisoning remains to be established.

REFERENCES

Bauer, C. A. (1971). *Funghi vivi, funghi che parlano*, Edit. Monauni, Trento, p. 62.
Beerman, W. (1971). *Chromosoma*, **34**, 152.
Benedict, R. G., and Brady, L. R. (1967). *Lloydia* **30**, 372.
Block, S. S., Stephens, R. L., and Murrill, W. A. (1955a). *J. Agr. Food Chem.* **3**, 584.
Block, S. S., Stephens, R. L., Barreto, A., and Murrill, W. A. (1955b). *Science* **121**, 505.
Boehringer, W. (1959). Dissertation, Universität Frankfurt a.M.
Bonelli, A., Genovese, E., and Napoli, P. A. (1967). *Boll. Soc. Ital. Biol. Sper.* **44**, 558.
Bucci, S., Mancino, G., Nardi, I., and Fiume, L. (1971). *Exp. Cell. Res.*, in press.
Burgermeister, W. (1969). Unpublished results.
Cessi, C., and Fiume, L. (1969). *Toxicon* **6**, 309.
Das, B. (1968). Personal communication.
Das, B. (1969). Personal communication.
der Decken, A., Löw, H., and Hultin, T. (1960). *Biochem. Z.* **332**, 503.
Dessy, G., and Francioli, M. (1938). *Boll. Ist. Sierotera Milan.* **17**, 779.
Dezelee, S., Sentenac, A., and Fromageot, P. (1970). *FEBS Letters* **7**, 220.
Fahrenholz, F., Faulstich, H., and Wieland, T. (1971). *Ann. Chem.* **743**, 83.
Faulstich, H. (1969). Unpublished results.
Faulstich, H., Wieland, T., and Jochum, C. (1968). *Ann. Chem.* **713**, 186.
Fiume, L. (1965). *Lancet* **1**, 1284.
Fiume, L. (1967). *Arch. Sci. Biol. (Bologna)* **51**, 85.
Fiume, L., and Laschi, R. (1965). *Sperimentale* **115**, 288.
Fiume, L., and Stirpe, F. (1966). *Biochim. Biophys. Acta* **123**, 643.
Fiume, L., and Wieland, T. (1970). *FEBS Letters* **8**, 1.
Fiume, L., LaPlaca, M., and Portolani, M. (1966). *Sperimentale* **116**, 15.
Fiume, L., Marinozzi, V., and Nardi, F. (1969). *Brit. J. Exptl. Pathol.* **50**, 270.
Floersheim, G. L. (1966a). *Biochem. Pharmacol.* **15**, 1589.
Floersheim, G. L. (1966b). *Physiol. Pharmacol. Acta* **24**, 219.
Frank, W. (1968). *Z. Naturforsch.* **23b**, 687.
Frimmer, M., and Weil, G. (1969a). *European J. Pharmacol.* **8**, 240.

Frimmer, M., Gries, J., Hegner, D., and Schnorr, B. (1967). *Naunyn-Schmiedebergs Arch. Exp. Pathol. Pharmakol.* **258**, 197.

Frimmer, M., and Weil, G. (1969b). *Klin. Wochschr.* **47**, 1116.

Gebert, U., Boehringer, H., and Wieland, T. (1967). *Ann. Chem.* **705**, 227.

Hegner, D., Lutz, F., Eckermann, V., Gries, J., and Schnorr, B. (1970). *Biochem. Pharmacol.* **19**, 487.

Hizukuri, S., and Larner, J. (1964). *Biochemistry* **3**, 1783.

Horgen, P. A., and Griffin, D. H. (1971). *Proc. Natl. Acad. Sci. U.S.* **68**, 338.

Incefy, G. S., and Kappas, A. (1971). *FEBS Letters* **15**, 153.

Ivanov, V. T., Miroshnikov, A. I., Abdullaev, N. D., Senyavina, L. B., Arkhipova, S. F., Uvarova, N. N., Kalilulina, K. Kh., Bystrov, V. F., and Ovchinnikov, Yu. A. (1971). *Biochem. Biophys. Res. Commun.* **42**, 654.

Jacob, S. T., Muecke, W., Sajdel, E. M., and Munro, H. N. (1970c). *Biochem. Biophys. Res. Commun.* **40**, 334.

Jacob, S. T., Sajdel, E. M., and Munro, H. N. (1970a). *Nature* **225**, 60.

Jacob, S. T., Sajdel, E. M., and Munro, H. N. (1970b). *Biochem. Biophys. Res. Commun.* **38**, 765.

Jahn, W. (1970). *Naunyn-Schmiedebergs Arch. Pharmakol. Exptl. Pathol.* **267**, 364.

Jost, J.-P., and Sahib, M. K. (1971). *J. Biol. Chem.* **246**, 1623.

Kedinger, C., Gniazdowski, M., Mandel, J. L., Jr., Gissinger, F., and Chambon, P. (1970). *Biochem. Biophys. Res. Commun.* **38**, 165.

Kobert, R. (1891). *Petersb. med. Wschr.* **16**, 463.

König, W., and Geiger, G. (1969). *Ann. Chem.* **727**, 125.

Küntzel, H., and Schäfer, K. P. (1971). *Nature New Biol.* **231**, 265.

Leloir, L. F., and Cardini, C. F. (1957). *J. Am. Chem. Soc.* **79**, 6340.

Leloir, L. F., and Goldenberg, S. H. (1960). *J. Biol. Chem.* **235**, 919.

Lindell, T. J., Weinberg, F., Roeder, R. G., and Rutter, W. J. (1970). *Science* **170**, 447.

Lynen, F., and Wieland, U. (1938). *Ann. Chem.* **533**, 93.

Mancino, G., Nardi, I., Corvaja, N., Fiume, L., and Marinozzi, V. (1970). *Exp. Cell. Res.* **64**, 237.

Marinozzi, V., and Fiume, L. (1971). *Exp. Cell Res.*, **67**, 311.

Matschinsky, F. (1959). Dissertation, Universität München.

Matschinsky, F., and Wieland, O. (1960). *Biochem. Z.* **333**, 33.

Matschinsky, F., and Wieland, O. (1961). Unpublished data.

Meihlac, M., Kedinger, C., Chambon, P., Faulstich, H., and Wieland, T. (1970). *FEBS Letters* **9**, 258.

Menon, I. A. (1971). *Can. J. Biochem.*, in press.

Meyer, U. (1966). Dissertation, Universität München.

Miller, F., and Wieland, O. (1967). *Arch. Pathol. Anat. Physiol.* **343**, 83.

Niessing, J., Schnieders, B., Kunz, W., Seifart, K. H., and Sekeris, C. E. (1970). *Z. Naturforsch.* **25b**, 1119.

Novello, F., and Stirpe, F. (1969). *Biochem. J.* **112**, 721.

Novello, F., and Stirpe, F. (1970). *FEBS Letters* **8**, 57.

Novello, F., Fiume, L., and Stirpe, F. (1970). *Biochem. J.* **116**, 177.

Palyza, V., and Kulhanek, V. (1970). *J. Chromatog.* **53**, 545.

Pfaender, P. (1966). Unpublished results.

Porter, K. R., and Bruni, C. (1959). *Cancer Res.* **19**, 997.

Prox, A., Schmidt, J., and Ottenheym, H. (1969). *Ann. Chem.* **722**, 179.

Puchinger, H. (1968). Dissertation, Universität Frankfurt a.M.

Puchinger, H., and Wieland, T. (1969a). *Ann. Chem.* **725**, 238.

Puchinger, H., and Wieland, T. (1969b). *European J. Biochem.* **11**, 1.

Rehbinder, D., Löffler, G., Wieland, O., and Wieland, T. (1963). Z. Physiol. Chem. 331, 132.

Rita, G. A., Zuretti, M. F., Baccino, F. M., and Dianzani, M. U. (1967). Enzyme Histochem., 1st. Lombardo Fondazione Baselli p. 3.

Roeder, R. G., and Rutter, W. J. (1969). Nature 224, 234.

Rott, R., and Scholtissek, C. (1970). Nature 224, 234.

Seifart, K. H., and Sekeris, C. E. (1969). Z. Naturforsch. 24b, 1538.

Seitz, U., and Seitz, U. (1971). Planta 97, 224.

Sekeris, C. E., Niessing, J., and Seifart, K. H. (1970). FEBS Letters 9, 103.

Siess, E., Wieland, O., and Miller, F. (1970). Arch. Pathol. Anat. Physiol. Abt. B6, 151.

Stirpe, F., and Fiume, L. (1967). Biochem. J. 105, 779.

Szabados, A. (1970). Dissertation, Universität München.

Tonelli, A. E., Patel, D. J., Goodman, U., Naider, F., Faulstich, H., and Wieland, T. (1971). Biochemistry 10, 3211.

Tosteson, D. C., Andreoli, T. E., Tiefenberg, M., and Cook, P. (1968). J. Gen Physiol. 51, 373.

Tyler, V. E., Jr., Brady, L. R., Benedict, G. R., Khanna, J. M., and Malone, M. H. (1963). Lloydia 26, 154.

Tyler, V. E., Jr., Benedict, G. R., Brady, L. R., and Robbers, J. E. (1966). J. Pharm. Sci. 55, 590.

Verne, J., Ceccaldi, P. F., and Hébert, S. (1950). Compt. Rend. Soc. Biol. 144, 645.

Vogel, G. (1968). Arzneimittel-Forsch. 18, 1063.

Weygand, F., and Mayer, F. (1968). Chem. Ber. 101, 2065.

Wieland, H., and Hallermayer, R. (1941). Ann. Chem. 548, 1.

Wieland, O. (1965). Clin. Chem. 2, 323.

Wieland, O., and Hasslinger, S. (1969). Unpublished data.

Wieland, O., and Szabados, A. (1968). Intern. Congr. Clin. Chem., 1966, Vol. 4, p. 59.

Wieland, T. (1957). Angew. Chem. 69, 44.

Wieland, T. (1967). Fortschr. Chem. Org. Naturstoffe 25, 214.

Wieland, T., and Buku, A. (1968). Ann. Chem. 717, 215.

Wieland, T., and Dölling, J. (1966). Naturwissenschaften 53, 526.

Wieland, T., and Dose, K. (1954). Biochem. Z. 325, 439.

Wieland, T., and Fahrmeir, A. (1970). Ann. Chem. 736, 95.

Wieland, T., and Gebert, U. (1966). Ann. Chem. 700, 157.

Wieland, T., and Grimm, D. (1965). Chem. Ber. 98, 1727.

Wieland, T., and Jeck, R. (1968). Ann. Chem. 713, 196.

Wieland, T., and Rehbinder, D. (1963). Ann. Chem. 670, 149.

Wieland, T., and Schöpf, A. (1959). Ann. Chem. 626, 174.

Wieland, T., and Wieland, O. (1959). Pharmacol. Rev. 11, 87.

Wieland, T., Schiefer, H., and Gebert, U. (1966). Naturwissenschaften 53, 39.

Wieland, T., Hasan, M., and Pfaender, P. (1968a). Ann. Chem. 717, 205.

Wieland, T., Lüben, G., Ottenheym, H., Faesel, J., de Vries, J. X., Konz, W., Prox, A., and Schmid, J. (1968b). Angew. Chem. 80, 209; Angew. Chem. Intern. Ed. Engl. 7, 204.

Wieland, T., Birr, C., and Flor, F. (1969a). Ann. Chem. 727, 130.

Wieland, T., Faesel, J., and Konz, W. (1969b). Ann. Chem. 722, 197.

Wieland, T., Lüben, G., Ottenheym, H., and Schiefer, H. (1969c). Ann. Chem. 722, 173.

Wieland, T., Faulstich, H., Burgermeister, W., Otting, W., and Möhle, W.; and Shemyakin, M. M., Ovchinnikov, Yu. A., Ivanov, V. T., and Malenkov, G. G. (1970). FEBS Letters 9, 89.

Wieland, T., Lapatsanis, L., Faesel, J., and Konz, W. (1971). Ann. Chem. 747, 194.

Wieland, T., Lüben, G., Ottenheym, H., and Schiefer, H. (1969c). Ann. Chem. 722, 173.

CHAPTER 11

Mushroom Toxins Other than *Amanita*

ROBERT G. BENEDICT

I. Introduction

During the past decade, some progress has been made in identifying toxic factors from basidiomycetes, but additional contributions would be welcome. In a comprehensive review on mushroom poisons, Tyler (1963) pointed out the need for a critical reevaluation of the existing literature in order to determine its validity. Even a casual examination of this article reminds the author of a quotation in *Nutritional Data* (1956), which states that "the casual risk of macrophagy is often greatly disproportionate to the gastronomic rewards."

With the exception of the deadly cyclopeptidic toxins from certain *Amanita* species, the writer will attempt to review the remaining toxins and irritants with regard to their lower plant sources, symptomatology, chemistry (if known), toxicology, and biosynthesis in culture. The reader will find an excellent review on the true *Amanita* toxins in the preceding

281

chapter. Unless otherwise noted, the author has utilized Singer (1962) as his taxonomic authority for species cited in this review.

A. STATISTICS ON MUSHROOM POISONING

Unlike their European and Far Eastern contemporaries, relatively few people in the United States collect wild mushrooms for culinary purposes. The records of Buck (1961, 1964, 1968) regarding fatalities from mycetisms (mushroom poisoning) indicate that only 73 reported deaths occurred in the United States between 1897 and November 1968 (see Table I). The medical records between 1897 and 1930 recorded only eight deaths and probably are far less complete than those from 1931 to the present.

Without doubt, some of the casualties in the 51 to 89 age bracket were people who foolishly believed that a mushroom's failure to tarnish a silver spoon during cooking was evidence of safe edibility. This old wive's tale is without any scientific basis whatsoever. According to Cann and Verhulst (1961), mushrooms account for less than 2% of the cases of accidental poisonings reported each year to the National Clearing House for Poison Control Centers in Washington, D.C.; mushrooms accounted for only 0.9% of approximately 30,000 accidental ingestions of poisonous substances during the year 1958. In 1967-1968, ingestions of mushrooms totalled 675 out of 83,704 cases (0.8%) with no fatalities; from January to June 1968, only 146 such incidences were reported from U.S. Poison Control Centers (Verhulst, 1969).

Although Great Britain had about one-fourth the population of the United States, the 14 deaths due to accidental ingestions of poisonous mushrooms between 1930 and 1945 (Dubash and Teare, 1946) were about equal to those in the United States for the same period. In countries like France, Poland, Russia, Italy, Germany, and Switzerland, where

TABLE I

REPORTED DEATHS FROM MYCETISMS[a] — UNITED STATES 1931-1968

Years	Children 10 months-11 years	Adults 19-50 years	51-89 years	Age unknown or not reported
1931-1939	1	3	2	6
1940-1957	5	4	15	1
1958-1968	8	10	9	0
Totals	14	17	26	7

[a]Phalloid type, 33 (see Section II, A and preceding chapter); *Helvella* sp., 7 (see Section II, E); miscellaneous, 4; unknown or not reported, 20.

edible mushrooms are avidly sought in season, poisoning cases are proportionately higher. In Switzerland, almost 5% of the 1980 cases of poisonings recorded between 1919 and 1958 were fatal, despite the help of the Swiss government's VAPKO (Vereinigung der amtlichen Pilzkontrollorgane der Schweiz), which advises amateur collectors on the safeness of their collections (Alder, 1961).

Imazeki, a Japanese mycologist, stated that 500 to 600 cases of mushroom poisonings occur in Japan each year, of which about 15 are fatal (Romagnesi, 1964). Most of these result from the ingestion of *Amanita verna* and *Amanita virosa*, both known to contain the deadly cyclopeptidic toxins.

B. TOXINS AND BASIDIOMYCETE TAXONOMY

Only one of two major subdivisions of the basidiomycetes need concern us in this review. The Homobasidiomycetes contain the Tremellales, or jelly fungi, which do not rate as edible or poisonous forms of consequence. Neither the Uredinales (Rusts) nor the Ustilaginales (Smuts), also classified in the Homobasidiomycetes, are of interest to the mycophagist. The majority of toxic Heterobasidiomycetes are found in the Agaricales. This review also will be concerned with the flesh pore fungi (Boletaceae) and the true pore fungi (Polyporales), although a majority of the latter are too woody to be of culinary interest. Others to be briefly considered as toxin sources are the Hydnaceae or hedgehog mushrooms, some of the Cantharellales or coral mushrooms, and a few toxic puffballs in the Lycoperdales. Last, but not least, will be a discussion of certain Ascomycetes in the family Helvellaceae, containing, among others, the genera *Morchella* and *Helvella*.

C. CLASSIFICATION OF MUSHROOM TOXINS AND IRRITANTS

The systems of Ford (1926) and Alder (1961) for classifying mushroom toxins were reviewed by Tyler (1963). Tyler adopted, with minor modifications, the three basic divisions of Alder and added one more. The author accepts the four basic divisions utilized by Tyler and has added one new division. The resulting five major divisions are:

1. Protoplasmic toxins
2. Toxins causing neurological manifestations
3. Gastrointestinal irritants
4. Disulfiram-like constituents
5. Toxins with antitumor activity

All of the compounds to be discussed in groups 1 to 4 are naturally occurring constituents of the fresh mushrooms involved and do not include unusual toxic reactions (hypersensitivities) due to mushroom proteins or to decomposition products caused by microbial action. Those in group 5 are special fermentation products, either not present or not biologically detectable in the mature carpophores involved.

II. Protoplasmic Toxins

All of the compounds in this section are capable of bringing about cellular degeneration in major organs of the body. The reader's attention is directed to the previous chapter for details on the toxic cyclopeptides from certain species of *Amanita*. Brief mention of some of these compounds here from certain *Galerina* species and from *Lepiota helveola* Bres. is made since these compounds are not limited to certain species of *Amanita*. Despite the knowledge released many years ago that *Galerina sulciceps* (Berk.) Sing., and, more recently, *Cortinarius orellanus* Bres., each contain potent, and probably unrelated, lethal factors, nothing is yet known about their chemical natures. The protoplasmic poison from certain species of *Helvella,* only recently identified as to structure, has been given the trivial name gyromitrin.

A. Cyclopeptides in *Galerina* sp.

The first fatalities from eating *Pholiota autumnalis* were recorded by Peck (1912). Singer (1962) lists this species as *Galerina autumnalis* (Peck) Sm. et Singer (*Naucoria,* Sacc., *Pholiota,* Peck 1908). Neither the toxins in *G. autumnalis* nor those in *G. venenata* A. H. Smith, implicated in a 1953 poisoning episode in Portland, Oregon, had been chemically classified until 1962. At that time, Tyler and Smith (1963) reopened the problem and tentatively identified both α- and β-amanitin from some of the 9-year-old herbarium specimens of *G. venenata* involved in the 1953 episode.

Later, Tyler *et al.* (1963) obtained both chromatographic and pharmacological evidence of the toxicity of *Galerina marginata* (Batsch ex Fr.) Kühner (*Pholiota,* Quél.) in addition to the two species already mentioned. This is not to imply that all species of *Galerina* are toxic; Tyler and Smith (1963) failed to find chromatographic evidence of toxins in *Galerina vexans* Smith et Singer and predicted that the toxigenic mushrooms of the genus *Galerina* may be restricted to members of certain stirps.

B. CYCLOPEPTIDES IN *Lepiota helveola* BRES. SENSU JOSSERAND

Certain species in the genus *Lepiota* also contain *Amanita* cyclopeptidic toxins. The most toxic one is *L. helveola* Bres. sensu Josserand, a European species not found in the United States. Mortara and Filipello (1967) pointed out that *Lepiota brunneoincarnata* Chodat and Martin and *Lepiota fuscovinacea* Moeller and Lange are also suspect. It is chemotaxonomically significant that Singer (1962), placed the former species in section Ovisporae, where one also finds *L. helveola,* and the latter in the closely related section Anomalie. The same authors cited the complete medical history of a nonfatal case of *L. helveola* poisoning. The patient had ingested mushrooms mixed with nontoxic *Marasmius oreades* (Bolt. ex Fr.) Fr. Their extensive review, including some previous cases of *L. helveola* poisonings, does not indicate the frequency of these poisonings in Europe, nor does it report any attempt to chemically analyze the mushroom toxins. There are indications, however, that *L. helveola,* on an equal weight basis, is probably just as toxic as *Amanita phalloides*.

C. LETHAL FACTOR FROM *Galerina sulciceps* (BERK) SING.

The mortality figures for *Amanita* toxin victims range from 34 to 63% (Alder, 1961). Poisoning by mushrooms containing these factors has always been considered the most lethal in the mushroom field, but this may no longer be true. A series of fatalities from mushroom poisonings between the years 1934 and 1937 were reported in Java by Boedijn (1938-1940). In each series, Boedijn identified the mushrooms as *Phaeomarasmius sulciceps* (Berk.) Scherffel. Singer (1962) presently lists this species as *Galerina sulciceps* (Berk.) Sing. (*Marasmius*) Berk. It is interesting that he classified it in section Naucoriopsis, where one also finds *Galerina marginata, G. autumnalis,* and *G. venenata,* but in a different stirps. In the 1934 outbreak, two of three natives died within 24 hours after eating the mushrooms; in October 1937, seven of nine victims died after the mushroom meal. No vomiting or diarrhea occurred, and the author did not know the fate of the other "very ill" patients. In November 1937, five victims who had consumed large quantities succumbed, one after 7 hours, one after 24 hours, two after 30 hours, and one after 51 hours. A sixth patient, who had eaten very little of the mushrooms survived. The symptoms included "spasms in the belly," nausea *without vomiting,* dizziness, palpitation of the heart, dyspnea, local anesthesia, "pins-and-needles" feeling, unconsciousness, and death. These symptoms are contrasted with a relatively long asymptomatic latent period (up to 24 hours) and a sudden violent vomiting and continuous diarrhea reported in the early phases of *Amanita* cyclopeptide poisonings. In the hope of eliminating the *Amanita* toxins as the etiological agent(s) in *G. sulciceps* poisoning, the reviewer (1967) contacted a mycologist at the herbarium in Bogor, Indonesia. All basidiomycetes retained there are preserved in alcohol, but specimens available from any new outbreaks of *G. sulciceps* poisoning will be dried and sent to the reviewer. The solution to an intriguing toxicological problem still awaits the availability of sufficient specimens and interested, qualified personnel.

D. ORELLANINE IN *Cortinarius orellanus* FR.

Cortinarius orellanus is an attractive mushroom with bright yellow to reddish yellow pigments. According to Desvignes (1965), the mushroom had not been suspect until 1952 to 1955, when observations in Poland by Grzymala showed that it contained a dangerous, slow-acting toxin capable of producing fatalities. An unusual feature is the extra long latent period before toxic manifestations begin. In 81 Polish cases studied by Grzymala, the latent period was 3 to 5 days for 34 patients, 7 days for 3, 8 to 10 days for 24, and 11 to 14 days for 20. Along with a dryness of the mouth and burning of the lips, an intense thirst developed, causing the patient to drink several liters of fluid per day. Gastric disturbances followed, with vomiting, intestinal pains, and a stubborn constipation or diarrhea. The patients shivered, felt a constant sensation of cold, and experienced violent and persistent headaches. In severe cases, there was insufficient renal function, oliguria, and sometimes anuria. Albuminuria occurred, blood urea levels rose, and an electrolytic imbalance existed. Signs of hepatic disturbances appeared, with pains, biliary vomitus, and a subicteric state. In terminal cases, neurological signs included drowsiness, loss of consciousness, and convulsions. The fatality rate varied between 10 and 20%. Autopsies showed renal lesions characteristic of toxic interstitial nephritis, often following a chronic nephritis if the disease had been prolonged.

The toxic substance is very stable to heat and prolonged desiccation. Grzymala (1962) Soxhlet-extracted the toxin in methanol and isolated a crystalline product, orellanine, the structure of which is still unknown or not yet reported. One gram of crystalline toxin was obtained from 100 gm dried mushroom, suggesting that each average-size specimen contains about 15 to 20 mg. Orellanine occurs as colorless to pale yellow crystals, only slightly soluble in water, but giving an acid reaction therein. It is more soluble in methanol and ethanol and very soluble in pyridine. The ultraviolet absorption spectrum in water presents maxima at 230–235 mμ, 300 mμ, and 340 mμ.

Toxicity tests have been conducted in cats by oral administration and parenterally in guinea pigs, mice, and rabbits. Karbler (quoted in Desvignes, 1965) reported the LD_{50} doses of orellanine for the cat, mouse, and guinea pig to be 4.9, 8.3, and 8.0 mg/kg, respectively.

Cortinarius orellanus also grows in central and southern France, and specimens collected there are just as toxic as their Polish counterparts. The mushroom has many close relatives in the Pacific Northwest, and disclosure of some chromatographic mobilities and a spray to detect the presence of orellanine might permit one to make comparative chemotaxonomic studies. Some taxonomic authorities disagree on the edibility of some cortinarii in taxonomic sections close to those in which one finds *C. orellanus;* for example, Singer (1962) writes, "All species belonging to subgenus *Dermocybe* are inedible or suspect, and so are at least some of those belonging to the subgenus *Cortinarius.*" However, A. H. Smith (1949) indicated that all of the above group (*Dermocybe*) are probably edible and that *C. cinnamomeus* (Fr.) S. F. Gray, the type species "is rather highly esteemed by some people."

E. GYROMITRIN IN *Helvella* SP.

Gyromitrin is another type of protoplasmic poison which is chemically unrelated to the cyclopeptidic toxins but capable of producing fatalities. This compound is found in some of the false morels, e.g., certain strains of *Helvella esculenta* Fr., *Helvella gigas* Krbh., and *Helvella underwoodii* Seaver. *Helvella infula* Fries, closely related to *H. esculenta*, is also suspect. Only seven fatalities from *Helvella* ingestions have been recorded in the United States (see Table I and Section I, A), but each year large quantities of these fungi are consumed in Europe, where a 2 to 4% fatality rate has been reported (Alder, 1961). The toxin is soluble in hot water and presumably may be separated from the fungus by discarding the parboil water. Two parboilings are recommended by some experts, but the author prefers to follow the simple directive, "Do not eat any species of *Helvella!*" Among the many well-illustrated mushroom handbooks to assist the nontaxonomist in distinguishing *Helvella* from the desirable morels are those of A. H. Smith (1967), McKenny (1962), Bandoni and Szczawinski (1964), and Krieger (1967).

The symptoms of poisoning by gyromitrin are similar to those caused by *Amanita* cyclopeptide ingestion. A latent period of about 6 to 10 hours is followed by a feeling of fullness in the stomach, then violent vomiting and watery diarrhea (Tyler, in McKinny, 1962). These conditions may continue for 1 to 2 days. Other symptoms are headache, lassitude, cramps, and severe pain in the hepatic and gastric region, followed by jaundice. In severe cases, general collapse, irregular pulse, difficulty in breathing, delirium, and convulsions occur. Liver damage and heart failure may cause death within 5 to 10 days.

Chemical characterization of gyromitrin was carried out by List and Luft (1967). Fresh lorchels *Helvella* (*Gyromitra*) *esculenta* were extracted with ethanol, and the extract, evaporated at low temperature, yielded a volatile, acid and base labile, autooxidizable compound melting at 5°C and having the empirical formula $C_4H_8N_2O$. Its proposed structure (I) was confirmed by nuclear magnetic resonance.

$$CH_3-\overset{\overset{\displaystyle H}{|}}{C}=N-N\overset{\displaystyle CH_3}{\underset{\displaystyle \underset{\displaystyle O}{\overset{\|}{C}}-H}{}}$$

Gyromitrin

(I)

This compound proved to be highly toxic for both guinea pigs and rabbits. Oral administration of an aqueous solution and even inspiration of vapors at normal temperatures and pressures were fatal. Since degassed aqueous extracts are nontoxic, it was concluded that volatile gyromitrin is the toxic component.

A few puzzling questions about the origin of gyromitrin are as yet unresolved. Is it formed only by the decomposition of protein in "overaged" specimens? If this is true, are the protein precursors present in only certain strains of *Helvella* species and not in others? Or is it inherently produced in young specimens, but only by select strains? Answers to these questions might account for the irregular occurrence of the toxin but would not explain the apparent resistance of some individuals to it. For example, a Michigan physician (Hendricks, 1940) treated several members of an Indian family that had consumed *H. esculenta*. The 69-year-old mother died 5 days after the mushroom meal; her 38-year-old daughter experienced violent vomiting, but had practically recovered by the next day. The victim's husband, age 80, son-in-law, 61, and two children, ages 2 and 4, suffered no ill effects whatsoever. In fact, the son-in-law said that he had eaten these fungi, in season, for at least 20 years, with no adverse reactions.

III. Toxins Causing Neurological Manifestations

These substances may be divided into four basic types: muscarine, isoxazole derivatives, substituted tryptamines, and unidentified hallucinogen(s) from *Gymnopilus spectabilis*. The activities of muscarine and some of the isoxazole derivatives are opposite and antagonistic, whereas those in the last three categories are similar in certain aspects and different in others.

A. MUSCARINE

1. EARLY STUDIES

The activity of muscarine as a toxic principle in the fly agaric *Amanita muscaria* (L. ex Fr.) Hooker was recognized more than 150 years ago, and the compound has been known by this trivial name since 1869. It is unfortunate that investigators had for so long relied on the fly agaric as the main source of supply of muscarine, since certain species of the genera *Inocybe* and *Clitocybe* are far richer sources (see Table II). In an earlier monumental effort to obtain enough muscarine for structural determination, Kögl *et al.* (1931) had cleaned and extracted 1250 kg of *A. muscaria*, peeling the outer, orange-red pigmented cap pellicles from 500 kg. All of this was done in one night's effort, and some of the workers suffered symptoms of muscarinic toxicity from this prolonged contact with the mushrooms.

The only naturally occurring sources of muscarine reported thus far are certain fleshy fungi, mainly species of *Amanita*, *Inocybe*, and *Clitocybe*. In his tabular discussion of 27 toxic species of *Amanita*, Tyler (1963) included *A. muscaria*, *A. pantherina* (D.C. ex Fr.) Secr., *A. parcivolvata* (Peck) Gilb. *Amanitopsis* Peck, *A. flavivolva* Murr., and *A. gemmata* (Fr.) Gill. as probable muscarine-containing species. The genus *Inocybe* is very large, and Malone *et al.* (1962) reported only one of thirty species to be nonmuscarinic; some of their results are shown in Table II.

Obviously *Inocybe napipes* is by far the most potent agaric listed. The percentage of muscarine detected by paper chromatographic analysis of extracts (Brown *et al.*, 1962) are compared with those obtained by the rat bioassay method of Malone *et al.* (1962). The latter detects the presence of other muscarine-like agents and/or physiological potentiators for *I. napipes* and *Inocybe lacera*. Most of the muscarine-containing species of *Clitocybe* are white and include *C. dealbata* (Table II), *C. rivulosa* (Pers. ex Fr.) Kummer, *C. truncicola* (Peck) Sacc., and *C. cerussata* (Fr.) Kummer and some of its varieties. Tyler (1963) also lists the reddish-orange to yellowish *C. illudens* (Schw.) Sacc., *C. subilludens* Murr., and *Pleurotus olearius* (D.C. ex Fr.) Gill., three closely related species corresponding to *Omphalotus olearius* (D.C. ex Fr.) Sing.

TABLE II

MUSCARINE CONTENT OF VARIOUS FLESHY FUNGI

Genus and species	Fresh weight and estimate through isolation (%)	Dry weight and rat bioassay (%)	Dry weight and chromatographic estimation (%)	References
Amanita muscaria[a]	0.0002–0.0003	—	—	Eugster (1956)
Inocybe patouillardi[a] Bres.	0.037	—	—	Eugster (1957a); Eugster and Müller (1959)
Inocybe nigrescens Atkinson	—	<0.001	<0.01	Malone et al. (1962); Brown et al. (1962)
Inocybe napipes Lange	—	2.10–3.15[b]	0.71[b]	Malone et al. (1962); Brown et al. (1962)
Inocybe picrosma Stuntz	—	0.005	<0.009	Malone et al. (1962); Brown et al. (1962)
Inocybe lacera (Fr.) Quél.	—	0.846–1.0[b]	0.08[b]	Malone et al. (1962); Brown et al. (1962)
Clitocybe dealbata (Sow. ex Fr.) Kummer	—	—	0.15 ± 0.04	Hughes et al. (1966)

[a]Contain about 90 to 93% moisture.
[b]More than one collection of fungi assayed.

2. CHEMICAL PROPERTIES

The lack of an efficient method of isolation, plus very low concentrations in the mushrooms, greatly delayed the progress of chemical studies on muscarine. From their large collection (over 1 ton) of *A. muscaria*, Kögl *et al.* (1931) recovered only 370 mg of muscarine reineckate (equivalent to 137 mg muscarine cation). When converted to the chloride, the final impure product probably did not have more than 25% muscarine. Some of the impurities in these earlier preparations were the related compounds muscaridine and unstable acetylcholine. Both of these impurities (see Fig. 1) are naturally occurring in certain mushrooms, and in 1957 acetylcholine was isolated from *A. muscaria* by Kögl and associates (Wilkinson, 1961).

Pure muscarine chloride was finally obtained from *A. muscaria* by Eugster and Waser (1954) and its composition determined as $C_9H_{20}O_2 \cdot N^+Cl^-$. The structure was finally determined some 3 years later by Jellinek, using X-ray crystallographic techniques in Kögl's laboratory (Kögl *et al.*, 1957b). The structure in Fig. 1 shows it to be the oxoheterocyclic quarternary salt 2-methyl-3-hydroxy-5-trimethylammoniummethyltetrahydrofuran chloride with *S* configuration at position 2, *R* at position 3, and *S* at position 5 in the ring. This structure was later verified by synthesis (Eugster, 1957b; Kögl *et al.*, 1957a). With three centers of asymmetry, four diasterioisomeric racemates and eight antipodes were anticipated, and nearly all have been obtained. For further particulars on the chemistry of these compounds, see Waser (1961), Eugster (1968), and Wieland (1968).

The L(+) isomer of muscarine as illustrated is by far the most active; it exhibits a highly specific muscarinic activity at postganglionic parasympathetic effector sites. As little as 0.01 μg/kg will decrease the blood pressure of cats, and similarly minute quantities (0.003-0.01 μg) will reduce

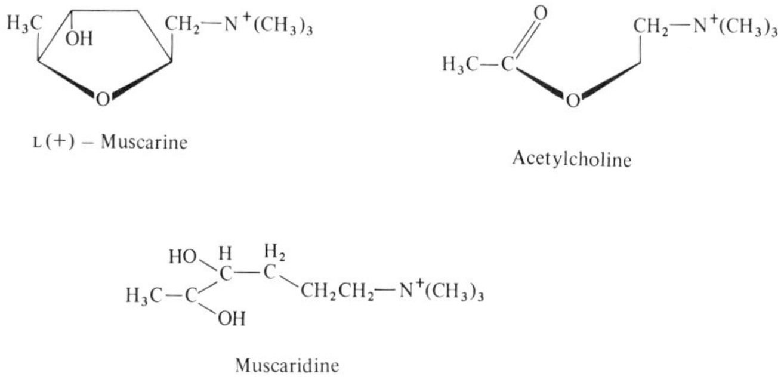

L(+) – Muscarine

Acetylcholine

Muscaridine

FIG. 1. Quaternary ammonium compounds in fleshy fungi.

the amplitude and rate of beat of the isolated frog heart of Straub. This action is immediately antagonized by atropine.

Statistics on muscarine poisoning in the United States are lacking; the majority of cases probably result from the ingestion of white *Clitocybe* species. In Europe, the brick red Risspilz (*Inocybe patouillardi* Bres. sensu *lateria* Ricken) is responsible for a number of cases each year, with a mortality rate up to 8% (Eugster, 1957a). According to Tyler (1963), symptoms of uncomplicated muscarinic poisoning appear quite rapidly, i.e., within 15 to 30 minutes. They are very characteristic — increased salivation, lachrymation, and perspiration, followed by vomiting and diarrhea. The pulse is slow and irregular, and breathing is asthmatic. Mild cases respond readily to atropine sulfate and other supporting therapeutic measures. Even in severe cases, death is uncommon, usually resulting from heart or respiratory failure.

The possibility is rather remote that muscarinic poisoning would result from ingestion of canned mushrooms processed in the United States. However, one such incidence did occur in the United States from the ingestion of canned mushrooms imported from Taiwan (Rose and Rieders, 1966). At a luncheon featuring Polynesian dishes, 55 of 86 women (wives of medical doctors) experienced bitemporal headache, malaise, and muscular tingling within 15 to 30 minutes after consuming imported mushroom soup. Careful laboratory investigation indicated that muscarine was the probable cause.

3. BIOSYNTHESIS

The choice of cultures for biosynthesis of muscarine in the laboratory is sharply limited. *Inocybe* species are strict mycorrhizally associated forms and usually do not grow in culture, even in the presence of excised coniferous root tissue. Swenberg *et al.* (1967a,b) utilized the strain of *Clitocybe rivulosa* available from the Centraalbureau voor Schimmelcultures, Baarn, The Netherlands. A medium containing beer wort, mannitol, succinate, and inorganic salts proved to be the best for muscarine production of many media on which the culture was grown. Stationary cultures were far superior to those in shaken flasks, and the small amount of muscarine formed occurred within the mycelium. Eighty flasks, harvested after 6 to 8 weeks at 24°C, yielded 491 gm of wet mycelium. Careful processing gave a total of only 38 mg of muscarine reineckate. Subsequent tests to determine dry weight, pH, and muscarine concentration (Fig. 2) showed that maximum growth was not reached until about 70 days, and muscarine levels peaked at close to 80 days. There was no sharp change in pH which could serve as a convenient indicator for optimum harvest time.

The structure of L(+)-muscarine and muscaridine (Fig. 1) suggests that

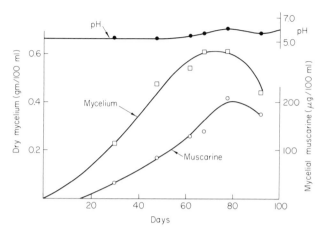

FIG. 2. Biosynthesis of muscarine in *Clitocybe rivulosa*. Courtesy of Swenberg *et al.* (1967b).

either a simple hexose or a 1-amino-3,6-deoxyhexose might serve as precursors for these compounds in nature. Brown *et al.* (1966) collected *Inocybe napipes* fruiting bodies at various stages of their development and determined that mature specimens, with dry weights four to six times those of young forms, contained approximately the same percentage of muscarine. It appeared that the muscarine content per carpophore had increased as the fruiting body matured. Fresh *I. napipes* could imbibe aqueous solutions through the base of their severed stipes, with subsequent diffusion throughout the fleshy part of the mushroom. This technique was employed to introduce D-glucose-U-^{14}C, and later, by the method of Brown *et al.* (1962), to chromatographically assay for its conversion to muscarine. Some radioactivity (0.46-1.9% of the administered isotope) was found to be incorporated into the lipid-containing extracts of the fruiting bodies. Since incorporation in the chromatographically purified muscarine totalled a very low 0.003 to 0.02%, the authors concluded that D-glucose is not a direct precursor of muscarine.

B. IBOTENIC ACID, MUSCIMOL, MUSCAZONE, AND TRICHOLOMIC ACID

1. USE IN FOLKLORE MEDICINE

The use of the fly agaric *Amanita muscaria* as a narcotic or intoxicant by the Koryak and neighboring tribes of Kamchatka was reported by early 18th century explorers who toured that part of Siberia. Orgies of tribesmen who participated in these ceremonies are vividly described by Ramsbottom (1953). The unknown active principles were assigned the name pilzatropine by Kobert in 1891. For a comprehensive historical review on pilzatropine, see the article by Tyler (1958).

The intense central nervous system stimulation brought on by the ingestion of *A. muscaria* or *Amanita pantherina* could not have been caused by muscarine, which has practically no central nervous system activity. Earlier reports that atropine, hyoscyamine, and scopalamine were responsible for the central nervous system activity could not be confirmed in subsequent studies by Brady and Tyler (1959), Salemink (1963), and Tabor and Vining (1963).

Investigations continued on the flyicidal and narcotic-intoxicant principles in selected *Amanita* and *Tricholoma* species, mainly in Japan, Switzerland, and Great Britain. The Japanese workers concentrated their efforts on the isolation and characterization of the flyicidal components, first from *Tricholoma muscarium* Kawamura (Takemoto, 1961) and later from *Amanita* species. The Swiss scientists turned their attentions to the narcotic-intoxicant components of *A. muscaria*. The ultimate discovery that three of these compounds are isoxazole derivatives and that a fourth is an oxazole was noteworthy because naturally occurring compounds of this type are rare.

Literature reports on the fly-killing activity of mushrooms list *A. muscaria* first and *Amanita pantherina* second. Some confusion still exists regarding another Japanese "flyicidal" source, *i.e., Amanita strobiliformis* (Paul.) Quél., collected for Takemoto *et al.* (1964a) by N. Matsumato. It is not possible from this designation to determine with certainty the identity of this species, and since Paulet is a pre-Fresian author, the citation given is obviously incorrect. Benedict *et al.* (1966b) surveyed 21 species of *Amanita* for isoxazoles (Table III) and found that these compounds were limited mainly to two species. Some mycological authorities—e.g., Dennis *et al.* (1960)—consider *A. strobiliformis* (Vitt.) Quél. to be conspecific with *A. solitaria* (Fr.) Secr. Collections of the latter species did not contain ibotenic acid or muscimol, but did contain other compounds which closely resembled the former in their mobilities in certain solvents and their distinctive color reactions with ninhydrin (Benedict *et al.*, 1966b) (see Section III, B, 3).

The distribution of *Tricholoma muscarium,* the only known source of tricholomic acid, appears to be limited to Japan (Kawamura, 1954). Unlike the other two isoxazole derivatives from mushrooms, tricholomic acid shows no special color changes with ninhydrin. However, a modification made by Benedict *et al.* (1966a) of the Scardi and Bonavita (1959) assay for the isoxazole antibiotic cycloserine (D-4-amino-3-isoxazolidone) enabled the detection of as little as 2 μg of tricholomic acid. Whereas the *T.*

muscarium specimens, kindly furnished by R. Imazeki of Japan were strongly positive, none of this compound was found during a chemotaxonomic study of 20 species of *Tricholoma* collected in western Washington.

Ibotenic acid and muscimol can be detected in dried *A. muscaria* and *A. pantherina* even after 5 and 7 years of storage at room temperature, respectively (see Table III). These two species are not available during the dry summer months in the Pacific Northwest, and muscazone could not be detected in either spring or fall collections from this area. Ibotenic acid is readily decarboxylated to muscimol during drying and even during two-dimensional chromatography (see Table III); hence, assay values for the two must be combined. Quantitative estimates of total isoxazoles (ibotenic acid and muscimol) from crude extracts are facilitated by a distinctive color change in the ninhydrin–isoxazole complex. After spraying chromatograms with ninhydrin and heating, a bright yellow color develops at the positions of the isoxazoles. Within a few hours, the color changes from yellow to purple. Dried *A. muscaria* averaged about 0.18% and *A. pantherina* contained up to 0.46% total isoxazoles (Benedict *et al.*, 1966b). Other quantitative figures are based mainly on recoveries during chemical purification processes, e.g., 0.025% for ibotenic acid from *A. muscaria* and 0.22% from *A. pantherina* (Takemoto *et al.*, 1964b). Both figures are calculated on a dry weight basis assuming 10% solids in fresh carpophores. Eugster (1968) reported that fresh *A. muscaria* collections from Switzerland varied markedly in isoxazole content, ranging from 0.03% in 1962 collections to a very high 0.1%, (approximately 1% on a dry weight basis) in summer specimens of 1966.

2. CHEMICAL PROPERTIES

Direct connections between the insecticidal isoxazoles and compounds from *A. muscaria* causing intoxications were not revealed until the publications of Eugster *et al.* (1965), Good *et al.* (1965), and Müller and Eugster (1965). The "prämuscimol" isolated by these workers from *A. muscaria* in 1960 was chemically indistinguishable from the ibotenic acid recovered from *A. strobiliformis* by Takemoto and associates. Various synonyms for these compounds had appeared in the literature, e.g., muscimol had been called β-toxin, pyroibotenic acid, agarin, and pantherine. Ibotenic acid had been designated as α-toxin or prämuscimol. Based on priority publication dates, Eugster and Takemoto (1966) accepted the trivial names muscimol, ibotenic acid, muscazone, and tricholomic acid. The structures of these compounds are shown in Fig. 3.

All of the compounds shown in Fig. 3 have been synthesized — mus-

TABLE III

VARIATION OF NARCOTIC-INTOXICANT COMPOUNDS IN *Amanita* SPECIES

Amanita species	Sources of plants	Ibotenic acid	Muscimol	Muscazone	References
A. muscaria	United States	+[a]	+	0	Benedict et al. (1966b)
A. muscaria	Switzerland	++	+	+[b]	Good et al. (1965)
A. muscaria var. formosa (Fr.) Sacc.	United States	+	+	0	Benedict et al. (1966b)
A. muscaria var. alba (Peck) Coker	United States	+	+	0	Benedict et al. (1966b)
A. pantherina	United States	++[a]	+	0	Benedict et al. (1966b)
A. pantherina	Japan	++	−[c]	−[c]	Takemoto et al. (1964b)
A. pantherina	Switzerland	0	0	0	Eugster and Takemoto (1966)
A. pantherina-gemmata intermediate form	United States	+	+	0	Benedict et al. (1966b)
A. strobiliformis	United States	0	0	0	Benedict et al. (1966b)
A. strobiliformis	Japan	+	−[c]	−[c]	Takemoto et al. (1964a)

[a]Survived long term storage at room temperature.
[b]Present in specimens collected in summer.
[c]Not reported.

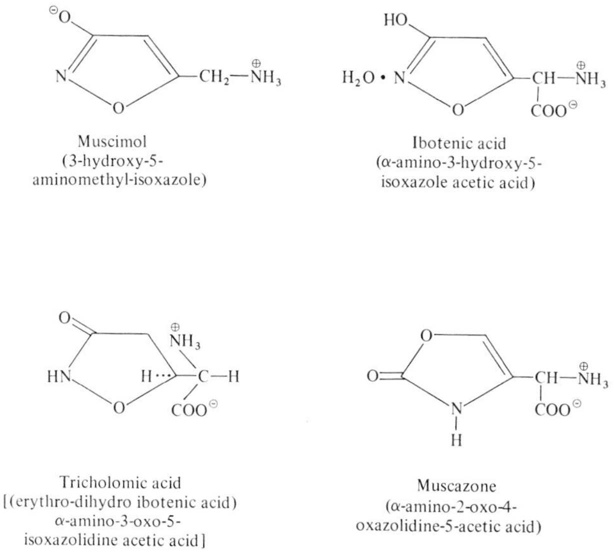

Muscimol
(3-hydroxy-5-
aminomethyl-isoxazole)

Ibotenic acid
(α-amino-3-hydroxy-5-
isoxazole acetic acid)

Tricholomic acid
[(erythro-dihydro ibotenic acid)
α-amino-3-oxo-5-
isoxazolidine acetic acid]

Muscazone
(α-amino-2-oxo-4-
oxazolidine-5-acetic acid)

Fig. 3. Flyicidal and narcotic intoxicants from mushrooms.

cimol by Gagneux *et al.* (1965a), tricholomic acid by Iwasaki *et al.* (1965), ibotenic acid by Gagneux *et al.* (1965b), and muscazone by Göth *et al.* (1967). An interesting photoreaction, similar to the Lossen rearrangement of hydroxamic acid derivatives, converts the isoxazole system of ibotenic acid into the 2-(3H)-oxazolone ring of muscazone (Göth *et al.*, 1967).

Since the concentrations of ibotenic acid and muscimol vary markedly according to geographic locations, season, and year of collection, one would expect considerable variation in the symptoms of persons who have accidentally eaten mushrooms containing these compounds. In general, these symptoms begin within 30 to 90 minutes after the meal. Vomiting does not necessarily occur, and the victim passes into a state resembling advanced alcoholic intoxication. Restlessness, confusion, visual disturbances, muscle spasms, and delirium follow. In a few severe cases involving *A. pantherina*, respiratory difficulties have necessitated tracheotomies. Following the excited state, the patient may pass into a deep sleep, and upon awakening, have little or no memory of his experience. The Swiss investigators believe that psychoactive compounds other than the isoxazoles under discussion may be present in *A. muscaria*, and different screening methods are being employed to search for these unknown compounds (Waser, 1967).

Proof that pure muscimol, in doses of 15 mg, exerts a marked psychoactive effect on a human is provided by Waser (1967), who tested the compound on himself. Forty minutes after ingestion, the psychotic manifestations included confusion, disarthria, disturbances of visual perception, illusions of color vision, myoclonia, disorientation in situation and time, weariness, fatigue, and sleep. Under psychiatric control, concentration tests indicated an improved performance with small doses (5 mg), but higher doses, 10 to 15 mg, resulted in diminished performance and learning, with an increased number of errors. Ibotenic acid, at 20 mg, produced slight face-flushing within 30 minutes, but no psychic stimulation. Neither heart rate nor blood pressure changes were noted. Lassitude and sleep were followed by migraine, with one-sided visual disturbance and an occipitally localized headache that continued in milder form for 2 weeks. Certain parallels are evident between the experiences of Waser with the pure drugs and previously listed symptoms following ingestion of the wild mushrooms.

According to Waser (1967), the sedative action of muscazone is far less than that of ibotenic acid or muscimol. The sedative–hypnotic actions of the above compounds were determined intraperitoneally in mice by observing their potentiating effect on the narcosis produced by a short-acting hypnotic (2-methoxy-4-allylphenoxyacetic acid diethylamide). Sedative action is noted after 4–8 mg/kg ibotenic acid or with only 1–2 mg/kg muscimol. Tricholomic acid would probably have little or no sedative action, since Takemoto (1961) reported that *Tricholoma muscarium* is occasionally used as a food by Japanese farmers without ill effects. Tricholomic acid may be the most effective fly killer among the mushroom isoxazoles; the above author described its toxic effect on flies as rather rapid, starting with paralysis of the legs and terminating in death due to "collapse or intense vacuole formation in the intestinal epithelium." The British investigators Bowden and Drysdale (1965) and Bowden *et al.* (1965) also used flies in their rather unique assays for muscimol (agarin) activity in crude preparations from *A. muscaria*.

A unique property of ibotenic acid and tricholomic acid, discovered accidentally in Japan, is that of flavor. They have approximately 20 times the flavor intensity of monosodium glutamate, and small quantities are said to act synergistically with nucleotide seasonings.

A number of derivatives of muscimol have been prepared by substitution at positions 3 or 5 (Belgian Patent, 1965), and these compounds are undergoing tests in psychiatric medicine. The results of these studies have not yet been released. To the authors' knowledge, biosynthetic studies on the isoxazoles from mushrooms have not yet been reported. Both *A. muscaria* and *A. pantherina* are mycorrhizal fungi, and growth in culture is relatively good but slow.

3. RELATED COMPOUNDS FROM *Amanita solitaria*

Amanita solitaria (Fr.) Secr. sensu D. E. Stuntz is uncommon in many areas of the Pacific Northwest. The chromatographic survey for isoxazoles in various *Amanita* species (Benedict *et al.*, 1966b) included collections of *A. solitaria* from several different areas of western Washington. Extracts were spotted on sheets and formed two-dimensionally, first in BAW (butanol-acetic acid-water, 4:1:1) and second in BuPyr (butanol-pyridine-water, 1:1:1). After processing, one particular ninhydrin-positive spot slowly changed from greenish yellow to purple and had an R_f in BAW close to that of ibotenic acid. However, this compound moved more slowly in BuPyr than did ibotenic acid. Extracts, semipurified by standard techniques to eliminate interfering amino acids, were chromatographically reexamined. A composite summary of the results is shown in Table IV.

Please note that the new compounds are chromatographically distinct from ibotenic acid, muscimol, and muscazone, and that the degradation of solitaric acid to solitarine follows the same pattern as that of ibotenic acid to muscimol. Studies on the isolation of these and other new compounds from *A. solitaria* are in progress (Chilton *et al.*, 1968).

C. SUBSTITUTED TRYPTAMINES

1. PSILOCYBIN AND PSILOCIN

Among many trips to determine the types of mushrooms consumed by mycophiles in various countries, the ethnomycologist team of Wasson and Wasson (1957) journeyed to remote parts of Mexico to learn about the sacred hallucinogenic mushrooms used in various religious ceremonies by native Indians. Little was known about the identity of the mush-

TABLE IV
CHROMATOGRAPHIC STUDIES ON *Amanita solitaria*

Component	R_f value	
	BAW	BuPyr
Ibotenic acid (added)	0.20	0.18
Muscimol (added)	0.39	0.31
Ibotenic acid[a]	0.20	0.34
Muscazone (added)	0.08	0.32
"Solitaric acid"	0.25	0.10
"Solitaric acid"[b]	0.25	0.20

[a]Ibotenic acid that had changed to muscimol after forming in BAW.
[b]Solitaric acid that had degraded to "solitarine" after forming in BAW.

rooms until 1956, when R. Heim accompanied R. G. Wasson to collect and study the most important types employed for religious purposes (Hofmann *et al.*, 1959). Heim found that they belonged primarily to the genus *Psilocybe* (11 species) but thought that one might be a *Stropharia* and another a species of *Conocybe* (Heim and Wasson, 1958).

Further studies (Hofmann *et al.*, 1959) showed that some of the species could be grown in the laboratory and that *Psilocybe mexicana* Heim was especially productive. When a sufficient number of laboratory-grown carpophores, either fresh or carefully dried, were ingested by the principal investigators, psychic manifestations occurred, showing that the laboratory specimens were active and that the compound(s) were not destroyed by drying. In addition, sclerotia-containing mycelia proved to be active, a significant finding in the quest for ample material for subsequent chemical studies.

Neither the dog nor mouse could be used as test animals in guiding extraction and concentration steps, but active fractions equivalent to 0.8 gm of dried mushroom did not incapacitate human volunteers and could readily be distinguished from blanks. Selective solvent extracts plus paper chromatographic techniques finally yielded a crystalline product which gave violet colors with both Keller's and van Urk's reagents, suggesting an indole derivative. The name psilocybin was applied to this compound, which had an R_f value of 0.10 in water-saturated butanol; a second active fraction with an R_f of 0.5 was called psilocin (see Fig. 4). The dried fruiting bodies of *Psilocybe mexicana* contain about 0.2 to 0.46% psilocybin and 0.05% psilocin. The mycelia or sclerotia-containing mycelia have 0.2 to 0.3% psilocybin, and none to trace amounts of psilocin.

Psilocybin is the first naturally occurring indolyl derivative with a substitution at position 4 on the aromatic ring. The proposed structures of the two compounds were later proven by synthesis (Hofmann *et al.*, 1959; Hofmann and Troxler, 1963). Various modifications have been made in

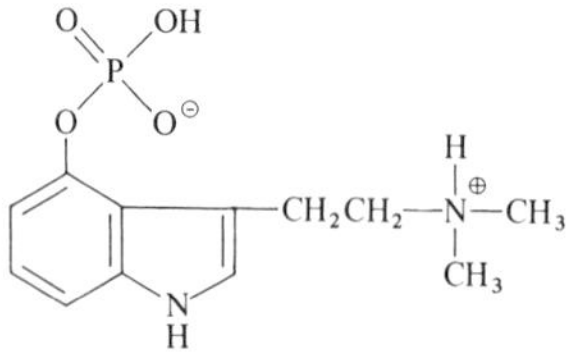

Psilocybin

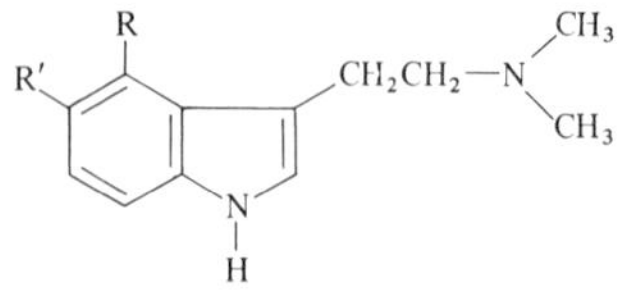

Psilocin: R' = H
        R = OH
Bufotenine: R' = OH
          R = H

FIG. 4. Hallucinogens from fleshy fungi.

their molecular structures to obtain more insight into relationships between structure and psychotropic action (Troxler *et al.*, 1959).

The presence of psilocybin and psilocin in cultures of *Stropharia cubensis*, listed by Singer as *Psilocybe cubensis* (Earle) Sing., which Heim collected in 1957 in both Thailand and Cambodia, showed that mushrooms containing these compounds were not geographically restricted to Mexico and surrounding countries. Since that time, several additional species have been reported in the United States and Canada (Benedict and Brady, 1969). Ola'h (1968) claimed that several species of *Panaeolus* also contain psilocybin.

Ingestion of 5 to 15 mg of psilocybin or 1 to 3 gm dry weight of a species containing the above levels of drug usually produces a striking hallucinogenic dysphoric state (McCawley *et al.*, 1962). The mood may be pleasant or filled with anxiety, depending upon the sensory impressions experienced. The hallucinations and illusions, although likely to be visual, may also be acoustic or gustatory. The subject's ability to comprehend suffers alterations, and he may experience depersonalization and misinterpretation of his environment. These circumstances may result in unmotivated compulsive movements and laughter, sometimes alternating with apathy and catatonia. The altered state may last for several hours.

The systemic effects of psilocybin in man and lower animals are similar to those resulting from sympathetic nervous system stimulation. Doses of 10 mg and 0.1 to 2 mg/kg in man and lower animals, respectively, produce hyperglycemia, temperature elevation, mydriasis, piloerection, contraction of the nictitating membrane, tachypnea, tachycardia, and a rise in blood pressure (Weidmann *et al.*, 1958; Delay *et al.*, 1959). The effect of psilocybin on the brain is poorly understood. Both compounds have identical physiological reactions; from *in vitro* mouse and rat tissue homogenate studies, it is postulated by Horita and Weber (1961) that psilocybin is rapidly dephosphorylated to psilocin in test animals and the latter compound is oxidized to an *o*-quinone type of compound.

Indiscriminant use of *Psilocybe* as a crude source of psychotomimetic agents is not recommended. Although he knew the identity in advance, Stein (1958) purposely ate about 5 gm of dried *Psilocybe cubensis* (greenhouse grown), and vividly described the unpleasant side effects endured for several hours thereafter. Even more serious is the apparent effect of crude psilocybin or psilocin on small children (McCawley *et al.*, 1962). Unidentified mushrooms collected in Milwaukie, Oregon, and Kelso, Washington, were served to children of ages 4, 4, 6, and 9 in two families. Each child ate approximately 1 gm of dried mushroom, later identified as *Psilocybe baeocystis* Singer and Smith. The parents ingested the same species and experienced anxiety and a "cheap drunk." The children developed fevers (102-106°F) and clonic-tonic type convulsions, usually intermittent and not precipitated by sudden sounds. Cerebral spinal fluid and blood pressures were normal prior to the convulsions. Despite a tracheostomy and intravenous administration of diphenylhydantoin to control the status convulsus, the 6-year-old girl died after 3 days of hospitalization. The only gross pathological lesions observed upon autopsy were cerebral edema and slight pulmonary edema. Analysis of mushrooms collected in the same areas indicated about 0.63% psilocybin and less than 0.1% psilocin present. A San Francisco physician informed the authors that he had treated two children with identical convulsion symptoms in a previous mushroom poisoning incident.

Not all hallucinogenic-containing species of *Psilocybe* are capable of synthesizing these substances in laboratory culture. Besides the previously mentioned studies on *Ps. mexicana*, Brack *et al.* (1961) grew *Psilocybe semperviva* Heim and Cailleux in surface culture and part of the psilocybin formed came from isotopically labeled tryptophan added to the medium. In submerged culture, *Psilocybe cubensis* NRRL A-9109 formed psilocybin but no psilocin in a succinate-glycine-glucose salts medium (Catalfomo and Tyler, 1964). Neither *Psilocybe cyanescens* Wakefield nor *Psilocybe pelliculosa* (Smith) Singer and Smith biosynthesized the drugs under the same conditions. Tryptophan added to replacement cultures of *Ps. cubensis* did not enhance psilocybin production and was soon degraded to kynurenine. Leung *et al.* (1965) found Catalfomo and Tyler's medium No. 1 adequate for psilocybin production with *Ps. baeocystis*, but two other nutrient preparations proved unsatisfactory. Recent work (Leung and Paul, 1969) on the biosynthesis of psilocybin and its analogs in culture (see Section III, C, 2) utilized seven carbon and four nitrogen sources. Under their conditions, hallucinogen production was highest at 2 weeks and declined sharply thereafter, although total growth continued up to 4 weeks. As single carbon sources, the highest yields of psilocybin occurred in glucose, i.e., glucose > galactose > trehalose, and control nitrogen (glycine plus ammonia nitrogen) was preferable to glycine > urea. Either source of control nitrogen alone produced no psilocybin analogs.

2. BAEOCYSTIN AND NORBAEOCYSTIN

Although pharmacological studies have not been reported on the above compounds, their close relationship to psilocybin and psilocin permits a brief discussion of current knowledge. In studies to determine conditions of culture to produce psilocybin with *Ps. baeocystis*, Leung *et al.* (1965) noted that extracts of mycelial pellets and wild carpophores contained a compound similar to, but not identical with, psilocybin. The new substance could be separated from the latter using two-dimensional chromatography. This chemical, named baeocystin, could not be detected in extracts of *Psilocybe strictipes* Smith, *Ps. caerulipes,* Peck, *Ps. pelliculosa,* Smith, or *Ps. atrobrunnea* (Lasch.) Gillet.

Crystalline baeocystin (mp 254-258°C dec.) had an ultraviolet spectrum identical with psilocybin, but different from psilocin. Mass spectrum and infrared comparisons suggested that baeocystin is the monomethyl analog of psilocybin (Leung and Paul, 1967). See Fig. 5 for the structure of this compound.

In the workup of a second, larger lot of fungal tissue (Leung and Paul, 1968), a new compound was isolated with a weak molecular ion peak of m/e 256 (low resolution mass spectrum) and a strong peak at m/e 176 due to the dephosphorylated species. This compound was named norbaeocystin.

The occurrence of these analogs of psilocybin appears to be limited to *Ps. baeocystis.* However, it had been known for many years that the "weed *Panaeolus*" (*P. venenosus,* Murr.), reputed to be poisonous, contained a psychotomimetic compound. Singer and Smith (1958), redescribed this species as *P. subbalteatus* (Berk. and Br.) Sacc. Later, Stein *et al.* (1959) reported that a 4-phosphorylindolyl derivative similar to psilocybin, but with a much higher melting point (> 250°C) is present in *P. venenosus.* They did not determine the chemical structure. Leung and Paul (1968) agree that this compound may be identical with baeocystin. In 1949, Stein had eaten *P. venenosus* mixed with *Agaricus bisporus,* and after an intensely painful diarrhea, a prolonged euphoric or excitatory period occurred.

A few studies are available on proposed pathways of biosynthesis of psilocybin and related compounds. Agurell (1966) and Agurell *et al.* (1966) have suggested that *Psilocybe cubensis* starts with tryptophan tryptamine → *N*-methyltryptamine → *N,N*-dimethyltryptamine → psil-

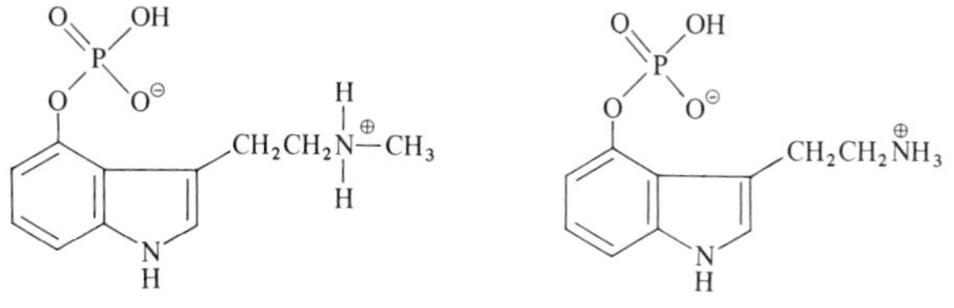

Baeocystin
(monomethyl analog of psilocybin)

Norbaeocystin
(demethyl analog of psilocybin)

FIG. 5. Metabolites from *Psilocybe baeocystis.*

ocin → psilocybin. Agurell and Nilsson (1968) obtained additional evidence that phosphorylation may also precede methylation, in that significant quantities of 4-hydroxytryptamine were converted to psilocybin and psilocybin-like compounds.

3. BUFOTENINE AND RELATED COMPOUNDS

In contrast to the relative abundance of 4-substituted dimethyl-tryptamine derivatives in *Psilocybe* sp., naturally occurring 5-substituted dimethyltryptamines are quite rare in fungi and thus far occur only in *Amanita citrina* (Schaeff. ex) S. F. Gray [*A. mappa* (Batsch ex Lasch. Quél.] and *Amanita porphyria* (A. & S. ex Fr.) Secr.

Bufotenine (5-hydroxydimethyltryptamine) (see Fig. 4) was first isolated from European collections of *A. citrina* by Wieland and Motzel (1953). Prior designation of this species as deadly poisonous proved to be erroneous; it contained no peptide toxins, but had often been mistaken for the deadly *Amanita phalloides*. Later, Tyler (1961) and Catalfomo and Tyler (1961) found bufotenine in American specimens of *A. citrina* and *A. porphyria*, respectively.

Extracts of the above species collected in Germany in 1963 by Tyler and Gröger (1964) contained a variety of Ehrlich-positive compounds in addition to bufotenine. Carefully prepared extracts, used in combination with a solvent system which gave unusually good separation on thin-layer plates, enabled the authors to detect 19 Ehrlich-positive compounds from *A. citrina* and 12 from *A. porphyria*. In several confirmatory tests, the chromatographic behavior of six of the unknowns did not differ from those listed in Table V. Note that active methylation systems are operative in both fungi and that the relative amount of bufotenine substantially exceeds that of its demethylated precursor, serotonin (5-hydroxytryptamine).

TABLE V
TRYPTAMINE DERIVATIVES IN EXTRACTS OF *Amanita* sp.

| | Relative concentration[a] | |
Compound	*A. citrina*	*A. porphyria*
Bufotenine *N*-oxide	4	2
Serotonin	2	1
N-Methylserotonin	2	1
Bufotenine	10	8
5-Methoxy-*N,N*-dimethyltryptamine	0.25	0.1
N,N-Dimethyltryptamine	0.25	0

[a]Expressed as 10 = maximum, 0 = none.

These fungi contain some of the same compounds as those reported in the tropical legume *Piptadenia* (*Anadenanthera*) *peregrina* (L.) Benth. by Stromberg (1954) and Fish *et al.* (1955) who had found dimethyltryptamine *N*-oxide, bufotenine, and bufotenine *N*-oxide. The seeds of this plant provide the most commonly known botanical source of snuffs prepared and inhaled by South American Indians to produce visions and hallucinations. The snuffs are inactive when taken orally, and this probably explains in part why *A. citrina* is mildly toxic (Hennig, 1958).

Considerable controversy exists relative to the action of bufotenine. Intravenous injection of 8 to 16 mg in human volunteers resulted in primary visual disturbances and changes in time and space perception, according to Fabing and Hawkins (1956). Turner and Merlis (1959) could not substantiate claims that bufotenine injections had any effect on the central nervous system, and Isbell, in Holmstedt and Lindgren (1967), reported that "inhalation of pure bufotenine in aerosol suspension in doses up to 100 mg (total dose) was without effect." Fischer (1968) joined the groups claiming that bufotenine is not a hallucinogenic agent, but does agree with Isbell that it is difficult to differentiate the alleged psychotomimetic action from the powerful cardiovascular effects of the compound.

Both *A. citrina* and *A. porphyria* are mycorrhizal-associated fungi. Western Washington strains of the latter species resisted all of the writer's efforts to subculture them. A German strain of *A. citrina* from a culture collection in Wiemar grew very slowly in surface culture on a malt extract-peptone-glucose broth utilized by Tyler and Gröger (1964). Six flasks yielded only 0.54 gm dry weight of mycelium in 85 days. A semiquantitative assay method indicated that the mycelium contained 0.03% bufotenine on a dry weight basis plus traces of other Ehrlich-positive compounds.

D. HALLUCINOGEN(S) FROM *Gymnopilus spectabilis* (FR.) SINGER

Gymnopilus spectabilis (Fr.) Singer is a bright yellow lignicolous fungus with caps sometimes exceeding a diameter of 8 to 10 inches and is found mainly on dead tree stumps. The first recorded note on its hallucinogenic activity is that in Imazeki's letter to Romagnesi (1964). At that time, efforts to isolate and identify the active compound(s) from Japanese collections were said to be in progress. Upon reading this paper, Walters (1965) recalled that hallucinations had occurred in an Ohio woman in 1942, following raw ingestion of a mushroom later identified by him as *Gymnopilus spectabilis*. This species has a bitter taste, and it is surprising that any culinary interest would be shown in it. In 1963, the reviewer noted that extracts of *Gymnopilus spectabilis* and *Polyporus schweinitzii* Fr. contained Ehrlich-positive compounds with similar mobilities and color changes.

Published case histories with *G. spectabilis* involvement appear limited to that of Buck (1967). A Massachusetts man who thought he had collected *Armillariella mellea* Karst, the edible honey agaric, ate two or three fried caps (approximate weight not estimated). Within 15 minutes he felt "woozy" and disconnected and experienced blurred vision and a shrinking of room size. All objects appeared unnaturally colored, i.e., grass and trees became "a vivid green, shot with purple." The sensations were not unpleasant. Thought collection proved difficult, and although questions asked of himself were readily answered, he could not find a book after he had put it down. Unsteadiness of foot and nausea accompanied the other symptoms. His wife and a neighbor ate lesser quantities and experienced uncoordination and cheap drunks. All three recovered within a few hours and gave coherent accounts of their experience. Boston Mycological Club members identified the mushroom as *G. spectabilis;* methanolic extracts appeared to contain indolyl compounds, with no probable evidence of either psilocybin or psilocin.

As of September 1969, the active hallucinogenic principle(s) in *G. spectabilis* had not, to the authors' knowledge, been revealed. However, a lead for clarifying the pharmacological activity in *G. spectabilis* has been provided by Hatfield and Brady (1968). They isolated and crystallized one of the yellow constituents in *Gymnopilus decurrens* Hesler, a chemotaxonomic relative of *G. spectabilis*. The *G. decurrens* metabolite proved to be bisnoryangonin [4-hydroxy-6-(4-hydroxystyryl)-2-pyrone] (see Fig. 6). Although previously synthesized as an analog of hispidin [4-hydroxy-6-(3,4-dihydroxystyryl)-2-pyrone] from *Polyporus hispidus* Fr. and *Polyporus schweinitzii* Fr., *bis*-noryangonin had not previously been reported from nature.

Yangonin is one of several closely related styrylpyrones found in *Piper* sp. (Kava) and *Aniba firmula* (Nees et Mart.) Mez. A beverage prepared

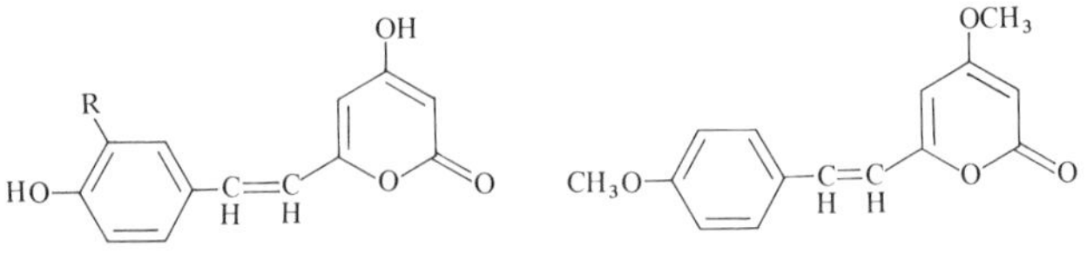

Hispidin: R = OH
bis-Noryangonin: R = H

Yangonin

FIG. 6. Styrylpyrones from plants and fleshy fungi.

from kava root has intoxicating properties, but pharmacological studies on the pure compounds do not account for all of the activity of the crude materials. For a comprehensive report on the chemistry and pharmacology of kava, see Klohs (1967) and Meyer (1967).

Recently, Hatfield and Brady (1969) isolated and characterized *bis*-noryangonin from *Gymnopilus spectabilis*. No evidence for indolyl compounds related to psilocybin, psilocin, or bufotenine could be found in this mushroom.

E. UNIDENTIFIED COMPOUND(S) FROM *Russula* SP.

Amongst the mushrooms reputed to bring about cerebral mycetisms among native tribesmen in Papua and New Guinea is a species of *Russula*. Singer (1958) reported that this mushroom is one of three inebriants in use among Mt. Hagen tribes. It had been reported earlier that it incites fits of frenzy and had been utilized by the natives "before going out to kill an enemy, or in times of anger, sorrow or excitement." The native name for this agaric was *nonda*.

Singer (1958) received mushroom specimens from some Australian investigators, none of whom had a mycological background. They referred to the phenomenon the mushrooms produced as the Wahgi River frenzies and to the fungi as "hysteria-producing." The mushrooms had been collected in Binj, the Territory of Papua, and New Guinea. Subsequent examination of the specimens permitted Singer to conclude that the *Russula* sp., although not close to any he knew, was a typical representative of that genus. He named it *Russula nondorbingi* Singer. spec. nov.

In further pursuit of knowledge of the "mushroom madness of Kuma," Heim and Wasson (1965) spent three weeks in the Waghi Valley in the fall of 1963. Contrary to the 1934 statement of Father W. A. Ross, they saw no signs of mushroom madness during their visit. However, there is general agreement that the feats of endurance performed during the attacks of madness by the "wild men" of the Kuma go beyond normal physical activity. For hours, they rush up and down mountain trails, brandishing weapons and shouting at the top of their lungs. Instead of one species of mushroom, as Singer (1958) was led to believe, several species in two genera (*Heimiella* and *Boletus*) are responsible, with no agreement among the natives as to which species are active. However, the authors found no record of a fatality or even of a serious injury from the menacing men. This reviewer wonders whether the suspect mushrooms actually contain any unusual physiological principle which might account for the reported feats of endurance. In the near future, Heim and Wasson hope to have more definite answers to questions they themselves have raised.

IV. Gastrointestinal Irritants

A. FROM *Agaricus* SP.

The genus *Agaricus* L. ex Fr. is very large and contains many edible species. However, some species are suspect and others cause gastrointestinal upsets, ranging from mild to severe. The latter type is usually accompanied by abdominal cramps and intense pain, but the possibility of a fatality resulting from the compound(s) is remote. This probably explains the lack of interest in determining the chemical nature of the irritants, which are presumed to be resin-like substances. Some strains of the suspect mushrooms contain little or no irritants.

Species of *Agaricus* reported to contain unidentified irritants by Tyler (1963) include *A. albolutescens* Zeller, *A. arvensis* Schaeff. ex Fr. var. *palustris* A. H. Smith, and probably *A. sylvicola* (Vitt.) Fr., which is closely related except for differences in the spores. Two species known to contain varying amounts of phenol or cresol-like compounds are *Agaricus hondensis* Murr. and *Agaricus placomyces* Peck. The phenolic odor of some collections of these fungi is unmistakable, and the phenolic component(s) may well be the toxic constituents of these mushrooms.

Even the well-known mushroom of commerce *Agaricus bispora* (Lange) Imbach contains unidentified components toxic to rats and mice. Kubin (1961) fed fresh cultivated mushrooms to young albino rats and various extracts and dialyzates to young albino mice. In an otherwise adequate diet, the mushrooms were the only source of protein. Death of the test animals occurred after an intense excitatory state. Carefully dried mushrooms had no ill effect.

B. FROM *Lactarius, Rhodophyllus,* AND *Tricholoma* SP.

Several species of *Lactarius* are known to cause nausea and abdominal pains following ingestion. One of these, *L. glaucescens* Crossl., is closely related to *L. piperatus* (L. ex Fr.) S. F. Gray. The latter species is edible after the "peppery" taste is removed by cooking. Charles (1942) recorded the first fatality, a 5-year-old child who succumbed from eating cooked *L. glaucescens*.

Lactarius torminosus (Schaeff ex Fr.) Gray, mycorrhizally associated with white birch trees, is said to be quite toxic when raw. Extracts of the raw fungus produced fatalities in both guinea pigs and rabbits and, according to Pilát and Usák (1954), are capable of causing death in children. Despite these ominous warnings, Russian mycophiles are said to collect *L. torminosus* for use fresh or for salting (like sauerkraut) or pickling, to be consumed with sour cream and vodka (Singer, 1962). Several other toxic species of *Lactarius* are listed by Tyler (1963) and include *L. helvus* (Fr.) Fr., *L. rufus* (Scap. ex Fr.) Fr., and *L. uvidus* (Fr. ex Fr.) Fr. Most of these are toxic only when eaten raw. *Lactarius helvus* collected in Finland and studied by Honkanen and Virtanen (1965) has an unusual ninhydrin-positive tropolone ring compound containing three carboxylic acid groups. There is no evidence at present that this compound is involved in the toxic reactions of this mushroom.

Rhodophyllus sp. This is a large and taxonomically difficult genus, and some mycologists prefer to separate it into several genera, e.g., *Entoloma, Leptonia, Nolanea,* and *Claudopus.* Tyler (1963) lists six species extracts of which are toxic to guinea pigs. One species is particularly dangerous, *R. sinuatus* (Bull ex. Fr) Sing., and is said to produce severe abdominal pains, vomiting, and diarrhea, and to incapacitate an individual for several days after symptoms begin. The agent in this fungus causes liver damage and has not been isolated. Deaths from this fungus are infrequent in adults, but children have been known to succumb.

Tricholoma sp. Like the toxic species of *Rhodophyllus*, several species of *Tricholoma* are suspect, and two of these, *Tricholoma pardinum* Quel. and *Tricholoma venenatum* Atk., produce violent gastronomic upsets. It is quite possible that the toxic component(s) of *Rhodophyllus sinuatus* and those of the *Tricholoma* species listed above are similar.

C. NORCAPERATIC ACID AND AGARICIC ACIDS

Cantharellus cibarius Fr., the orange chanterelle, is widely collected for food by mycophiles in the Pacific Northwest. The closely related fungus *Gomphus floccosus* (Schw.) Singer, formerly *Cantharellus floccosus* Schw., is sometimes mistaken for *C. cibarius,* and in some individuals ingestion of *G. floccosus* causes mild to severe gastric upsets. Miyata *et al.* (1966) found that the mushroom contained substantial quantities (4.5% on a dry weight basis) of an acid related to agaricic acid. According to Karrer (1958), this acid was first reported from *Polyporus officinalis* Fr. in 1845 by von Martius. Thoms and Vogelsang (1907) determined the structure of agaricic acid as shown in structure II.

$$
\begin{array}{cc}
\text{COOH} & \text{COOH} \\
| & | \\
\text{CH}-\text{C}_{16}\text{H}_{33} & \text{CH}-\text{C}_{14}\text{H}_{29} \\
| & | \\
\text{OH}-\text{C}-\text{COOH} & \text{OH}-\text{C}-\text{COOH} \\
| & | \\
\text{CH}_2 & \text{CH}_2 \\
| & | \\
\text{COOH} & \text{COOH}
\end{array}
$$

Agaricic acid	Norcaperatic acid (α-tetradecylcitric acid)
(II)	(III)

The structure of the acid from *C. floccosus* is closely related to caperatic acid from the lichen *Parmelia caperata* L., first investigated by Helle (1897). Caperatic acid has a $COOCH_3$ instead of $COOH$ as one of the terminal carboxyl groups.

Henry and Sullivan (1969) recently found norcaperatic acid in *Gomphus kauffmanii* (A. H. Smith) Corner at levels of 4.4%. Several species of *Cantharellus* did not contain this compound; hence, these findings may have chemotaxonomic significance. Carrano and Malone (1967) studied the pharmacological effects of norcaperatic and agaricic acids in rats. Both acids produced delayed onset, dose-related effects of mydriasis, central nervous system depression, and skeletal muscle weakness. In addition, both were competitive inhibitors of the enzyme aconitase, with norcaperatic acid being the more active of the two. They postulated that the potentially toxic properties of norcaperatic acid are probably involved with its close structural relationship to citric acid.

V. Disulfiram-Like Action of *Coprinus atramentarius* (Bull ex Fr.) Fr.

Coprinus atramentarius is an edible mushroom, and when not combined with an alcoholic beverage, produces no gastronomic or other discomforts. However, if alcoholic beverages are consumed with the mushroom, or even up to within 48 hours after the mushroom meal, the person generally experiences one or more of the following symptoms: profound flushing of the face, a metallic taste in the mouth, peresthesia of the extremities, palpitation and tachycardia, and a feeling of swelling in the hands. The initial symptoms are usually followed by nausea and vomiting. To the unsuspecting person, it is a rather shocking experience. Reynolds and Lowe (1965) have an excellent account of four cases of this type wherein the mushrooms were eaten the night before and illness followed consumption of beer on the next day. The considerable effort put forth to find the responsible component in the mushroom has not borne fruit. It is believed that the raw mushroom contains the precursor which is converted to the active compound during the cooking process. The possibility that the guanidine in *C. atramentarius* is converted to cyanamid appears rather remote. Winterstein *et al.* (1913) isolated guanidine from *Boletus edulis*, a choice edible bolete lacking the same effect as *C. atramentarius*. Simandl and Franc (1956) had reported the isolation of tetraethylthiuram disulfide from *C. atramentarius*, but Weir and Tyler (1960) were unable to confirm these findings. In Sweden, Barkman and Perman (1963) found that rabbits could serve as satisfactory test animals for assaying mushroom sensitivities and predicted that this animal would be most useful in attempts to isolate the compound that induces supersensitivity to ethanol. According to Romagnesi (1964) as instructed by Imazeki, *Clitocybe claviceps* (Pers. ex Fr.) Kummer, combined with sake or beer, has the same effect on Japanese mycophiles as does *C. atramentarius* plus an alcoholic beverage.

An incident involving the potentiation by ethanol of unidentified gastrointestinal irritants in narrow capped morels (*Verpa bohemica?*) was reported by Groves (1964). The morels had been picked, cleaned, and refrigerated on a Saturday. The family of four ate the morels at noon the next day and later in the afternoon attended a birthday party. The two daughters, ages 14 and 17, consumed only soft drinks, whereas the parents each had a mixed drink containing rye whiskey. Within 15 to 20 minutes both parents became violently ill, with diarrhea and vomiting as the main symptoms. Complete recovery occurred by 9:00 P.M. Neither of the girls was affected, and no *Helvella* species mushrooms had been mixed in with the specimens.

VI. Toxins with Antitumor Activity

The first indications that certain basidiomycetes contained antitumor principles were noted by Lucas *et al.* (1957). They tested extracts of fruiting bodies of *Boletus edulis* Bull. ex Fr. and other holobasidiomycetes against sarcoma 180 in mice. The results were promising, but attempts to cultivate *B. edulis in vitro* proved laborious and slow. Later on, the same group (Lucas *et al.*, 1958-1959) investigated several species of the puffball *Calvatia,* including *C. bovista* (Pers.) Kambly et Lee, *C. cyathiformis* (Bosc.) Morg., *C. craniformis* (Schw.) Fr., and *C. gigantea* (Pers.) Lloyd. In the latter cases the authors produced the antitumor principle in laboratory culture. Roland *et al.* (1960) published their findings on a purified fraction from the carpophores of *C. gigantea* which acted upon 14 of 24 different tumors in mice, rats, and hamsters. Fifteen hundred pounds of fresh fungi processed at Armour and Company, Chicago, yielded about one pound of calvacin.

1. CALVACIN

According to Sternberg *et al.* (1963), calvacin is a nondialyzable, basic mucoprotein of high molecular weight with good antitumor activity in experimental animals. Unfortunately, it is cumulative in effect, and extremely low dosages cause a prolonged intoxication, characterized by anorexia and extreme weight loss. Among the diverse lesions produced in experimental animals are myofibrillar necrosis in cardiac and skeletal muscle, pulmonary hemorrhages, hepatic necrosis, biliary obstruction, renal tubular necrosis, and fibrinoid degeneration. Doses as low as 2 mg/kg/day produced severe anaphylactoid reactions in dogs. Since calvacin is a mixture of antigenic substances, it was predicted that clinical testing would be difficult and hazardous due to the cytotoxic actions of the drug.

2. LAMPTEROL

Nakanishi *et al.* (1963) isolated a crystalline antitumor substance from *Lampteromyces japonicus* (Kawamura) Sing., a fleshy fungus limited geographically to Japan (Singer, 1962). The compound lampterol melted at 124–127°C and had the empirical formula $C_{15}H_{28}O_4$. Its activity against Ehrlich mouse ascitic tumors was quoted at 120 μg/kg, and its toxicity as 50 mg/kg on intraperitoneal injection in mice.

Later work by McMorris and Anchel (1965), who characterized two antibacterial substances, illuden S and illuden M from *Clitocybe illudens* (Schwein.) Sacc., now listed as *Omphalotus olearius* (D. C. ex Tr.) Sing., showed that lampterol and illuden S are identical compounds. The structures of these novel sesquiterpenoids are shown in Fig. 7.

An active basidiomycete screening program, started at Wyeth in 1961 (Gregory *et al.*, 1966), covered the fermentation products of more than 7000 cultures of basidiomycetes; 50 cultures, representing 20 different genera, formed compounds having an inhibitory effect on sarcoma 180, mammary adenocarcinoma 755, or leukemia L-1210. One of the most active materials came from aqueous mycelial extracts of the wood rot fungus *Poria corticola* (Fr.) Cke., using an isolate obtained from the culture collection at Baarn, Holland. Ruelius *et al.* (1968) established fermentation conditions enabling the production of consistently high titers of the active compound poricin.

3. PORICIN

This compound is an acidic protein with an isoelectric point near 3.7. Purification was accomplished by ethanol precipitation plus a sequence of

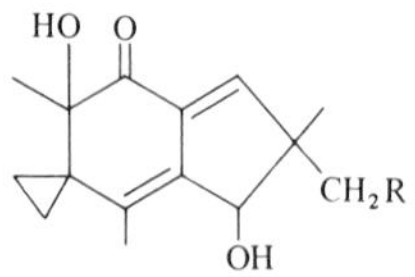

"Lampterol"
(illuden S): R = OH
Illuden M: R = H

FIG. 7. Fungal sesquiterpenoid triols.

chromatographic procedures. Unfortunately, the toxicity of the compound did not decrease as purification proceeded. Poricin has been assigned the number 107,455 by the Cancer Chemotherapy National Service Center.

Highly purified preparations from Sephadex columns inhibited the growth of sarcoma 180 at 0.7-1 μg/mouse/day. The inhibition was such that little or no tumor could be seen at autopsy. It is very toxic, however, and some groups of mice would die shortly before an experiment was to be terminated. Schillings and Ruelius (1968) crystallized the active principle and found that it contained 18 common amino acids with a molecular weight (by gel filtration) of $\sim$ 100,000.

4. FLAMMULIN

Only one other proteinaceous antitumor substance from mushrooms has thus far been isolated and characterized. It is a basic protein called flammulin from *Flammulina velutipes* (Curt. ex Fr.) Sing. Komatsu *et al.* (1963) and Watanabe *et al.* (1964) found that flammulin had a molecular weight of about 24,000 and contains 19 amino acid residues. It is active against Ehrlich ascites tumors when administered in daily doses of 2.5 μg/mouse. Toxicity data were not cited.

VII. Miscellaneous Unidentified Factors

An indication that many mushrooms have interesting unidentified compounds with physiological activity is provided by Malone *et al.* (1967). Of 66 different species of higher fungi whose 70% ethanol solubles had been hippocratically screened in rats, the following activities are recorded: three, parasympathomimetics; eight, psychotropics; five, metabolic poisons; six, mild central nervous system depressants; fourteen, diuretics; two, skeletal muscle relaxers; three, tranquilizers; and one contained a sympathetic stimulant. Two fungi had apparent dual activity.

Naematoloma fasciculare (Hud. ex Fr.) Karst., a lignicolous mushroom, is sometimes mistaken for the edible *Naematoloma sublateritium* (Fr.) Karst. or *Naematoloma capnoides* (Fr.) Karst. The former has a very bitter taste, whereas the latter two species are mild. Singer (1962) noted that cases of poisoning from eating *N. fasciculare* had been reported in Italy, Japan, and Russia. Herbich *et al.* (1966) studied some fatalities caused from eating this fungus. Some 9 hours may elapse before gastrointestinal symptoms begin, but during this symptom-free period extensive liver damage occurs which may later bring death to the patient. Upon autopsy, the liver shows marked pathological changes, including liver cell necrosis, bleeding in the liver parenchyma, and subacute liver atrophy. Kidney functions are also impaired, with extensive fatty degeneration as the main lesions in these organs. Similar degeneration is noted in the heart muscles and the ganglion cells of the brain.

Ito *et al.* (1967) have grown *N. fasciculare* in shaken flask culture on a glucose-cottonseed meal medium and extracted from the broth naematolin $C_{17}H_{24}O_5$, mp 145, $[\alpha]_D^{20}$ 360 ($c = 1.0$, $CHCl_3$). The compound had no effect on bacteria at 1 mg/ml, but was cytotoxic against HeLa S_3 cells in tissue culture at 6.25 μg/ml. It inhibited polio virus by the plaque method and showed coronary vasodilation action in isolated guinea pig heart at 100 μg/heart.

For many years, *Paxillus involutus* (Batsch ex Fr.) Fr. has had a reputation for being poisonous if not thoroughly cooked before eating. Singer (1962) states that it is the *babye oohky* of the Russian farmer, who knows it to be a poor choice, but salts it in large numbers in years when more preferable mushrooms are not available. Bschor and Mallach (1963) made a thorough study of four cases of poisoning from eating insufficiently cooked *P. involutus* mushrooms. Only one of the victims survived. The symptoms of poisoning began rapidly and included intense stomach pains and a severe acute circulatory collapse. Blood vessels and capillaries of the stomach became greatly distended. Some evidence of fatty degeneration of the liver, kidneys, heart, and skeletal muscle were noted. The presence of fat emboli indicated a strong disturbance to the emulsion stability of the physiological blood fat.

In hippocratic screen studies conducted on rats by Malone *et al.* (1967), *P. involutus* extracts had a diuretic effect and served also as a peripheral vasodilator and metabolic poison. Although evidence is lacking that this compound is involved in the toxic syndromes induced by heat-labile factors in *P. involutus*, Edwards *et al.* (1967) recently isolated and characterized involutin, a diphenylcyclopentenone, from this mushroom. The compound is 5-(3′,4′-dihydroxyphenyl)-3,4-dihydroxy-2-(*p*-hydroxy-phenyl)cyclopent-2-en-1-one.

Mature specimens of *Chlorophyllum molybditis* (Meyer ex Fr.) Mass. are white and rather distinctive with their slate-green gills and unusual greenish spore deposits. Floch *et al.* (1966) reported that 0.5 ml press juice, given intraperitoneally, killed mice in 4 hours. The active factor is soluble in ethanol or chloroform, but not in ether. Heating at 100°C destroys it. Some people claim they can eat the mushroom without fear; others are said to experience mild to severe gastric upsets plus vomiting, diarrhea, and mental haziness.

Clitocybe toxica sp. n. collected and studied in South Africa by Sapeika and Stephens (1965) contains a potent, heat-labile substance which inhibits cytochrome oxidase. Very small amounts are lethal for mice and rats. The authors proved that the action of the toxin is not due to a muscarinic or cyanide effect.

Clavaria sp. Two species of *Clavaria* (coral fungi) have been reported (Tyler, 1963) to produce stomach pains and diarrhea in humans. These are *C. formosa*, Pers. ex Fr. and *C. gelatinosa* Coker.

In a recent presentation, Fildago and Kauffman Fildago (1969) described a coral fungus in the *Ramaria formosa* group as *Ramaria flavobrunnescens* Atk. Corner, (which R. H. Peterson believes is uncorrectly identified and may be *Ramaria subformosa*). This fungus is probably mycorrhizally associated with eucalyptus trees in certain parts of South America. It is accidentally consumed during grazing and causes fatalities in both cattle and sheep. The nature of the toxin is unknown, but it is apparently heat labile, since cooking renders the mushroom harmless. In one 1965 outbreak involving a herd of 650 cattle, 150 were found ill and 35 died. After a 1969 episode in which many sheep died, serious studies were started on this fungus. The effect is cumulative, although the toxin levels are probably quite low. After feeding lambs about 0.5 lb (wet weight) of the above *Ramaria* sp. daily, symptoms began on the seventh day, and by the twenty-seventh day the animals were dead. Gross external symptoms in cattle included loss of hair and gradual loss of sight.

Many mycophiles collect morels (*Verpa* sp. and *Morchella* sp.) in the spring of the year, and it is agreed that few edible fleshy fungi have flavors more desirable than these. However, A. H. Smith (1967) points out that caution should be exercised not to overeat, especially if the species is not recognized. In his own experience, Smith noted a definite lack of muscular coordination about 4 hours after eating about a pint of cooked *Verpa bohemica* (Smith, 1967). The obvious danger to one driving an automobile during such a period cannot be overemphasized.

Boletus sp. Although most species of *Boletus* are considered edible, those in which the tubes have red mouths are to be avoided. This group contains *B. eastwoodiae* (Murr.) Sacc. and Trotter, *B. frostii* Russel with sp. *floridanus* Sing., *B. luridus* Schaeff. ex Fr., *B. satanus* Lenz, and probably many others in Singer's Section 6 Luridi, which includes the above species. The symptoms of poisoning are said to vary and range all the way from mild gastronomic upsets to paralysis. Some specimens may contain muscarine, but this suggestion has not been verified. Keinholz (1934) tasted and swallowed a small piece of *B. eastwoodiae* and experienced severe stomach cramps and diarrhea. The symptoms were promptly relieved with 1/50 grain atropine.

Scleroderma sp. Most puffballs with white interiors are edible and choice if they are prepared before any yellowing has occurred. Puffballs whose interiors are dark violaceous to purplish, including many species of *Scleroderma*, should be avoided. Stevenson and Benjamin (1961) reported one clear case of poisoning in a physician who ate *S. aurantium*. They also cited a case involving *S. cepa* Pers. This was eaten raw, and within 30 minutes the patient reported stomach pains, weakness, nausea without dizziness, and a tingling sensation over his entire body. Muscular rigidity, accompanied by stomach cramps, profuse sweating, and facial pallor were promptly relieved after forced emesis.

Finally, in a brief return to the gill mushrooms, the reviewer should mention a writer who suggested that one only need to peel the skin to detoxify certain mushrooms. The following statement is quoted from Moench (1963): "*Amanita rubescens* and *Amanita pantherina* may be eaten and taste very good, if the skin is first removed." The first-mentioned species is probably harmless, but the reviewer would not agree that the recommended treatment would eliminate the headaches from the second species!

ACKNOWLEDGMENTS

The author wishes to thank the editors of this treatise for the invitation to contribute, and he is also indebted to J. L. McLaughlin, College of Pharmacy, University of Washington, for critical review of the chapter.

REFERENCES

Agurell, S. (1966). *Acta Pharm. Suecica* 3, 71.
Agurell, S., and Nilsson, J. L. G. (1968). *Tetrahedron Letters* No. 9, 1063.
Agurell, S., Blomkvist, S., and Catalfomo, P. (1966). *Acta Pharm. Suecica* 3, 37.
Alder, A. E. (1961). *Deut. Med. Wochschr.* 86, 1121.
Bandoni, R. J., and Szczawinski, A. F. (1964). "Guide to Common Mushrooms of British Columbia," British Columbia Provincial Museum Handbook No. 24, Victoria, B.C.
Barkman, R., and Perman, E. S. (1963). *Acta Pharmacol. Toxicol.* 20, 43.
Belgian Patent 665,249. (1965). Nouveau dérivé de l'isoxazole et sa preparation.
Benedict, R. G., and Brady, L. R. (1969). *Proc. 3rd Intern. Ferment. Symp., Rutgers University, 1968* p. 63. Academic Press, New York.
Benedict, R. G., Brady, L. R., Stuntz, D. E., and Tyler, V. E., Jr. (1966a). Unpublished data.
Benedict, R. G., Tyler, V. E., Jr., and Brady, L. R. (1966b). *Lloydia* 29, 333.
Boedijn, K. B. (1938-1940). *Bull. Jard. Botan. Buitenzorg.* [3] 16, 76.
Bowden, K., and Drysdale, A. C. (1965). *Tetrahedron Letters* No. 12, 727.
Bowden, K., Drysdale, A. C., and Mogey, G. A. (1965). *Nature* 206, 1359.
Brack, A., Hofmann, A., Kalberer, F., Kobel, H., and Rutschmann, J. (1961). *Arch. Pharm.* 294, 230.
Brady, L. R., and Tyler, V. E., Jr. (1959). *J. Am. Pharm. Assoc., Sci. Ed.* 48, 417.

Brown, J. K., Malone, M. H., Stuntz, D. E., and Tyler, V. E., Jr. (1962). *J. Pharm. Sci.* 51, 853.

Brown, J. K., Tyler, V. E., Jr., and Brady, L. R. (1966). "Biochemie und Physiologie der Alkaloide," p. 593. *Abhandl. Deut. Akad. Wiss. Berlin.*

Bschor, F., and Mallach, H. J. (1963). *Arch. Toxikol.* 20, 82.

Buck, R. W. (1961). *Mycologia* 53, 537.

Buck, R. W. (1964). *Clin. Med.* 71, 1353.

Buck, R. W. (1967). *New Engl. J. Med.* 276, 391.

Buck, R. W. (1968). Personal communication.

Cann, H. M., and Verhulst, H. L. (1961). *Am. J. Diseases Children* 101, 128.

Carrano, R. A., and Malone, M. H. (1967). *J. Pharm. Sci.* 56, 1611.

Catalfomo, P., and Tyler, V. E., Jr. (1961). *J. Pharm. Sci.* 50, 689.

Catalfomo, P., and Tyler, V. E., Jr. (1964). *Lloydia* 27, 53.

Charles, V. K. (1942). *Mycologia* 34, 112.

Chilton, W. S., Tsou, G., Kirk, L., and Benedict, R. G. (1968). *Tetrahedron Letters* No. 60, 6283.

Delay, J., Pichot, P., Lemperiere, T., Nicholas-Charles, P., and Quentin, A. M. (1959). *Presse Med.* 67, 1368.

Dennis, R. W. G., Orton, P. D., and Hora, F. B. (1960). *Brit. Mycol. Soc. Trans. Suppl.*, p. 1.

Desvignes, A. (1965). *Alimentation et la Vie* 53, 155.

Dubash, J., and Teare, D. (1946). *Brit. Med. J.* 1, 45.

Edwards, R. L., Elsworthy, G. C., and Kole, N. (1967). *J. Chem. Soc.,* C p. 405.

Eugster, C. H. (1956). *Helv. Chim. Acta* 39, 1002.

Eugster, C. H. (1957a). *Helv. Chim. Acta* 40, 886.

Eugster, C. H. (1957b). *Helv. Chim. Acta* 40, 2562.

Eugster, C. H. (1968). *Naturwissenschaften* 55, 305.

Eugster, C. H., and Müller, G. (1959). *Helv. Chim. Acta* 42, 1189.

Eugster, C. H., and Takemoto, T. (1966). *Helv. Chim. Acta* 50, 126.

Eugster, C. H., and Waser, P. G. (1954). *Experientia* 10, 298.

Eugster, C. H., Müller, G. F. R., and Good, R. (1965). *Tetrahedron Letters* No. 23, 1813.

Fabing, H. D., and Hawkins, J. R. (1956). *Science* 123, 886.

Fildago, O., and Kauffman Fildago, M. E. P. (1969). *Abstr. 11th Intern. Botan. Congr., Seattle, 1969* p. 59. Univ. of Washington Press, Seattle, Washington.

Fischer, R. (1968). *Nature* 220, 411.

Fish, M. S., Johnson, J. M., and Horning, E. C. (1955). *J. Am. Chem. Soc.* 77, 5892.

Floch, H., Labarbe, C., and Roffi, J. (1966). *Rev. Mycol.* [N.S.] 31, 317.

Ford, W. W. (1926). *J. Pharmacol. Exptl. Therap.* 29, 305.

Gagneux, A. R., Häfliger, F., Meier, R., and Eugster, C. H. (1965a). *Tetrahedron Letters* No. 25, 2081.

Gagneux, A. R., Häfliger, F., Good, R., and Eugster, C. H. (1965b). *Tetrahedron Letters* No. 25, 2077.

Good, R., Müller, G. F. R., and Eugster, C. H. (1965). *Helv. Chim. Acta* 48, 927.

Göth, H., Gagneux, A. R., Eugster, C. H., and Schmid, H. (1967). *Helv. Chim. Acta* 50, 137.

Gregory, F. J., Healy, E. M., Agersborg, H. P. K., Jr., and Warren, G. H. (1966). *Mycologia* 58, 80.

Groves, J. W. (1964). *Mycologia* 56, 779.

Grzymala, S. (1962). *Bull. Soc. Mycol. France* 78, 394.

Hatfield, G. M., and Brady, L. R. (1968). *Lloydia* 31, 225.

Hatfield, G. M., and Brady, L. R. (1969). *J. Pharm. Sci.* **58**, 1298.

Heim, R., and Wasson, R. G. (1958). "Les Champignons hallucinogenes du Mexique," Editions du Muséum National d'Histoire Naturelle, Paris. **6**, 123.

Heim, R., and Wasson, R. G. (1965). *Botan. Museum Leaflets Harvard Univ.* **21**, 1.

Helle, O. (1897). *Chem. Ber.* **30**, 357.

Hendricks, H. V. (1940). *J. Am. Med. Assoc.* **114**, 1625.

Hennig, B. (1958). "Michael/Hennig Handbuch für Pilzfreunde," Vol. 1, p. 132. Fischer, Jena.

Henry, E. D., and Sullivan, G. (1969). *Abstr. 10th Ann. Meeting, Am. Soc. Pharmacognosy, Oregon State Univ.* p. 11.

Herbich, J., Lohwag, K., and Rotter, R. (1966). *Arch. Toxikol.* **21**, 310.

Hofmann, A., and Troxler, F. (1963). U. S. Patent 3,075,992 (to Sandoz).

Hofmann, A., Heim, R., Brack, A., Kobel, H., Frey, A., Ott, H., Petrzilka, T., and Troxler, F. (1959). *Helv. Chim. Acta* **42**, 1557.

Hölmstedt, B., and Lindgren, J-E. (1967). *U.S., Public Health Serv., Publ.* **1645**, 369.

Honkanen, E., and Virtanen, A. I. (1965). *Acta Chem. Scand.* **18**, 1319.

Horita, A., and Weber, L. (1961). *Biochem. Pharmacol.* **7**, 47.

Hughes, D. W., Genest, K., and Rice, W. B. (1966). *Lloydia* **29**, 328.

Ito, Y., Kurita, H., Yamaguchi, T., Sato, M., and Okuda, T. (1967). *Chem. & Pharm. Bull. (Tokyo)* **15**, 2009.

Iwasaki, H., Kamiya, T., Oka, O., and Ueyanagi, J. (1965). *Chem. & Pharm. Bull. (Tokyo)* **13**, 753.

Karrer, W. (1958). "Konstitution und Vorkommen der Organische Pflanzenstoffe," p. 353. Birkhäuser, Basel.

Kawamura, S. (1954). "Icones of Japanese Fungi," Vol. IV, p. 449. Kazamashobo, Tokyo.

Keinholz, J. R. (1934). *Mycologia* **26**, 275.

Klohs, M. W. (1967). *U. S., Public Health Serv., Publ.* **1645**, 126.

Kögl, F., Duisberg, H., and Erxleben, H. (1931). *Ann. Chem.* **489**, 156.

Kögl, F., Cox, H. C., and Salemink, C. A. (1957a). *Ann. Chem.,* **608**, 81.

Kögl, F., Salemink, C. A., Schouten, H., and Jellinek, F. (1957b). *Rec. Trav. Chim.* **76**, 109.

Komatsu, N., Terakawa, H., Nakanishi, K., and Watanabe, Y. (1963). *J. Antibiotics (Tokyo)* **A16**, 139.

Krieger, L. C. C. (1967). "The Mushroom Handbook." Dover, New York.

Kubin, G. (1961). *Nahrung* **5**, 41.

Leung, A. Y., and Paul, A. G. (1967). *J. Pharm. Sci.* **56**, 146.

Leung, A. Y., and Paul, A. G. (1968). *J. Pharm. Sci.* **57**, 1667.

Leung, A. Y., and Paul, A. G. (1969). *Lloydia* **32**, 66.

Leung, A. Y., Smith, A. H., and Paul, A. G. (1965). *J. Pharm. Sci.* **54**, 1576.

List, P. H., and Luft, P. (1967). *Tetrahedron Letters* No. 20, 1893.

Lucas, E. H., Byerrum, R. U., Clarke, D. A., Reilly, H. C., Stevens, J. A., and Stock, C. C. (1957). *Antibiot. Chemotherapy* **7**, 1.

Lucas, E. H., Byerrum, R. U., Clarke, D. A., Reilly, H. C., Stevens, J. A., and Stock, C. C. (1958-1959). *Antibiot. Ann.* p. 493.

McCawley, E. L., Brummett, R. E., and Dana, G. W. (1962). *Proc. West. Pharmacol. Soc.* **5**, 27.

McKenny, M. (1962). "The Savory Wild Mushroom." Univ. of Washington Press, Seattle, Washington.

McMorris, T. C., and Anchel, M. (1965). *J. Am. Chem. Soc.* **87**, 1594.

Malone, M. H., Robichaud, R. C., Tyler, V. E., Jr., and Brady, L. R. (1962). *Lloydia* **25**, 231.

Malone, M. H., Tyler, V. E., Jr., and Brady, L. R. (1967). *Lloydia* **30**, 250.

Meyer, H. J. (1967). *U. S., Public Health Serv., Publ.* **1645**, 133.

Miyata, J. T., Tyler, V. E., Jr., Brady, L. R., and Malone, M. H. (1966). *Lloydia* **29**, 43.

Moench, G. L. (1963). *J. Am. Med. Assoc.* **186**, 1025.

Mortara, M., and Filipello, S. (1967). *Minerva Med.* **58**, 3628.

Müller, G. F. R., and Eugster, C. H. (1965). *Helv. Chim. Acta* **48**, 910.

Nakanishi, K., Tada, M., Yamada, Y., Ohashi, M., Komatsu, N., and Terakawa, H. (1963). *Nature* **197**, 292.

"Nutritional Data." (1956). 3rd ed. H. J. Heinz Co., Pittsburgh, Pennsylvania.

Ola'h, G. M. (1968). *Compt. Rend.* **267**, 1369.

Peck, C. H. (1912). *Bull. N. Y. State Museum* **157**, 5.

Pilát, A., and Ušák, O. (1954). "Mushrooms." H. W. Bijl, Amsterdam.

Ramsbottom, J. (1953). "Mushrooms and Toadstools," p. 44. Collins, London.

Reynolds, W. A., and Lowe, F. H. (1965). *New Engl. J. Med.* **272**, 630.

Roland, J. F., Chmielewicz, Z. F., Wiener, B. A., Gross, A. M., Boening, O. P., Luck, J. V., Bardos, T. J., Reilly, H. C., Sigiura, K., Stock, C. C., Lucas, E. H., Byerrum, R. U., and Stevens, J. A. (1960). *Science* **132**, 1897.

Romagnesi, M. H. (1964). *Bull. Soc. Mycol. France* **80**, IV.

Rose, E. K., and Rieders, F. (1966). *Ann. Internal Med.* [N.S.] **64**, 372.

Ruelius, H. W., Janssen, F. W., Kerwin, R. M., Goodwin, C. W., and Schillings, R. T. (1968). *Arch. Biochem. Biophys.* **125**, 126.

Salemink, C. A., ten Broeck, J. W., Schuller, P. L., and Veen, E. (1963). *Planta Med.* **11**, 139.

Sapeika, N., and Stephens, E. L. (1965). *S. African Med. J.* **39**, 749.

Scardi, V., and Bonavita, V. (1959). *Clin. Chim. Acta* **4**, 161.

Schillings, R. T., and Ruelius, H. W. (1968). *Arch. Biochem. Biophys.* **127**, 672.

Simandl, J., and Franc, J. (1956). *Chem. Listy* **50**, 1862.

Singer, R. (1958). *Mycopathol. Mycol. Appl.* **9**, 275.

Singer, R. (1962). "The Agaricales in Modern Taxonomy," 2nd ed. Cramer, Weinheim.

Singer, R., and Smith, A. H. (1958). *Mycopathol. Mycol. Appl.* **9**, 280.

Smith, A. H. (1949). "Mushrooms in their Natural Habitat," p. 485. Sawyer's Inc., Portland, Oregon.

Smith, A. H. (1967). "The Mushroom Hunter's Field Guide." Univ. of Michigan Press, Ann Arbor, Michigan.

Stein, S. I. (1958). *Mycopathol. Mycol. Appl.* **9**, 263.

Stein, S. I., Closs, G. L., and Gabel, N. W. (1959). *Mycopathol. Mycol. Appl.* **11**, 205.

Sternberg, S. S., Philips, T. S., Cronin, A. P., Sodergren, J. E., and Vidal, P. M. (1963). *Cancer Res.* **23**, 1036.

Stevenson, J. A., and Benjamin, C. R. (1961). *Mycologia* **53**, 438.

Stromberg, V. L. (1954). *J. Am. Chem. Soc.* **76**, 1707.

Swenberg, M-L. L., Kelleher, W. J., and Schwarting, A. E. (1967a). *Science* **155**, 1259.

Swenberg, M-L. L., Kelleher, W. J., and Schwarting, A. E. (1967b). *Abstr. 154th Meeting Am. Chem. Soc., Chicago* p. Q10.

Tabor, G., and Vining, L. C. (1963). *Can. J. Botany* **41**, 639.

Takemoto, T. (1961). *Japan. J. Pharm. Chem.* **33**, 252.

Takemoto, T., Yokobe, T., and Nakajima, T. (1964a). *Yakugaku Zasshi* **84**, 1186.

Takemoto, T., Nakajima, T., and Sakuma, P. (1964b). *Yakugaku Zasshi* **84**, 1233.

Thoms, H., and Vogelsang, J. (1907). *Ann. Chem.* **357**, 145.

Troxler, F., Seeman, F., and Hofmann, A. (1959). *Helv. Chim. Acta* **42**, 2073.

Turner, W. J., and Merlis, S. (1959). *A.M.A. Arch. Neurol. Psychiat.* **81**, 121.

Tyler, V. E., Jr. (1958). *Am. J. Pharm.* 130, 264.

Tyler, V. E., Jr. (1961). *Lloydia* 24, 71.

Tyler, V. E., Jr. (1963). *Progr. Chem. Toxicol.* 1, 339.

Tyler, V. E., Jr., and Gröger, D. (1964). *Planta Med.* 12, 397.

Tyler, V. E., Jr., and Smith, A. H. (1963). *Mycologia* 55, 358.

Tyler, V. E., Jr., Brady, L. R., Benedict, R. G., Khanna, J. M., and Malone, M. H. (1963). *Lloydia* 26, 154.

Verhulst, H. L. (1969). Personal communication.

Walters, M. B. (1965). *Mycologia* 57, 837.

Waser, P. G. (1961). *Pharmacol. Rev.* 13, 465.

Waser, P. G. (1967). *U. S., Public Health Serv., Publ.* 1645, 437.

Wasson, V. P., and Wasson, R. G. (1957). "Mushrooms, Russia and History," Vols. I and II. Pantheon Books, New York.

Watanabe, Y., Nakanishi, K., Komatsu, N., Sakabe, T., and Terakawa, H. (1964). *Bull. Chem. Soc. Japan* 37, 747.

Weidmann, H., Taeschler, M., and Konzett, H. (1958). *Experientia* 14, 378.

Weir, J. K., and Tyler, V. E., Jr. (1960). *J. Am. Pharm. Assoc., Sci. Ed.* 49, 426.

Wieland, T. (1968). *Science* 159, 946.

Wieland, T., and Motzel, W. (1953). *Ann. Chem.* 581, 10.

Wilkinson, S. (1961). *Quart. Rev. (London)* 15, 153.

Winterstein, E., Reuter, C., and Karolev, R. (1913). *J. Chem. Soc. Org. Abstr.* 104, 433.

CHAPTER 12

Ergot

DETLEF GRÖGER

I. Introduction

The variety of natural products that are synthesized by bacteria and fungi seems not to be less than those synthesized by higher plants. A good example of the eminent biochemical capabilities of microorganisms are the different structures of N-heterocyclic compounds found in microorganisms. Among those metabolic products are "true" alkaloids. According to their importance as poison and remedy undoubtedly the constituents of ergot are the best known and the most thoroughly investigated. Different species of the genus *Claviceps* are capable of forming alkaloids. Ergot fungi are parasitizing mainly on rye and different wild grasses. The best-known species is *Claviceps purpurea* (Fr.) Tulasne in which the sclerotia are designated *Secale cornutum* in pharmacopeias.

The active principles of ergot are poisonous when they are ingested in large quantities. The admixture of ergot in flour and bread often occurred widely in Europe centuries ago and caused epidemics in many regions. The damaging effects of ergot on health remained obscure for a long time, although *Claviceps purpurea* is one of the oldest known producers of mycotoxins. The following characterization of mycotoxicoses according to Feuell (1966) proves true also for gangrenous and convulsive ergotism:

1. The disease is not transmissible.
2. Drug and antibiotic treatment have little or no effect on the disease.
3. The outbreak is usually associated with a specific foodstuff.

In contrast to many other mycotoxicoses today the typical forms of ergotism are practically eliminated. During its long fascinating history ergot changed from a feared poison to a desired starting material for the production of important remedies. Continuous chemical investigation of the active principles paved the way for their multi-fold use in medicine. Progress in ergot research especially in the last 20 years will be discussed in this chapter. For earlier literature on *Claviceps* the reviews of Tschirch (1923), Barger (1931), Stoll (1952), Guggisberg (1954), Hofmann (1964), and Bové (1970) should be consulted.

II. Historical Background

The occurrence of ergotism as well as the use of ergot in medicine by the Ancients is uncertain. There are some hints in the old literature, but final proof is still lacking. Large-scale epidemics raged in the Middle Ages, and these were very often designated "holy fire" or *ignis sacer* in those parts of west and middle Europe where rye was used for bread. Two characteristic forms, gangrenous and convulsive ergotism, are known to have occurred but at that time the causative agent was not recognized. A review on ergot epidemics from ancient to modern times has been given by Kobert (1889). In the *Kreuterbuch* of Adam Lonicer (1582) ergot was described for the first time as a drug, and there are some remarks on its use by women.

In Caspar Bauhin's *Theatrum Botanicum* (1658) the first illustration of ergot designated *Secale luxurians* appeared. Von Münchhausen (1764) was the first to recognize ergot as a fungus and to not consider ergot a diseased rye grain. At the beginning of the 19th century ergot was introduced into official medicine (Stearns, 1808), and the first chemical investigation was carried out (Vauquelin, 1816). The life cycle of *Claviceps purpurea* was described by Tulasne (1853) in a classic paper in France. Also in France the first crystalline alkaloid preparation *ergotinine cristallisée* was isolated by the pharmacist Tanret (1875). The first chemically pure, homogeneous alkaloid, which exhibits all the typical biological properties of ergot, was ergotamine (Stoll, 1918). The fundamental principles of the parasitic cultivation of ergot were achieved by von Békésy (1935) and W. Hecht (1941). This method was improved in the laboratories of the Sandoz Company of Basel (Stoll and Brack,

1944a,b). Lysergic and isolysergic acid were obtained as degradation products of ergot alkaloids by Jacobs and Craig (1934). Stoll and Hofmann (1943a) described the separation of the ergotoxine complex into its components, ergocristine, ergocornine, and ergokryptine. In the same year the synthesis of lysergic acid diethylamide (LSD) was published (Stoll and Hofmann, 1943b). After World War II the first representative of a new class of ergoline derivatives agroclavine was isolated from saprophytic cultures (Abe, 1951a,b). The first total synthesis of lysergic acid was accomplished by Kornfeld *et al.* (1954), and a few years later Hofmann *et al.* (1961) described the synthesis of ergotamine. A phytochemical sensation was the isolation of ergoline derivatives from higher plants (Hofmann and Tscherter, 1960). The large-scale production of simple lysergic acid derivatives in submerged culture is described by Arcamone *et al.* (1961a). In the past 10 years the biosynthesis of ergoline derivatives was investigated in many laboratories (for a review see Gröger, 1969), while the elucidation of the structure of various ergot pigments was also performed in the last decade (as reviewed by Franck, 1969).

III. Biology

A. Some Observations on the Life Cycle of Ergot

The life cycle of ergot fungi was given in a general outline about 100 years ago by Tulasne (1851, 1853) and Kühn (1865). These authors related the formation of stromata and perithecia on sclerotia with the development of the sphacelia stage in the inflorescence of rye plants after infection with ascospores and the subsequent production of honeydew containing a large number of conidiospores. They established a definite connection between the sphacelia and subsequent formation of sclerotia. Before the characteristic fructifications as well as the sphacelia mycelium were regarded as separate fungi. Also the origin of honeydew was long obscure. The formation of asci proceeds by a primitive form of a sexual process in the capitulum. Between an antheridium and a ascogonium plasmogamy takes place (Killian, 1919). Every ascus contains eight filamentous ascospores. A detailed description of the development of asci in ergot fungi has not yet been given. Kirchhoff (1929) made the observations that germinating spores (ascospores of the primary infection and conidiospores of the secondary infection) grow around the ovary and enter the latter at its base. From this point the hyphae of the sphacelia stage spreads out. Ergot fungi can develop only in female sex organs of grasses. Therefore the parasitism of ergot was designated by Gäumann (1951) as an highly organospecific infectious disease of grasses.

Seedlings of higher plants form roots for their attachment in soil. Von Békésy (1956a) found that the fixing of sclerotia of *C. purpurea* in the ground takes place by the formation of a necklike ring of hyphae at the base of the stalk. Apparently the hyphae are important not only for the attachment of sclerotia but also serve to provide the water supply of the fruiting bodies. The conditions for germination of ergot sclerotia were thoroughly investigated by Kirchhoff (1929) and Krebs (1936). After a chilling period of 3 to 4 weeks at -1-$3°$ C, the sclerotia start to germinate over the next two months at a temperature of 10-$15°$ C. Von Békésy (1956b) in his germination experiments inserted the sclerotia between long-stemmed moss plants (*Dicranium scoparium*) and got, in this way, the necessary moisture while avoiding decay. He observed that various races of ergot reacted differently after storage in the cold. Furthermore the pigmentation of the stipes and capitula of the stromata varied considerably. Mitchell and Cooke (1968a,b) performed similar investigations using sclerotia of *C. purpurea* from a number of grass species. It was found that storage at low temperature markedly influenced the physiology of the sclerotia. After the chilling and the pregermination periods during the subsequent germination, a great increase in water uptake and respiration rate occurs. But before this increase a degradation of lipids takes place. Fat-hydrolyzing enzymes are probably mobilized during or immediately after activation treatment. The relation between the formation of fruiting bodies and the respiration rate with regard to fumaric acid metabolism was investigated by Garay (1958).

Microscopic morphological studies on the development of sclerotia beginning from the sphacelia stage were carried out by Milovidov (1954). Kybal (1964) assumes that sclerotia are formed by the germination of honeydew spores. Probably this process takes place step by step in the phase of intensive daily growth. The conidiospores are able to form either sphacelial or sclerotial hyphae. The question of whether the sphacelia isolate different types of spores with different genetic predispositions has been brought up. *Claviceps* is a homothallic fungus. This was stated by McCrea (1931) and was verified later by von Békésy (1956b). The germination physiology of conidiospores of saprophytic and parasitic origin has been investigated in detail (Garay and Kökenyesi, 1955; Garay, 1956a,b; Gláz, 1955; Lewis, 1959; Rozumek, 1963). It was found that the germination rate is markedly influenced by the density of spore suspensions, and honeydew even in a dilution of 1:5000 promoted germination.

The cytology of *C. purpurea* was investigated by different authors (Thielmann, 1955; Kybal *et al.*, 1957; Milovidov, 1954, 1957; Jung and

Rochelmeyer, 1960). To some extent the results reported have been conflicting. It is clear that the mass of conidiospores of honeydew and saprophytic cultures are uninucleate. Young sporeforming saprophytic mycelium and sphacelia mycelium are always uninucleate. In older mycelium, cells with several nuclei can be observed. The occurrence of multinucleate cells in saprophytic cultures is regarded as a sign of beginning sclerotinization because cells of normal sclerotia are multinucleate. The formation of multinucleate cells can take place, for instance, through anastomoses. The importance of sexual cycles for a generative renewing of vegetative selected strains is diminished by the copulation of mycelia (Jung and Rochelmeyer, 1960).

B. TAXONOMIC PROBLEMS AND DISTRIBUTION OF *Claviceps* Species

According to Gäumann (1964) the family Ostropaceae with six genera and Clavicipitaceae with ten genera belong to the order Clavicipitales. Other authors include the Clavicipitaceae in the order Hypocreales, i.e., Sphaeriales. The ripe asci are cylindrical, being oblong with a typical apical swelling. The ascospores are slender hyaline septated and break down easily in different segments.

Gäumann (1964) differentiated three lines of evolution within the Clavicipitaceae: *Hypomyces—Ascopolyporus* line; *Epichloe—Claviceps* line; and the genus *Cordyceps*. Important characteristics in the consideration of the taxonomy of *Claviceps* species include the morphology and pigmentation of stromata and perithecia as well as the structure of asci and the length of ascospores. Most important is the type of sexual reproduction, which has been studied only in *C. purpurea* and *Claviceps paspali* (Oddo and Tonolo, 1967). *Claviceps paspali* forms, in contrast to *C. purpurea*, ascogonia but no antheridia, and autogamy takes place. Also the two species differ in type of ascospores, germination of ascospores, and production of alkaloids. In addition to these characteristics it was observed that sclerotia of *C. paspali* and pseudostroma of some *Balansia* species have a similar structure. It has been proposed that *C. paspali* Stev. et Hall be regarded as a member of a new genus—*Mothesia* (intermediate between *Balansia* and *Claviceps*) (Oddo and Tonolo, 1967). This new genus would include *C. paspali* Stev. et Hall, type species; *Claviceps gigantea* Fuentes, de la Isla, Ullstrup, Rodriguez; *Claviceps grohii* Groves; *Claviceps* sp. from *Pennisetum typhoideum* Rich.; and *Claviceps* sp. from *Sorghum* sp. Further investigations are necessary to substantiate the validity of this new genus.

Critical reviews on the taxonomy of the genus *Claviceps* and the differentiation of single species have been given by Langdon (1954) and Skalický and Starý (1962). These authors have not always held the same opinion on taxonomy. Since the compilation of Langdon (1954), five new species have been described: *Claviceps phalaridis* Walker (1957); *Claviceps zizaniae* (Fyles) Pantidou (1959); *Claviceps gigantea* Fuentes *et al.* (1964); *Claviceps fusiformis* Loveless (1967a); and *Claviceps cyperi* Loveless (1967b) (Table I).

TABLE I

Claviceps SPECIES *sensu* LANGDON TO 1969[a]

Species	Synonyms
C. annulata Langdon	
C. cinerea Griff.	
C. cynodontis Langdon	
C. cyperi Loveless	
C. diadema (Möller) Diehl	
C. digitaria Hansford	*C. balansoides* Möller
C. flavella (Berk. und Curt.) Petch	*C. pallida* Pat.
C. fusiformis Loveless	*C. patouillardiana* P.Henn.
C. gigantea Fuentes, de la Isla Ullstrup, Rodriguez	
C. glabra Langdon	
C. grohii Groves	
C. hirtella Langdon	
C. inconspicua Langdon	
C. litoralis Kawatani	
C. lutea Möller	
C. maximensis Theis	
C. nigricans Tul.	
C. orthocladae (P.Henn.) Diehl	
C. paspali Stev. & Hall.	*C. rolfsii* Stev. & Hall.
C. phalaridis Walker	
C. platytricha Langdon	*C. microcephala* (Wallr.)Tul.
C. purpurea (Fr.) Tul.	*C. sesleriae* (Stäger)
C. pusilla Ces.	*C. setulosa* Quel.
C. queenslandica Langdon	
C. ranuncoloides Möller	
C. sulcata Langdon	
C. tripsaci Stev. & Hall.	
C. uleana P. Henn	
C. viridis Padwick und Azmat.	
C. yanagawensis Togashi	
C. zizaniae (Fyles) Pantidou	

[a]According to Skalický and Starý (1962) there exist also the following species: *C. philippii* Rehm and *C. wilsoni* Cooke. The following species are considered autonomous (not synonymous): *C. patouillardiana* P. Henn.; *C. rolfsii* Stev. & Hall.; *C. setulosa* Quel. *C. litoralis* kawatani is synonymous with *C. purpurea* (Fr.) Tul.

Lists of the host plants to various *Claviceps* species have been published repeatedly (Atanasoff, 1920; Barger, 1931; Kawatani, 1952; 1953; Grasso, 1954, 1957; Lindquist and Carranza, 1960; Brady, 1962). A critical and comprehensive report of parasitism by *Claviceps* species with regard to the phylogenetic relationship of the hosts was given by Brady (1962). On members of Juncaceae one *Claviceps* species was in fact found only in the sphacelial stage (*Claviceps junci* = *Sphacelia junci*). Three *Claviceps* species are parasitic on members of Cyperaceae; *Claviceps grohii* on *Carex* species; *Claviceps cyperi* on *Cyperus esculentus* L. and *Cyperus rotundus* L.; and *Claviceps nigricans* on the genera *Scirpus* and *Eleocharis*. Most of the *Claviceps* species (N28) have been reported on Gramineae.

The morphotypes of ergot are primarily influenced by the host plants. *Claviceps purpurea* and related species grow on grasses of the subfamily Pooideae = Festucoideae, Oryzoideae, and Bambusoideae. Ergot fungi parasitizing members of the subfamily Panicoideae form yellow stromata, and their sclerotia are mostly globose and grayish yellow. *Claviceps* species that grow on the Panicoideae have a certain relationship to the genus *Balansia*. Ergot fungi of the *C. paspali* type are widely distributed in eurytropical and eurysubtropical areas, while *C. purpurea* is typical in amphiboreal areas (Skalický and Starý, 1962; Tonolo, 1967). *Claviceps purpurea* was found recently for the first time on *Festuca obturbans* in East Africa (Brack *et al.*, 1963) and on a *Paspalum* species in South America (Kobel and Stopp, 1967).

Within the various *Claviceps* species different "races" have been described. Following host specificity Stäger (1922) has defined so-called "biological" races. The occurrence of specific races, related to parasitism on particular host plants, was challenged by the experiments of several authors (von Békésy, 1956b; Meinicke, 1956; Kybal and Brejcha, 1955; Campbell, 1957). According to Skalický and Starý (1962), biotypes of *C. purpurea* exist; and they divide them into physiological, phenological, and geographical races and climate types. In the case of biochemical races the alkaloid content of the sclerotia is a criterion. In wild ergot one finds alkaloid-free and alkaloid-yielding sclerotia. In selected ergot for parasitic cultivation the chemical races can be designated according to their main alkaloid or alkaloids. For example Kybal and Brejcha (1955) reported the existence of five different races of ergot, and Gröger (1956, 1957) isolated strains that produce one or two main alkaloids — ergotamine, ergosine, and ergocristine — as well as an ergocornine/ergokryptine race. Further possibilities for classification exist, if the ability of an ergot strain to produce alkaloids in saprophytic culture is considered. One of the pioneers of ergot research classified the ergot fungi into 16 races (Abe, 1963; Abe and Yamatodani, 1964). The most important ones are (a) *Agropyron*-type fungi, (b) *Elymus*-type fungi, (c) *Pennisetum*-type fungi, (d) *Secale*-type fungi, and (e) *Paspalum*-type fungi.

Three chemical races were found in *C. paspali* (Kobel *et al.*, 1964): (a) Lysergic acid methylcarbinolamide race, (b) 6-methyl-$\Delta^{8,9}$-ergolene-8-carboxylic acid race, and (c) ergometrine race. More than 90% of the isolated *C. paspali* strains produce lysergic acid methylcarbinolamide under submerged conditions. Race (b) grows apparently only on *Paspalum* species in Portugal and the south of France. The ergometrine race was found in Argentina. A form of the *Secale*-type (*Claviceps purpurea*) was isolated from Triticale (an artificial hybrid between rye and wheat) in Spain (Tonolo, 1966). This race is remarkably different from the "normal" *C. purpurea* types. Under submerged conditions plectenchymatic cells are formed but no conidia. Under certain special nutritional conditions ergotamine is synthesized in high yield. The ergot fungi of the *C. purpurea* type and the *C. paspali* type show other biochemical differences. Only *C. paspali* strains are able to produce 2,3-dihydroxybenzoic acid.

According to Gröger (1968) the genus *Claviceps* may be divided biochemically into three groups:

1. Ergot fungi that are able to form the ergoline ring system starting from tryptophan and mevalonic acid. These strains have not (or have lost) the ability to oxidize elymoclavine under allylic rearrangement to lysergic acid. They correspond to the *Pennisetum* and *Agropyron* type.

2. Ergot fungi of the *C. paspali* type that can produce clavine alkaloids and simple lysergic acid derivatives but no peptide alkaloids.

3. *Claviceps purpurea* and related species that are able to synthesize clavine alkaloids, simple lysergic acid derivatives, and peptide alkaloids.

IV. Constituents of Ergot

A. SOME REMARKS

Probably only a few other drugs have been so thoroughly investigated chemically as has ergot. Since the time of Vauquelin (1816) — a French pharmacist — a great many natural compounds, representing different structural types, were isolated from sclerotia and saprophytic cultures of various ergot species. Ergot was called a "treasure house" of pharmacologically active compounds. The most important substances in toxicology and medicine are the alkaloids. For a long time the ergoline derivatives have been regarded as specific for the genus *Claviceps*.

Recently ergot alkaloids were isolated from fungi outside the genus *Claviceps* and from higher plants. The variable composition of the alkaloid mixture in the drug and the complex chemical structure of the bases and their sensitivity have impeded the chemical evaluation of ergot alkaloids. The chemistry of ergot has been critically reviewed by Hofmann (1964) and Stoll and Hofmann (1965).

B. ALKALOIDS

1. STRUCTURAL TYPES

Ergot alkaloids are 3,4-substituted indole (I) derivatives (Fig. 1). The tetracyclic ring system has been designated ergoline (II) (Jacobs and Gould, 1937). On the basis of their structures ergot alkaloids may be divided into two groups: (a) Acid amide type derivatives of lysergic acid (6-methyl-8-carboxyl-$\Delta^{9,10}$-ergolene) and the stereoisomeric isolysergic acid (VI), general formula (III). Lysergic acid derivatives are especially found in officinal ergot (sclerotia of *C. purpurea*). Simple lysergic acid derivatives (XV, XVI) are formed in large amounts in saprophytic cultures of *C. paspali* (Figs. 2 and 3). (b) The so-called clavine alkaloids or clavines. These are simple hydroxy and dehydro derivatives of 6,8-

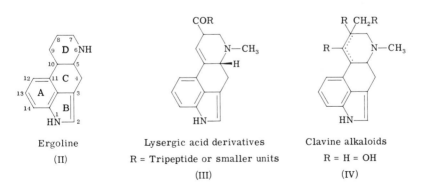

FIG. 1. General formulas of the ergoline derivatives.

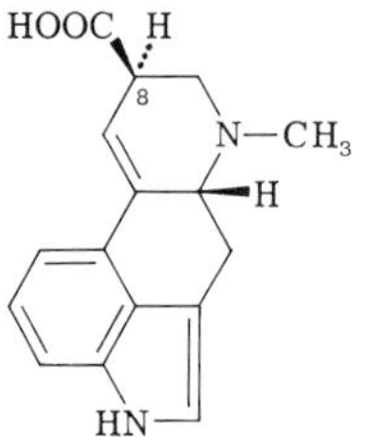

Lysergic acid
(V)

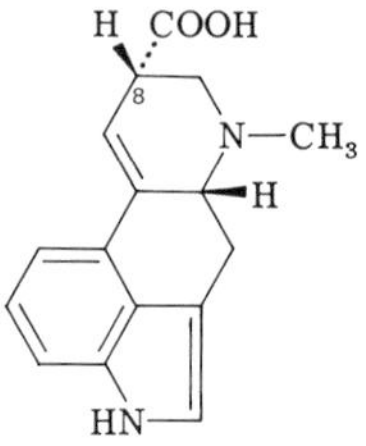

Isolysergic acid
(VI)

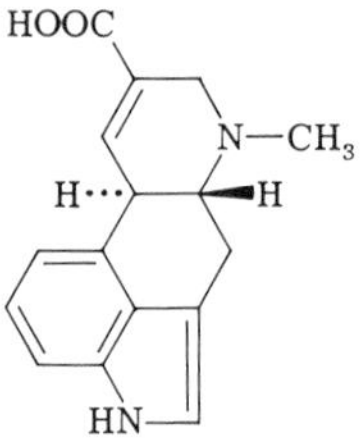

6-Methyl-$\Delta^{8,9}$-ergolene-8-
carboxylic acid $\Delta^{8,9}$-
lysergic acid
(VII)

FIG. 2. Lysergic acid formulas.

dimethylergoline, (general formula (IV). To this group belong also the isomeric chanoclavines, in which ring D is open between N-atom 6 and C-atom 7.

The clavine alkaloids occur mainly in sclerotia of *Claviceps* species that parasitize wild grasses in the Far East and Africa. The saprophytic cultures of these strains also can produce clavines. Trace amounts of this ergoline derivatives can be found in sclerotia and saprophytic cultures of *C. purpurea* and *C. paspali* (Fig. 4).

2. LYSERGIC ACID DERIVATIVES

A survey of natural occurring lysergic acid derivatives is given in Table II and Fig. 3. The structural element common to all alkaloids of this group is lysergic acid, which was first obtained by alkaline hydrolysis of ergotinine (Jacobs and Craig, 1934). The natural dextrorotatory lysergic acid was named *d*-lysergic acid and its iso form *d*-isolysergic acid is a stereoisomer that differs only in the configuration in the center of asymmetry at C-8. The hydrogen atoms at C-5 and C-8 of (V) are in the trans position, and, in the case of isolysergic acid, in the cis position. The carboxyl group is in an equatorial position in lysergic acid and more distant to the nitrogen atom in ring D than the carboxyl group of isolysergic acid in the axial position. Ring D has a pseudo-chair form in both epimers (Hofmann, 1964). The absolute configuration was determined by Leemann and Fabbri (1959) and Stadler and Hofmann (1962). *d*-Lysergic acid lactam can be degraded by oxidation to a D-aspartic acid derivative. According to the Cahn-Ingold nomenclature, *d*-lysergic acid possess the (5*R*:8*R*) configuration. From submerged cultures of a Portuguese *C. paspali* strain a structural isomer of lysergic acid can be isolated. In this compound the double bond of ring D is located in the 8,9-position (VII). This acid is very unstable and readily transformed into lysergic acid

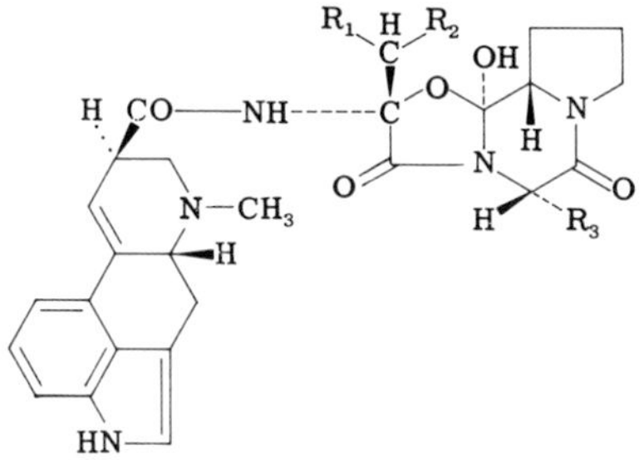

$R_1 = R_2 = H$	$R_1 = R_2 = CH_3$	$R_1 = H; R_2 = CH_3$	R_3
Ergotamine (VIII)	Ergocristine (IX)	Ergostine (X)	 $\langle$benzyl$\rangle$—CH$_2$—
Ergosine (XI)	α-Ergokryptine (XII)		$-CH_2-\overset{H}{\underset{CH_3}{C}}{<}^{CH_3}_{CH_3}$
	β-Ergokryptine (XIII)		$-\overset{H}{\underset{C}{C}}{<}^{CH_3}_{CH_2-CH_3}$
	Ergocornine (XIV)		$-\overset{H}{\underset{C}{C}}{<}^{CH_3}_{CH_3}$

Peptide type alkaloids

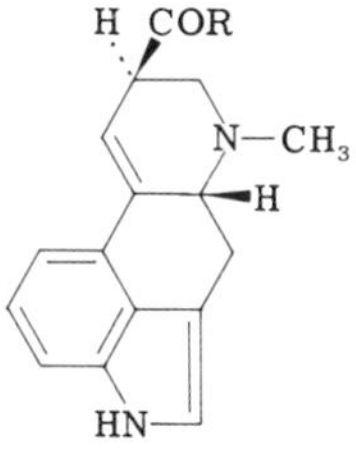

(XV) R = NH$_2$: Ergine

(XVI) R = NH·$\overset{H}{\underset{OH}{C}}{<}^{CH_3}$:   Lysergic acid methylcarbinolamide

(XVII) R = NH$\overset{H}{\underset{CH_2OH}{C}}{<}^{CH_3}$:   Ergometrine (ergobasine)

(XVIII) R = NH$\overset{H}{\underset{COOCH_3}{C}}{<}^{CH(CH_3)_2}$:   Lysergic acid-L-valinemethylester

FIG. 3. Natural ergot alkaloids derived from lysergic acid.

(Kobel *et al.*, 1964). Simple acid amide type derivatives of lysergic acid can be obtained from *C. paspali* strains: lysergic acid methylcarbinolamide (XVI) and lysergic acid amide = ergine (XV). Compound XVI readily splits off acetaldehyde and is transformed into XV. *d*-Lysergic acid-L-valinemethylester (XVIII) was isolated from mother liquors of an alkaloid production. In many ergot drugs traces of XVIII could be detected (Schlientz *et al.*, 1963). Ergometrine (XVII) has been well

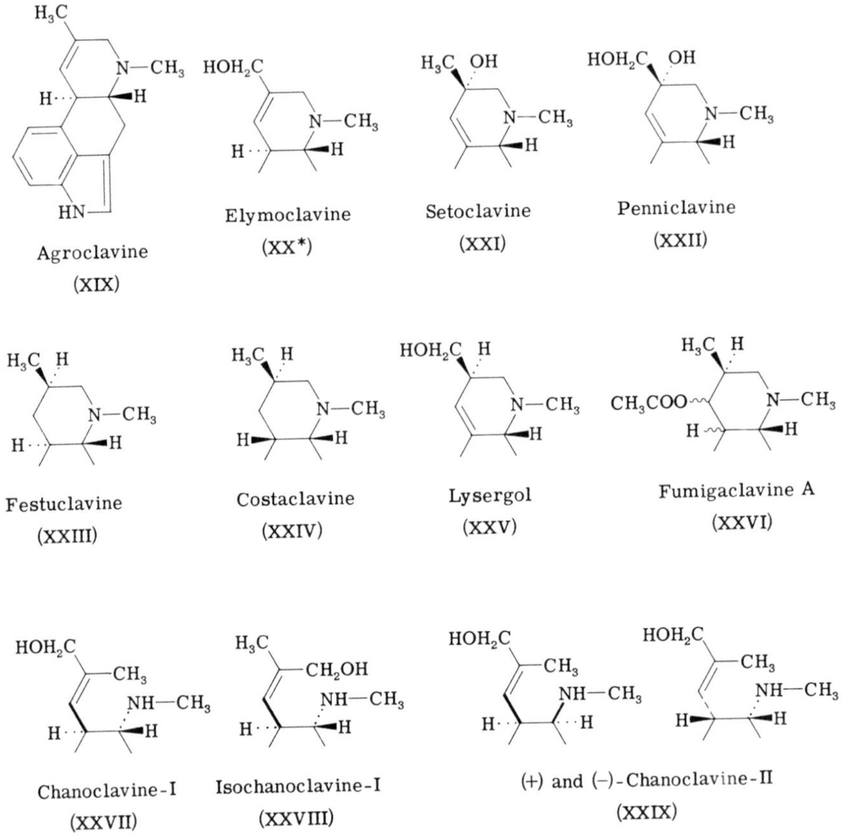

Agroclavine
(XIX)

Elymoclavine
(XX*)

Setoclavine
(XXI)

Penniclavine
(XXII)

Festuclavine
(XXIII)

Costaclavine
(XXIV)

Lysergol
(XXV)

Fumigaclavine A
(XXVI)

Chanoclavine-I
(XXVII)

Isochanoclavine-I
(XXVIII)

(+) and (–)-Chanoclavine-II
(XXIX)

* Starting with elymoclavine, only ring D is depicted.

FIG. 4. Clavine alkaloids.

known for a long time. It is detectable in many ergot drugs of different
origin and is also synthesized in saprophytic cultures of *C. purpurea* and
C. paspali.

The main alkaloids of *C. purpurea* are the peptide alkaloids (Fig. 3). On
hydrolysis they decompose to lysergic or isolysergic acid, one equivalent
of ammonia, one α-keto acid, and two amino acids. Proline is common to
all peptide alkaloids, and it is obtained in the L-form upon mild alkaline
hydrolysis. The α-keto acids yielded are in the ergotamine group, pyruvic
acid; in the ergotoxine group, dimethylpyruvic acid; and in the ergoxin
group, α-keto butyric acid. The various alkaloids differ from each other
in the amino acids obtained upon hydrolytic cleavage. These are L-

TABLE II

ALKALOIDS OF THE LYSERGIC AND ISOLYSERGIC ACID SERIES

Alkaloid	Formula	Reference
1. Ergotamine group		
Ergotamine	$C_{33}H_{35}O_5N_5$	Stoll (1918)
Ergotaminine		
Ergosine	$C_{30}H_{37}O_5N_5$	S. Smith and Timmis
Ergosinine		(1937)
2. Ergotoxine group		
Ergocristine	$C_{35}H_{39}O_5N_5$	Stoll and Burckhardt
Ergocristinine		(1937)
α-Ergokryptine	$C_{32}H_{41}O_5N_5$	Stoll and Hofmann (1943a)
α-Ergokryptinine		
β-Ergokryptin	$C_{32}H_{41}O_5N_5$	Schlientz *et al.* (1967)
β-Ergokryptinine		
Ergocornine	$C_{31}H_{39}O_5N_5$	Stoll and Hofmann (1943a)
Ergocorninine		
3. Ergoxine group		
Ergostine	$C_{34}H_{37}O_5N_5$	Schlientz *et al.* (1964)
Ergostinine		
4. Ergobasine group		
Ergobasine	$C_{19}H_{23}O_2N_3$	Dudley and Moir (1935); Kharasch and Legault (1935); Stoll and Burkhardt (1935); Thompson (1935)
Ergobasinine		
5. Miscellaneous		
Ergosecaline	$C_{24}H_{28}O_4N_4$	Abe *et al.* (1959)
Ergosecalinine		
Ergine	$C_{16}H_{17}O\ N_3$	Arcamone *et al.* (1960)
Erginine		
D-Lysergic acid and isolysergic acid methylcarbinolamide	$C_{18}H_{21}O_2N_3$	Arcamone *et al.* (1960)
$\Delta^{8,9}$-Lysergic acid	$C_{16}H_{16}O_2N_2$	Kobel *et al.* (1964)
D-Lysergyl-L-valine-methylester	$C_{22}H_{27}O_3N_3$	Schlientz *et al.* (1963)

phenylalanine, L-leucine, or L-valine. In the case of ergosecaline, the structure for which is not yet entirely elucidated, the cleavage products lysergic acid, NH_3, pyruvic acid, and valine can be detected (Abe *et al.*,

1959). The sequence of the various building blocks of the peptide alkaloids and their type of linkage were mainly elucidated by Stoll and Hofmann's group (Stoll and Hofmann, 1965). On the basis of cleavage experiments two unusual structural features in the peptide moiety of peptide alkaloids were observed, namely, a α-hydroxy-α-amino acid grouping and the so-called cyclol structure.

Some years ago a new alkaloid, ergostine (X), was isolated from the "ergotamine" fraction of Swiss ergot of rye. A total synthesis for ergostine and its isomer ergostinine has been developed. This alkaloid pair is the first member of a third group of peptide alkaloids, which is called the ergoxine group (Schlientz et al., 1964). An isomer of ergokryptine was also detected by Schlientz et al. (1967). This alkaloid has been named β-ergokryptine (XIII). The earlier described alkaloid of the ergotoxine group is now designated α-ergokryptine (XII). Both isomers differ in the peptide portion. α-Ergokryptine yields L-leucine upon hydrolysis, and β-ergokryptine L-isoleucine. The yield of β-ergokryptine as well as α-ergokryptine varies considerably depending on the origin of the drug.

Several types of isomerization reactions have been observed in the series of lysergic acid derivatives. The lysergic acid $\rightleftarrows$ isolysergic acid isomerization has been known for a long time. This epimerization at carbon atom 8 proceeds via an enolic symmetric intermediate. The conversion is favored in polar solvents. The position of the equilibrium depends on the pH of the solution and the nature of the substituent at the carboxyl radical of lysergic acid. The derivatives of d-lysergic acid are designated by the suffix -in and all derivatives of d-isolysergic acid end in -inine. Lysergic acid-derived alkaloids are pharmacologically highly active compounds and are levorotatory (in $CHCl_3$). Isolysergic acid derivatives are strongly dextrorotatory and exhibit only a weak pharmacological action.

The aci-conversion was detected by Schlientz et al. (1961a). The formation of aci-alkaloids occur in the peptide alkaloids and the corresponding dihydro derivatives. Isomerization takes places in acidic solutions, and the alkaloids obtained are designated by the prefix "aci." In the course of the aci-conversion the peptide moiety is affected, but the detailed mechanism is not yet known. Aci-alkaloids are pharmacologically inactive.

After intense irradiation with UV light of acid aqueous solutions of lysergic acid-derived alkaloids lumi-derivatives are formed. By this reaction one molecule of water is added at the double bond in the 9-10 position. At C-10 a new center of asymmetry is created, and two stereoisomer irradiation products can be isolated (Fig. 5).

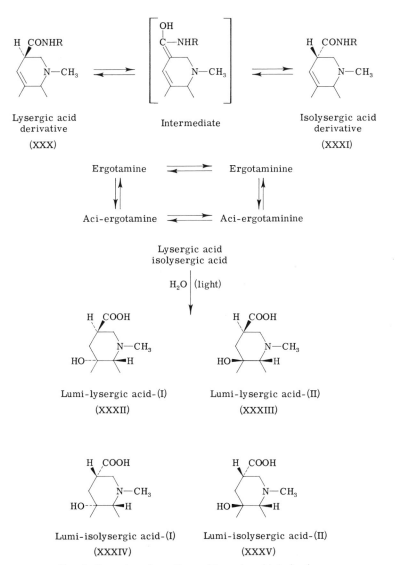

FIG. 5. Some transformations of lysergic acid derivatives.

3. SYNTHESIS OF ERGOLINE DERIVATIVES

The first synthesis of the tetracyclic ergoline ring system (Fig. 6) was achieved by Jacobs and Gould (1937). In this synthesis 3-nitro-α-naphthoic acid (**XXXVI**) was used as a starting material. After reduction of the nitro group, according to the chinolin synthesis of Skraup, ring D was

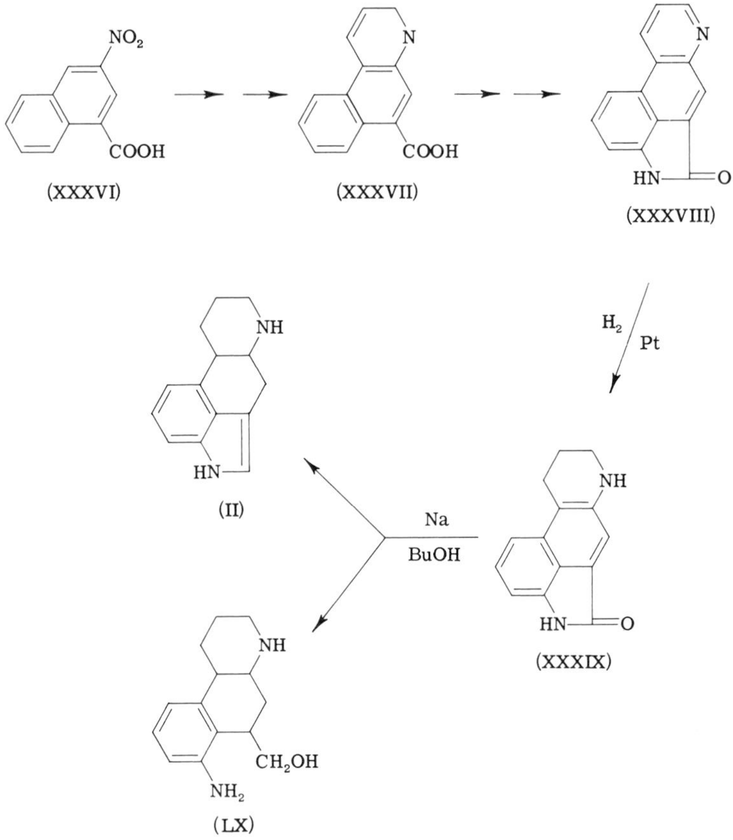

FIG. 6. Synthesis of ergoline.

added. In several steps a naphthostyril derivative was obtained that already contained rings A, B, C, and D (XXXVIII). Partial hydrogenation and reduction of XXXIX with sodium in butanol yielded ergoline (II) and an amino alcohol (XL). A naphthotyril derivative as starting material was used by Uhle and Jacobs (1945) for their synthesis of racemic dihydrolysergic acid. Stoll and Rutschmann (1950) synthesized the four isomeric racemic dihydro-nor-lysergic acids, while the optically active dihydrolysergic acids were first obtained by Stoll et al. (1950). The racemates were resolved into optical antipodes by using the corresponding L-norephedrides.

   The first synthesis of lysergic acid was announced by the group of Kornfeld and Woodward (Kornfeld et al., 1954). Earlier experiments of several authors using the naphthostyril- or benz- [c,d] indoline system as starting material failed, due to the inability to introduce the double bond in the 9,10-position of ring D. Kornfeld *et al.* (1954) started with an *N*-acyl-2,3-dihydroindole derivative and added rings C and D by classic methods. The dehydrogenation of the 2,3-dihydroindole system to the indole system was one of the latest steps in this particular synthesis. A new synthesis of lysergic acid has been described recently by Julia *et al.* (1969). First, in the 3-position of 5-bromisatine 6-methyl-nicotinic acid methylester was added to form an oxindole system. The crucial step was the formation of ring C between C-5 of an tetrahydro-nicotinic acid system and the 4-position of the indole part, which was achieved by aryne cyclization.

A partial synthesis of ergometrine and its stereoisomers, which can also be performed on a technical scale, was developed by Stoll and Hofmann (1943b,c). Racemic lysergic acid hydrazide was resolved with the aid of the di-(*p*-toluyl)-tartaric acids into two optical antipodes and converted to the corresponding azides. The azides were condensed with L-2-aminopropanol-(1) and D-2-aminopropanol-(1). By rearrangement of the resulting isolysergic acid derivatives into the lysergic acid derivatives all eight optically active condensation products were isolated. Two of them occur in nature.

For the total synthesis of the peptide portion of the peptide alkaloids, special methods for the formation of the labile α-hydroxy-α-amino acid grouping and the cyclol structure had to be developed. Procedures for the synthesis of α-hydroxy-α-amino acid derivatives and their incorporation into di- and tripeptides were first announced by the group of Shemyakin *et al.* (1959). The construction of a cyclol system was not achieved by these investigators. This problem was solved successfully by Hofmann *et al.* (1961, 1963). It was observed that cyclol formation occurred spontaneously if certain structural conditions are fulfilled. The α-hydroxy-α-amino acid grouping was added to the cyclol group (Fig. 7). (3S:9S) Phenylalanylproline lactam (XLII) was used as starting material for ergotamine synthesis. In this compound two of four centers of asymmetry of the peptide portion possess the correct configuration. The lactam was reacted with methylbenzyloxymalonic acid semiester acid chloride in pyridine. Acylated diketopiperazine was obtained and was treated with palladium–hydrogen to remove the benzyl group. The resulting compound with a free hydroxyl group cyclized spontaneously to the cyclol form. Since a racemic acid chloride had been used, two diastereomer products were obtained, which could be resolved by fractional crystallization. By means of a Curtius degradation in both isomers the carbethoxy group was converted to an amino group. The resulting amino cyclols could be crystallized in the form of its hydro-

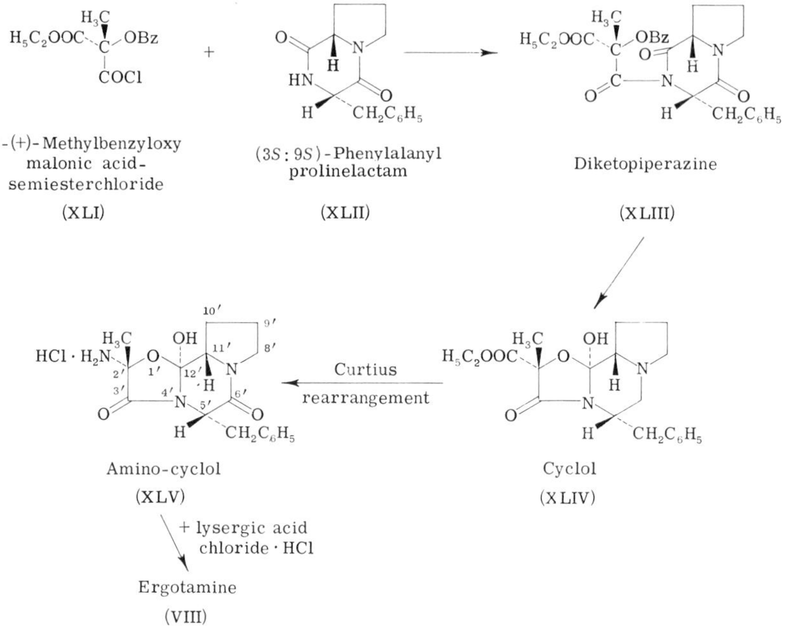

FIG. 7. Total synthesis of ergotamine.

chlorides. Both the amino cyclol compounds were reacted with d-lysergic acid chloride hydrochloride. With one of the peptide portion isomers an alkaloid was obtained that was identical in every respect to the naturally occurring ergotamine. The other amino cyclol gave a stereoisomer of ergotamine after condensation with the lysergic acid moiety. Apparently it differs from ergotamine only in the spatial arrangement at the C-2′ atom of the peptide portion. In a second improved procedure the acid chloride of the S-(+)-methylbenzyloxymalonic acid semiester (XLI) was used and in this way (XLIII → XLIV) only one cyclolization product could be obtained. During the formation of the cyclol system (XLIV) a new center of asymmetry at C-12′ is formed. It was observed that the ring closure takes places in a stereo-specific manner and the hydroxy group is located in the α-position. Other peptide alkaloids were synthesized by the same procedure (Stadler et al., 1969). By varying the amino acids used, alkaloids were obtained that had not yet been found in nature.

4. CLAVINE ALKALOIDS

The carboxyl group at the C-8 atom of the ergoline ring system, which is typical for lysergic acid derivatives, is reduced in the clavine

TABLE III
CLAVINE ALKALOIDS

Alkaloid	Formula	Reference
Agroclavine	$C_{16}H_{18}N_2$	Abe (1951a)
Elymoclavine	$C_{16}H_{18}O\ N_2$	Abe et al. (1952)
Molliclavine	$C_{16}H_{18}O_2N_2$	Abe and Yamatodani (1955)
Penniclavine	$C_{16}H_{18}O_2N_2$	Stoll et al. (1954)
Festuclavine	$C_{16}H_{20}N_2$	Abe and Yamatodani (1954)
Pyroclavine	$C_{16}H_{20}N_2$	Abe et al. (1956)
Costaclavine	$C_{16}H_{20}N_2$	Abe et al. (1956)
Setoclavine	$C_{16}H_{18}O\ N_2$	Hofmann et al. (1957)
Isosetoclavine	$C_{16}H_{18}O\ N_2$	Hofmann et al. (1957)
Isopenniclavine	$C_{16}H_{18}O_2N_2$	Hofmann et al. (1957)
Chanoclavine-I	$C_{16}H_{20}O\ N_2$	Hofmann et al. (1957)
(Secaclavine)		Abe et al. (1956)
Lysergine	$C_{16}H_{18}N_2$	Abe et al. (1961)
Lysergene	$C_{16}H_{16}N_2$	Abe et al. (1961)
Lysergol	$C_{16}H_{18}O\ N_2$	Abe et al. (1961); Hofmann and Tscherter (1960)
Fumigaclavine A	$C_{18}H_{22}O_2N_2$	Spilsbury and Wilkinson (1961)
Fumigaclavine B	$C_{16}H_{20}O\ N_2$	Spilsbury and Wilkinson (1961)
Isochanoclavine-I	$C_{16}H_{20}O\ N_2$	Stauffacher and Tscherter (1964)
Chanoclavine-II	$C_{16}H_{20}O\ N_2$	Stauffacher and Tscherter (1964)
Dihydrolysergol-(I)	$C_{16}H_{20}O\ N_2$	Agurell and Ramstad (1965)
Isolysergol	$C_{16}H_{18}O\ N_2$	Agurell (1966a)
Norsetoclavine	$C_{15}H_{16}O\ N_2$	Ramstad et al. (1967)
Elymoclavine-O-β-D-fructoside	$C_{22}H_{28}N_2O_6$	Floss et al. (1967c)
Cycloclavine	$C_{16}H_{16}N_2$	Stauffacher et al. (1970)

series to a hydroxymethyl or methyl group. The double bond in ring D is located in the 8,9-, or 9,10-position. In some alkaloids ring D is saturated. They correspond to the dihydro derivatives of lysergic acid not found in nature. Rings C and D are in general trans linked (Fig. 4). The structural and configurative relationships of clavine alkaloids have been elucidated by conversion within the clavine series as well as by their connection with derivatives of lysergic and dihydrolysergic acid of known configuration. The chemistry of clavine alkaloids was comprehensively described by Hofmann (1964). By catalytic hydrogenation of agroclavine (XIX) the diastereomer alkaloids festuclavine (XXIII) and pyroclavine were obtained (Yamatodani and Abe, 1956; Schreier, 1958). These stereoisomeric alkaloids were also obtained from the lactam of d-dihydrolysergic acid-I. In this way agroclavine (XIX) was unequivocally connected with the d-dihydrolysergic acid series. Furthermore heating of agroclavine in a sodium butylate solution yielded only lysergine. Oxidation of the same clavine with potassium dichromate in dilute sulfuric acid produced setoclavine (XXI) and isosetoclavine (XLVII) (Fig. 8) (Hofmann *et al.*, 1957). Catalytic hydrogenation of elymoclavine (XX) gave a mixture of *d*-dihydrolysergol-I and *d*-dihydroisolysergol-I (XLVIII) (Yamatodani and Abe, 1955). From this result the trans linkage of rings C and D in XX may be deduced, and it is evident that the spatial configuration at C-5 is the same as in lysergic acid. Oxidation of elymoclavine (XX) yielded penniclavine (XXII) and isopenniclavine (Yamatodani and Abe, 1955; Hofmann *et al.*, 1957). A mixture of agro-clavine, lysergine, pyroclavine, festuclavine (XXIII), costaclavine (XXIV), and the isomeric *d*-dihydrolysergols-I was obtained by the reduction of elymoclavine (XX) with sodium in butanol (Yamatodani and Abe, 1956, 1957).

The clavine alkaloids, found in *Aspergillus*, fumigaclavine A (XXVI) and fumigaclavine B are easily transformed by acetylation and de-

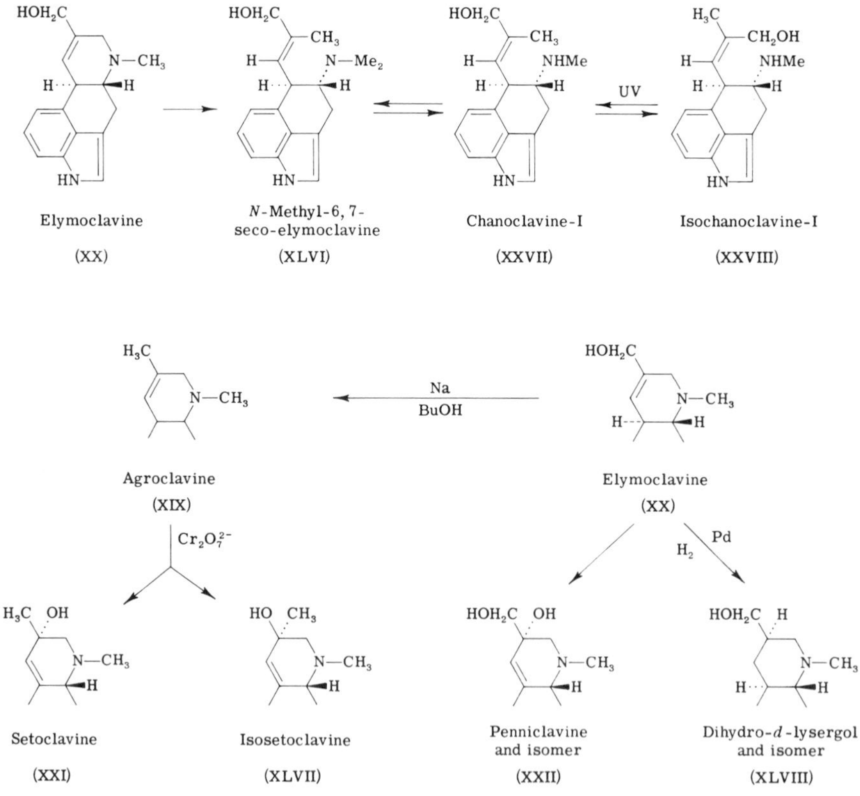

FIG. 8. Some chemical transformations in the clavine series.

acetylation into each other. Heating of fumigaclavine B with NaOH yielded lysergine (Spilsbury and Wilkinson, 1961).

The stereoisomeric chanoclavines occupy an exceptional position among the ergoline derivatives (Fig. 4). In these clavines ring D is open. The structural elucidation of chanoclavine-I (XXVII) was described by Hofmann *et al.* (1957). The position of the double bond in chanoclavine-I (XXVII), which is not in conjugation with the indole system, was determined by its UV spectrum. Chanoclavine-I easily formed an *O,N*-diacetyl derivative and gave, after catalytic hydrogenation, a small amount of festuclavine. Since festuclavine is correlated with *d*-dihydrolysergic acid, the spatial configurations at C-5 and C-10 in chanoclavine can be determined. Some isomers of chanoclavine-I have been isolated by Stauffacher and Tscherter (1964). By comparison of the NMR spectra of the single alkaloids as well as their acetyl derivatives the configuration of the various isomers could then be deduced. Further evidence for the stereochemistry of chanoclavine-I and iso-chanoclavine-I (XXVIII) by chemical means was given by Acklin *et al.* (1966). Upon UV irradiation both alkaloids were transformed, one into the other, and chanoclavine-I could be correlated with elymoclavine (XX).

5. ERGOLINE DERIVATIVES OUTSIDE THE GENUS *Claviceps*

A surprising phytochemical finding was the detection of ergoline derivatives in higher plants, in the genera of the family of twining plants (Convolvulaceae). From the ritual drug *ololiuqui*, used by the Aztec and other Mexican Indians, Hofmann and Tscherter (1960) were able to isolate ergine (XV), isoergine, and chanoclavine (XXVII) in crystalline form. Later the isolation of elymoclavine (XX), lysergol (XXV), and ergometrine (XVII) was achieved (Hofmann, 1961, 1963). Seeds of various Convolvulaceae species are designated *ololiuqui*. These include *Rivea corymbosa* (L.) Hall.f. and *Ipomoea violacea* L. (*badoh negro*). Other authors have screened many species of the genera *Argyreia, Ipomoea,* and *Rivea* and found in many cases traces of ergoline derivatives. Of special interest is the isolation of the peptide alkaloids ergosine (XI) and ergosinine from *Ipomoea argyrophylla* Vatke (Stauffacher *et al.*, 1965). Recently, cycloclavine, the first member of a new type of clavine, was detected in *Ipomoea hildebrandtii* Vatke (Stauffacher *et al.*, 1970). This subject has been reviewed by Der Marderosian (1967).

Spilsbury and Wilkinson (1961) found festuclavine (XXIII), and fumigaclavine A and B in *Aspergillus fumigatus* Fres. Fumigaclavine was also discovered in cultures of *Rhizopus nigricans*. Yamano *et al.*

(1962) confirmed these results and have also found, in different strains of *Aspergillus fumigatus,* fumigaclavine C, agroclavine, chanoclavine, and elymoclavine. By paper chromatography, various clavine alkaloids were identified in some species of *Aspergillus* and *Penicillium* (Abe *et al.,* 1967). Costaclavine (XXIV) was isolated from *Penicillium chermesinum* Biourge (Agurell, 1964). *Penicillium concavo-rugulosum* was investigated by Abe *et al.* (1969). In addition to chanoclavine-I, two new alkaloids rugulovasine A and rugulovasine B were found. The structures of the "rugulovasines" have not yet been fully elucidated. Apparently in these ergolines ring D is open, and they contain an unsaturated lactone grouping. Lysergic acid derivatives have hitherto not been found in fungi outside the genus *Claviceps.* On the other hand some clavine alkaloids are not formed in strains of *Claviceps.*

6. BIOSYNTHESIS OF ERGOT ALKALOIDS

In most of the indole alkaloids the indole ring system is substituted in positions 2 and 3. To the few exceptions belong the ergoline alkaloids in which the 3 and 4 position of the indole ring is connected with other rings. By the use of isotope technique and saprophytic cultures the building blocks of the ergoline skeleton were elucidated, and a first insight into the mechanism of the formation of the tetracyclic ring system was gained (Fig. 9). The interrelations of ergot alkaloids have been intensively investigated. Current activities in different laboratories are concerned with the enzymology and the *in vivo* formation of the cyclic peptides of the ergotamine and ergotoxine groups. Some reviews have been published on this topic (Agurell, 1966b; Ramstad, 1968; Voigt, 1968; Gröger, 1969; Floss, 1969). The present status of research is as follows. The ergoline ring system in ergot fungi and higher plants is constructed from tryptophan (XLIX) and mevalonic acid (L). The *N*-methyl group of the ergot alkaloids is derived from methionine via a transmethylation reaction (Baxter *et al.,* 1961). L-Tryptophan is the immediate precursor of the ergoline nucleus. In the course of incorporation of tryptophan into ergoline at the α-carbon atom of the side chain of the amino acid an inversion of configuration takes place. The α-hydrogen at this particular carbon atom is retained during this process (Floss *et al.,* 1964). For the first time in alkaloid biochemistry it has been shown that, in ergoline biosynthesis, an isoprene unit can participate in the formation of alkaloids (Gröger *et al.,* 1960; Birch *et al.,* 1960). The connection of the isoprene unit (isopentenylpyrophosphate$\rightleftarrows$ dimethylallylpyrophosphate) with tryptophan occurs at the 4-position of the indole nucleus by way of 4-dimethylallyltryptophan (LI). The above-mentioned compounds are incorporated into clavine alkaloids (Plieninger

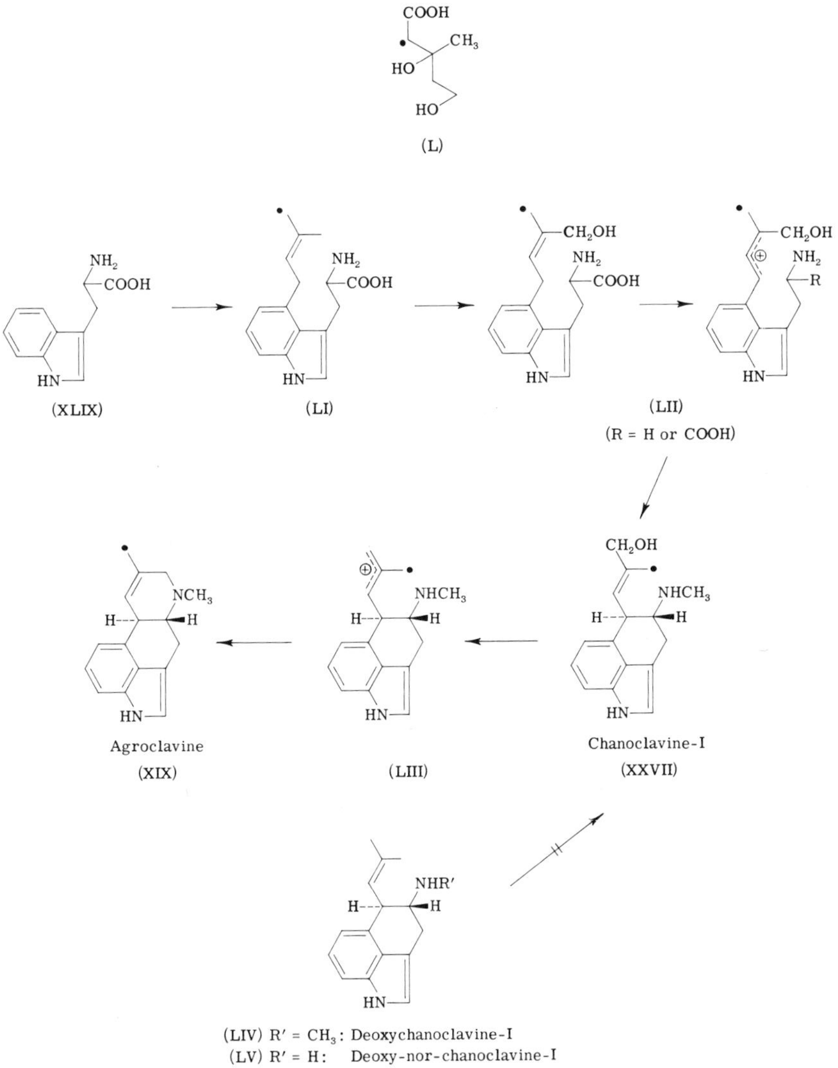

• Position of label after feeding mevalonate-2-^{14}C.

FIG. 9. Biosynthetic pathway of the ergoline ring system.

et al., 1959, 1967; Plieninger, 1966). Recently the natural occurrence of 4-dimethylallyltryptophan in saprophytic cultures of *Claviceps* was shown (Robbers and Floss, 1969a; Agurell and Lindgren, 1968).

The mechanism of formation of agroclavine, the first alkaloid with a complete tetracyclic ring system in a biosynthetic pathway, has been repeatedly investigated (Arigoni, 1965; Fehr *et al.*, 1966; Floss, 1967a; Floss *et al.*, 1967a, 1968; Voigt *et al.*, 1967). Deoxychanoclavine-I (LIV) and deoxy-norchanoclavine-I (LV) are not incorporated into other ergolines (Arigoni, 1965). Therefore one of the methyl groups of the dimethylallyl residue must be hydroxylated before ring C is closed. From experiments with (4*R*)-mevalonic-2-^{14}C-4-T it follows that hydroxylation occurs at the *cis* methyl group of the dimethylallyl residue that is not derived from the C-2 of mevalonic acid. After administration of mevalonic-2-^{14}C the radioactivity is always found in the methyl group of chanoclavine-I and other chanoclavine isomers. Consequently during the formation of ring C in chanoclavine-I a cis–trans isomerization at the exocyclic double bond must have taken place in which the hydroxymethyl group occurs in the trans position. Experiments with different chanoclavine isomers have revealed that only chanoclavine-I is incorporated with high efficiency into tetracyclic ergolines, e.g., agroclavine and elymoclavine. It is well documented that, after administration of mevalonic acid-2-^{14}C or chanoclavine-I-7-^{14}C, agroclavine and elymoclavine are labeled at C-17. Therefore during the cyclization of chanoclavine-I to agroclavine another cis–trans isomerization of the double bond occurs. Thus the apparently "normal" labeling of tetracyclic ergot alkaloids from mevalonic acid-2-^{14}C in the trans carbon of the isoprenoid moiety is an accident caused by two isomerizations. The cyclization of chanoclavine-I proceeds with 100% retention of the hydrogen at C-10 but with only 70% retention of the hydrogen at C-9.

Conversions within the clavine type alkaloids and their relations to the lysergic acid derivatives have been intensively investigated (Fig. 10) (Agurell and Ramstad, 1962b; Ramstad, 1968; Abe, 1966, 1969). Agroclavine is oxidized to elymoclavine which is the progenitor of lysergic acid (Mothes *et al.*, 1962). In various clavine alkaloids different reactions may occur, e.g., isomerization, hydroxylation, or hydrogenation. The formation of the diastereoisomeric setoclavines and penniclavines starting from agroclavine or elymoclavine seems to be well understood. These hydroxylations are carried out not only by ergot fungi but also by a variety of other fungi and even by plant homogenates. The reaction is catalyzed by peroxidase. Initially C-10 is attacked to form 10-hydroxy-agroclavine or 10-hydroxyelymoclavine; under mild conditions a rearrangement takes place to yield the corresponding 8-hydroxy-alkaloids (Ramstad, 1968). The oxidation of the methyl group at C-8 of agroclavine to elymoclavine and the further conversion of the C-17 hydroxymethyl group to lysergic acid has only been observed in members of the genus *Claviceps*. The C-methyl oxidation of agroclavine is not catalyzed by peroxidase (Jindra *et al.*, 1968). The hydroxyl oxygen of chano-

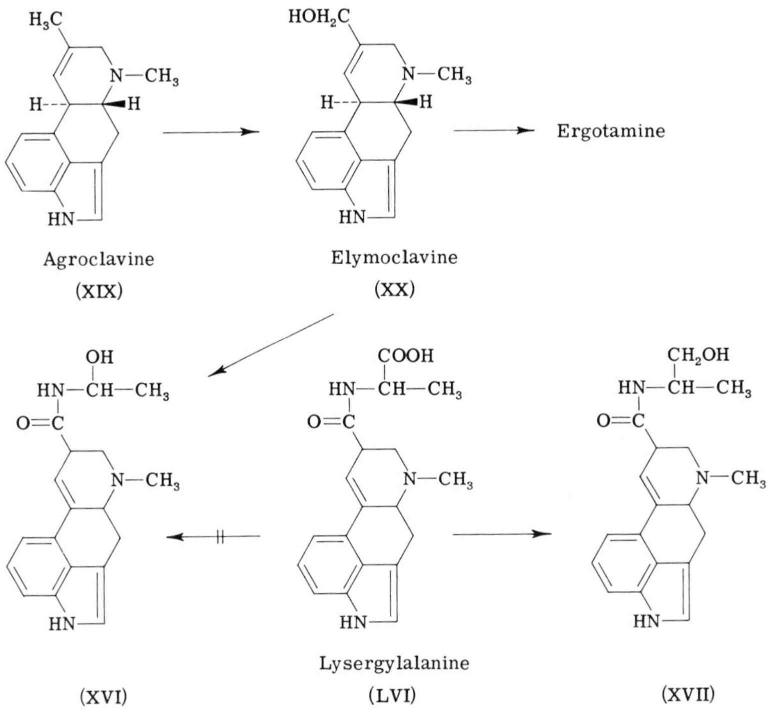

H₃C

N—CH₃

H-- ◄H

HN

Agroclavine

(XIX)

HOH₂C

N—CH₃

H-- ◄H

HN

Elymoclavine

(XX)

⟶ Ergotamine

OH
|
HN—CH—CH₃
|
O=C
|

N—CH₃

HN

(XVI)

COOH
|
HN—CH—CH₃
|
O=C
|

N—CH₃

HN

Lysergylalanine

(LVI)

CH₂OH
|
HN—CH—CH₃
|
O=C
|

N—CH₃

HN

(XVII)

FIG. 10. Biosynthetic relationships among ergot alkaloids.

clavine-I and of elymoclavine is derived from molecular oxygen and not from water (Floss *et al.*, 1967b). The conversion of ergoline derivatives by using cellfree preparations of different *Claviceps* strains have been recently described (Abe, 1969). After incubation of *d,l*-tryptophan-3-^{14}C with a cellfree system of a particular *Claviceps* strain the formation of labeled setoclavine, isosetoclavine, and, to a lesser degree, lysergene was detected. According to Abe, lysergene and not agroclavine would be the primary alkaloid with a tetracyclic ring system in ergot alkaloid biosynthesis.

Lysergic acid is a precursor for lysergic acid amide and lysergic acid methylcarbinolamide (Agurell, 1966b) as well as ergometrine (Minghetti and Arcamone, 1969). In no instance was lysergic acid amide found to be converted to the corresponding α-hydroxyethylamide (Agurell, 1966b). It has been established that the side chain of the lysergic acid methyl-carbinolamide arises from pyruvate-^{14}C or alanine-^{14}C. Alanine-^{15}N serves also as donor of the amide nitrogen (Castagnoli and Tonolo, 1967; Gröger *et al.*, 1968a). Alanine is also specifically incorporated into ergometrine but not alaninol or α-methylserine (Nelson and Agurell, 1969). Lysergylalanine (LVI) has been found to be an intermediate in ergometrine biosynthesis (Basmadjian et al., 1969). The formation of the α-hydroxy-α-amino acid grouping in the peptide alkaloids is obscure. Phenylalanine is incorporated into the peptide portion of ergotamine and ergocristine (Vining and Taber, 1963a) and proline into the peptide moiety of ergocornine and ergokryptine (Gröger and Erge, 1970). Abe (1969) prepared a cellfree system capable of carrying out the incorporation of labeled prolyldiketopiperazines into peptide-type alkaloids.

7. CHEMICAL ASSAY OF ERGOT

Many of the ergot alkaloids and their semisynthetic derivatives possess properties of therapeutic or toxicological value. Therefore numerous methods have been described for the qualitative and quantitative estimation of these compounds. The medically used lysergic acid derivatives are sensitive substances easily decomposed by oxidation and exposure to light. Furthermore they have a tendency to form isomers. In the conversion to isolysergic acid derivatives the pharmacological effect is practically lost. These tendencies must be considered in the chemical assay of the active principles in crude drugs and pharmaceutical preparations.

Ergot alkaloids give characteristic color reactions with sulfuric acid, Keller reagent, and van Urk's reagent (p-dimethylaminobenzaldehyde in sulfuric acid). The van Urk color reaction (van Urk, 1929) was improved by M. J. Smith (1930) and later standardized for quantitative measurements by the addition of ferric chloride (Allport and Cocking, 1932). Lysergic acid and isolysergic acid derivatives and most of the clavines give an intense blue color. The color test depends upon the indole nucleus of the ergolines. The mechanism of this reaction has been studied by Pöhm (1953) and Dibbern and Rochelmeyer (1963). Most of the quantitative methods are based on the van Urk–Smith reaction. Ergolines are also quantitatively estimated by titration, ultraviolet absorbance, and the extremely sensitive spectrophotofluorometric methods (Foster, 1955; Gyenes and Bayer, 1961; Hofmann, 1964). The IR spectra of numerous ergolines have been recorded (Hofmann, 1964), and Mrtek et al. (1968) now published the NMR data of various clavine alkaloids. The estimation of the total alkaloid content seems an insufficient parameter for the evaluation of drug samples or pharmaceuticals. It is often necessary to estimate the different components in a complex mixture selectively. An initial separation may be obtained by fractionation into water-soluble and water-insoluble alkaloids. Individual alkaloids or closely related groups of alkaloids may be best separated with the aid of column, paper, or thin-layer chromatography. On aluminum oxide columns it is possible to elute separately with chloroform/methanol mixtures (a) the dextrorotatory alkaloids of the ergotamine and ergotoxine group as well as the ergotoxine complex, (b) ergotamine and ergosine, and (c) ergometrine (Hofmann, 1964).

A paper chromatographic method for the separation of ergolines was first used by Foster *et al.* (1949). Using unbuffered paper and a mixture of butanol:water:acetic acid (4:1:5) as a solvent system, he achieved the separation of ergometrine and ergometrinine. Later, numerous procedures using paper chromatography for the assay of ergot alkaloids were described. The different alkaloids of the pharmaceutically important ergotamine and ergotoxine group and the corresponding dihydroalkaloids were successfully separated by the use of phthalic acid dimethylester-impregnated Whatman No. 1 paper and acidified formamide: water mixtures as the mobile phase (Stoll and Rüegger, 1954). This system can also be used for the estimation of the aci-alkaloids (Schlientz *et al.,* 1961b). Similarly good results were obtained by employing formamide-impregnated paper (Pöhm and Fuchs, 1954; Macek and Vaněček, 1955). When the mobile phase was varied, the separation of peptide alkaloids, dihydrogenated alkaloids, and clavine type alkaloids was achieved. The paper chromatographic separation of the water-soluble alkaloids and the clavines was thoroughly investigated by Voigt (1959), Yamatodani (1960), and Agurell and Ramstad (1962a).

Thin-layer chromatographic procedures for the resolution and quantitative determination of ergot alkaloids have also been employed. The literature has been reviewed by Agurell (1965) and Santavý (1967). Silica gel G, alumina, and formamide-impregnated cellulose powder are used as adsorbents for the thin-layer technique. The pioneering work of Rochelmeyer's group should be mentioned here (Rochelmeyer, 1958; Teichert *et al.,* 1960a,b; Klavehn and Rochelmeyer, 1961; Klavehn *et al.,* 1961). Mixtures of chloroform:ethanol (95:5 and 90:10), are useful as solvent systems for the separation on silica gel G for ergometrine as well as alkaloids of the ergotamine and ergotoxine group and a series of clavine alkaloids. A mobile phase of ethylacetate:ethanol:dimethylformamide (85:10:5) is particularly effective for the separation of various clavine alkaloids as well as ergometrine/ergometrinine on silica gel G. The chromatographic behavior on thin-layer chromatograms of peptide alkaloids and ergometrine and the corresponding dextrorotatory isomers has been published in detail (McLaughlin *et al.,* 1964). Agurell (1965) has discussed the correlation between the structure of clavines and their R_M-values. The detection of ergolines can be achieved by fluorescence in UV light (lysergic acid derivatives and various clavines) or by spraying the chromatograms with an acidified *p*-dimethylaminobenzaldehyde solution.

Methods for the quantitative measurements of single alkaloids after chromatographic separation have been repeatedly described. A direct densitometric method on thin-layer plates for the determination of ergine, isoergine, and some clavines was used by Genest (1965). A spectrophotofluorometric procedure for the assay of LSD in the presence of heroin, other narcotics, and certain restricted drugs has been developed by Genest and Farmilo (1964). In most cases the alkaloids are colorimetrically estimated with van Urk's reagent after extraction from paper or thin-layer adsorbents (Klavehn *et al.*, 1961; McLaughlin *et al.*, 1964; Hofmann, 1965; Röder *et al.*, 1967).

C. ERGOT PIGMENTS

The sclerotia of various *Claviceps* species, and especially *C. purpurea*, contain a number of yellow and red-violet pigments. Crystalline pigments were first isolated about 90 years ago by Dragendorff and Podwyssotzky (1877). Later many authors were concerned with the isolation, purification, and structure elucidation of colored material obtained from ergot. Important progress has been made mainly in the last decade. The difficulties in chemical investigation may be explained by the fact that the pigments often possess very similar structures or occur as diastereomers and that, in many cases, the classic separation methods fail. According to the excellent work of the groups of deMayo (Canada), Whalley and Robertson (England), and, especially, Franck (Germany) the results may be summarized as follows (Franck, 1965, 1969). The pigments are divided according to their color and acidity into two groups: (1) orange-red pigments, strongly acid ($pK_a = 4.2$), anthraquinone carboxylic acids and (2) light-yellow pigments, weakly acid ($pK_a = 8.5$-9.4) ergochromes (Fig. 11).

To the first group belong the hydroxyanthraquinone carboxylic acids, endocrocin (LXVII), and clavorubin (LXVIII) (Franck and Reschke, 1959, 1960; Franck and Zimmer, 1965). Clavorubin differs only from endocrocin, which is also synthesized by other microorganisms, by an additional hydroxy group in the 5-position. The majority of ergot pigments are of weakly acid, pale yellow substances. All the earlier described pigment preparations were mostly mixtures and were this type. According to a proposal by Franck *et al.* (1965) the pigments of the second group are designated ergochromes. Ergochromes are dimerization products of four monomeric xanthones, whose relative and absolute configuration have been elucidated. By using two letters and two numbers it

(A)

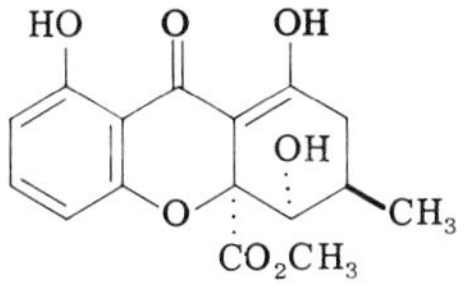

(LVII) A – A = Ergochrome AA [4, 4']
(secalonic acid A)

(LVIII) B – B = Ergochrome BB [4, 4']
(secalonic acid B)

(B)

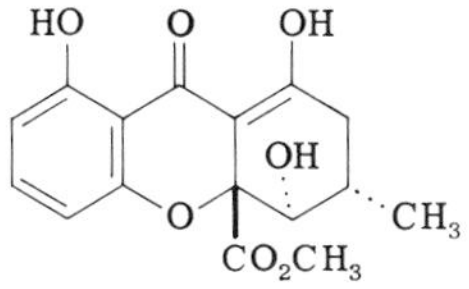

(LIX) C – C = Ergochrome CC [2, 2']
(ergoflavin)

(LX) A – B = Ergochrome AB [4, 4']
(secalonic acid)

(C)

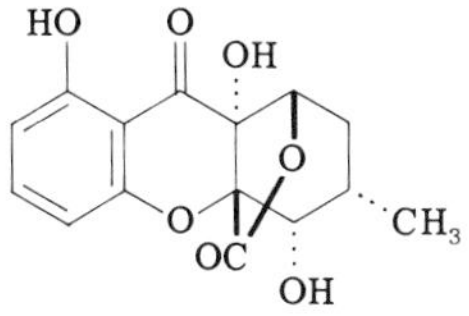

(LXI) A – C = Ergochrome AC [2, 2']
(ergochrysin A)

(LXII) B – C = Ergochrome BC [2, 2']
(ergochrysin B)

(D)

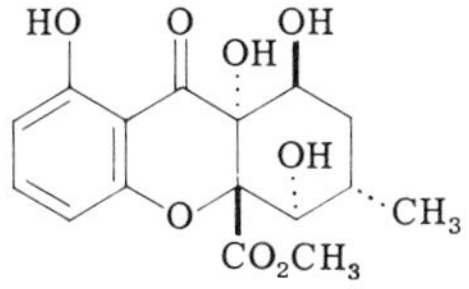

(LXIII) A – D = Ergochrome AD [2, 2']

(LXIV) B – D = Ergochrome BD [2, 2']

(LXV) C – D = Ergochrome CD [2, 2']

(LXVI) D – D = Ergochrome DD [2, 2']

Ergochromes

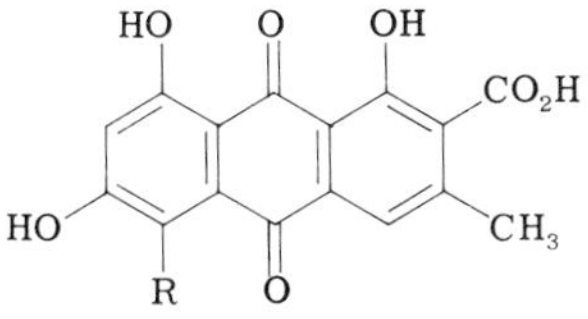

(LXVII) R = H : Endocrocin

(LXVIII) R = OH: Clavorubin

FIG. 11. Pigments of ergot.

is possible to characterize the dimers and the type of linkage. For some pigments, trivial names have been retained.

Ergoflavin (LIX) was crystallized in pure form by Freeborn (1912). By a combination of classic chemical degradation and X-ray analysis the complete structure was determined (Apsimon *et al.*, 1963; McPhail *et al.*, 1963). Ergoflavin [ergochrome CC (2,2')] contains, as opposed to the secalonic acids, no carbomethoxy groups, no enol groupings, but two γ-lactone groups. Like ergoflavin, secalonic acid A and secalonic acid B are symmetric ergochromes (ergochrome AA, ergochrome BB).

Secalonic acid C, ergochrysin A, and ergochrysin B proved to be un-
symmetric dimerization products of the xanthone moieties A, B, and C.
In all studies (e.g., IV, IR, NMR, and mass spectra) they behaved like a
mixture of two of the above-mentioned symmetric pigments. Therefore
they have the structure ergochromes AB, AC, and BC. The remaining
ergochromes AD, BD, CD, and DD are combinations of a further xanthone derivative D with the already known monomeric parts A, B, and
C. The structures could be confirmed by conversion into known ergochromes and by partial synthesis. Ergochrome AD, for example, can be
converted, by heating in acetic acid, into ergochrysin A (ergochrome
AC). The mass spectrum revealed that the conversion was accompanied
by the loss of 32 mass units corresponding to the cleavage of methanol.
Furthermore in the IR spectra the intensity of the absorption band of
the ester carbonyl was reduced by half, and the γ-lactone band of the
ergochrysins appeared. So it was deduced that ergochrome AD differs
from ergochrysin A by methanolysis of the γ-lactone grouping, and the
monomer possesses the structure of xanthone D. According to the
NMR spectra the two monomer units can be coupled at the phenyl
residue in the o- or p-position to the phenolic hydroxyl. By further comparison with test substances it was concluded that all ergochromes that
possess a C- or D-unit are o,o'- (2,2'-)linked, but the secalonic acids
A, B, and C must have a p,p'-(4,4'-)linkage. Ergochromes were isolated
from ergot of a different origin. The amount of the individual pigments
in the mixture varies. Ergochromes and anthrachinones are also formed
by some strains in still culture under saprophytic conditions. With the
aid of saprophytic cultures the biosynthetic pathway of this new class
of natural products was elucidated (Franck et al., 1966, 1968; Gröger
et al., 1968b).

The common occurrence of anthraquinones and xanthone pigments
as well as the results of model experiments led to speculations on a new
type of biogenetic correlation. Both tricyclic ring systems consist of
15-16 C-atoms with a similar arrangement of the substituents. Endocrocin
is built up from acetate units via acetyl-CoA or malonyl-CoA (Gatenbeck, 1958). Ergochromes may be derived by anthrachinones according
to the scheme in Fig. 12: by oxidative fission of anthraquinones (LXIX),
benzophenone-carboxylic acids (LXX) are formed, which, in turn, are
converted by ring closure to xanthone derivatives (LXXI). Ergochromes
may be considered to be seco-anthraquinones. This hypothesis was
substantiated by feeding experiments using labeled acetate and emodin.
Acetic acid is incorporated via the hypothetical heptaketopalmitic acid
(LXXIII) and endocrocin or a related anthraquinone into ergochrome,
so that after administration of acetate-2-[14]C a labeling pattern should

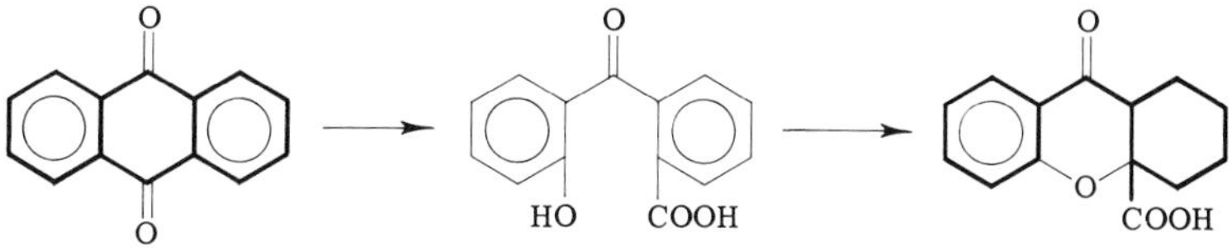

(LXIX) (LXX) (LXXI)

FIG. 12. A possible biogenetic pathway of the ergochromes.

result as illustrated in Fig. 13. By a specific degradation the expected distribution of radioactivity in the pigments was observed. Of special interest was the result that the carboxyl-C of ergochromes is derived from a Me-group of acetic acid. Further confirmation came from experiments with emodin-T and emodin-U-^{14}C. Emodin-U-^{14}C as well as ergochrome BB, which was isolated after the feeding of this anthraquinone, was degraded. An analogous labeling pattern was observed in the ergochrome and fed emodin so a direct conversion can be assumed. Model experiments revealed that the oxidative ring fission occurs not in the anthraquinone itself but rather in the anthrone and its hydroperoxide.

8 $\overset{*}{C}H_3COOH$

(LXXII)

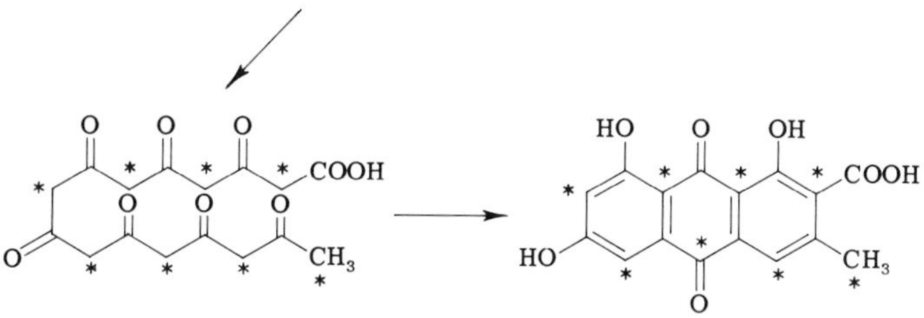

(LXXIII) (LXVII)

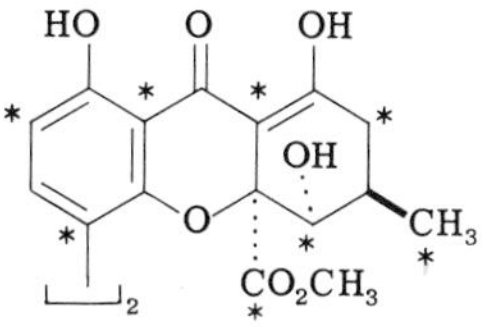

(LVII)

FIG. 13. Incorporation of acetate-2-^{14}C into ergochrome AA.

D. METABOLITES OF VARIOUS STRUCTURES

In the course of the numerous chemical investigations of ergot in the last few years a variety of compounds were isolated from the genus *Claviceps* for the first time. Arcamone *et al.* (1961b) detected 2,3-dihydroxybenzoic acid [DOB (LXXIV)] in the culture broth of lysergic acid methylcarbinolamide-producing strains of *C. paspali*. Apparently DOB is only synthesized by this particular *Claviceps* species and might be a chemotaxonomic marker in the genus *Claviceps*. In contrast to higher plants the formation of DOB by *C. paspali* involves degradation of tryptophan by the kynurenine pathway. Anthranilic acid can be regarded as a direct precursor of LXXIV (Tyler *et al.*, 1964; Gröger *et al.*, 1965). A monooxygenase reaction seems to be the route by which 2,3-dihydroxybenzoic acid is formed from anthranilic acid (Floss *et al.*, 1969).

The nucleotides of *C. paspali* were investigated by Ballio *et al.* (1966). In this study guanine propionate was identified. This purine was hitherto found only as a riboside in *Fusarium* sp. and was detected after incubation of extracts of *Eremothecium ashbyi* with glucose and guanine. On its function in the metabolism of fungi, nothing is known.

The composition of the lipids from different *Claviceps* species was recently investigated using thin-layer, column, and gas chromatography

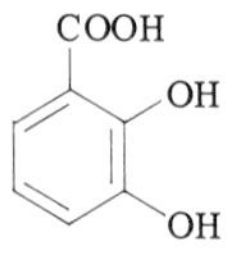

2, 3-Dihydroxy-
benzoic acid
(LXXIV)

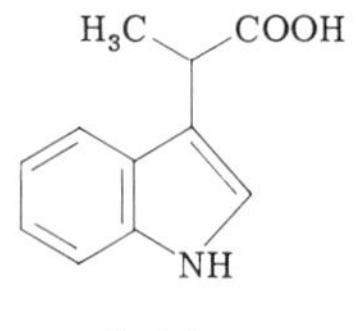

Indole-
isopropionic acid
(LXXV)

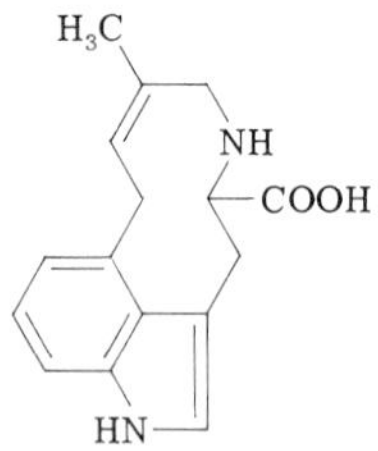

Clavicipitic acid
(LXXVI)

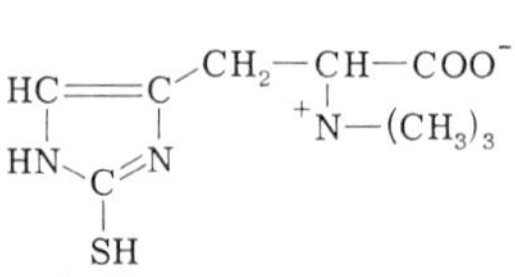

Ergothioneine
(LXXVII)

FIG. 14. Various metabolites of ergot.

(Thiele, 1964; Morris, 1967). 12-Hydroxystearic acid and (+)-threo-9,10-dihydroxystearic acid were the unusual fatty acids found. Ricinoleic acid is biosynthesized in *C. purpurea* by hydration of linoleic acid and not by oxidative hydroxylation of oleic acid (Morris *et al.*, 1966). Unlike the ricinoleic acid found in castor oil, the hydroxyl groups of this fatty acid in ergot are acylated with various long-chain fatty acids. These compounds are called estolides. In addition to the amino acid metabolism the occurrence and formation of amines in sclerotia (Stein von Kamienski, 1958; Steiner and Hartmann, 1958; Kordts and Voigt, 1961) and saprophytic cultures (Hartmann, 1965) were thoroughly investigated. The sclerotia of *C. purpurea* are rich in amines and seem to contain amines of greater diversity than most plants. The following amines have been detected: methyl-, ethyl-, trimethyl-, n- and i-propyl, i-butyl-, i-amyl-, n-hexyl-, and β-phenylethylamine. Alkaloid-free and alkaloid-yielding ergot contain the same amines. A correlation between alkaloid and amine content was not observed. Saprophytic cultures mainly form i-amylamine and n-hexylamine, and their concentration steadily increases during the growth phase. Under conditions of autolysis no amine formation was detectable. i-Amylamine and n-hexylamine originate by decarboxylation of L-leucine and D,L-α-aminoheptanoic acid, respectively. The last-mentioned amino acid belongs to the group of the few known aliphatic C_7 amino acids and was found in nature for the first time in saprophytic cultures of *Claviceps* (Steiner and Hartmann, 1964).

Two indole derivatives of a new type, paspalin and paspalicin, were isolated from the mycelium of *C. paspali* by Fehr and Acklin (1966). Paspalin contains an indole ring system and a C_{20}-unit. Apparently this moiety is an alicyclic terpenoid system, because mevalonic acid is incorporated here. The final structure remains to be elucidated. Robbers and Floss (1969b) obtained a 4-substituted indolic amino acid from a clavines-synthesizing *Claviceps* strain. From NMR data and feeding experiments with DL-tryptophan-(alanine-2-^{3}H-carboxyl-^{14}C) and mevalonic acid-(2-^{14}C-5-^{3}H) the unusual structure of clavicipitic acid (LXXVI) with a ten-membered ring system was deduced. It seemed worthwhile to see if, by decarboxylation and ring closure, the tetracyclic ring system could be formed. Yamano (1961) isolated *R*-indoleisopropionic acid (LXXV) from sclerotia parasitizing on *Festuca arundinacea*. Only a few strains of *Claviceps* can produce this indolic acid under saprophytic conditions. Biosynthetically LXXV is derived from L-tryptophan and the intact methyl group of L-methionine. Anthranilic acid and indole are also good precursors of indoleisopropionic acid and are even more efficiently incorporated than either tryptophan or methionine. Apparently problems of compartmentation of the amino acids or of different rates of uptake are involved here. Indoleacetic acid is not incorporated into LXXV. Side-chain methylation of an aromatic amino acid or a derivative thereof has been observed in this compound for the first time (Hornemann et al., 1969).

Ergothioneine (LXXVII), which is known to be present in microorganisms and animal red blood cells, was first isolated from ergot (Tanret, 1909). The biosynthesis of LXXVII in ergot was studied by Wildy and Heath (1957) and Heath and Wildy (1958). The administration of ^{35}S-methionine yielded ^{35}S-ergothioneine. Histidine served as a direct precursor for the betaine of mercaptohistidine. In *Neurospora* L-2-thiolhistidine is not incorporated into ergothioneine but hercynine is an effective precursor of LXXVII. Apparently the methylation of histidine precedes the attachment of sulfur to the imidazole ring. The same biogenetic pathway can also be assumed for ergot.

Two groups have reported on the isolation and formation of glucans in ergot: Perlin and Taber (1963) and Buck et al. (1968). The glucans are branched polysaccharides, whose main chain consists of 1-3 linked β-D-glucopyranosyl units.

V. Production of Ergot Alkaloids

A. GENERAL INFORMATION

In principle there are various possibilities for the production of ergot alkaloids.

1. Isolation from the drug (sclerotia of *Claviceps purpurea*).
2. Extraction from saprophytic cultures of different ascomycetes, mainly *Claviceps* species.
3. Partial and total synthesis.

The most economical method for the production of peptide alkaloids seems to have been, until now, extraction from the crude drug. One is no longer dependent on the collecting of wild ergot. Ergot can be produced parasitically by infection of rye plants with suitable strains. For this purpose strains selected with respect to the quantity and quality of the alkaloid content are employed. Nevertheless the parasitic technique has some drawbacks. One crop a year may be obtained and the yield is extremely dependent upon the weather. Therefore over many years the search for methods of producing ergoline alkaloids in saprophytic culture has continued. From the commercial standpoint it is very important to find high-yield strains and suitable culture conditions for submerged cultivation. This problem seems to have been satisfactorily solved for clavine alkaloids and simple lysergic acid derivatives. The total synthesis of peptide alkaloids is economically not feasible, because no methods for a practical synthesis of lysergic acid have been described. The partial synthesis of ergolines of the amide type, e.g., ergometrine, has been commercially available for many years (Hofmann, 1964). In the future, the partial synthesis of peptide alkaloids will become more important. Biologically produced lysergic acid is linked with the synthesized peptide portion of the classic alkaloids. Furthermore by variation of the peptide portion alkaloids that do not occur in nature may be produced, and these may have new useful therapeutic effects. Biological methods for ergot alkaloid production will be discussed in detail.

B. PARASITIC PRODUCTION

The general principles for the parasitic cultivation of ergot were stated by von Békésy (1935, 1938), W. Hecht (1944), and Stoll and Brack (1944a,b). A prerequisite for the large-scale field production was the artificial inoculation of rye spikes, which was described first by von Békésy. Inoculation with spore suspensions may be performed using puncture boards, inoculation pistols, or machines. The older methods of infection were critically reviewed by von Békésy (1938). Ergot cultivation is carried out mainly in Hungary, Austria, Switzerland, Czechoslovakia, and Germany. The inoculation of rye plants by the injection technique can begin two or three weeks before flowering and continue some time after flowering had begun. An early date of inoculation renders an intense honeydew production that is necessary for a good secondary infection.

In the ergot crop the number of sclerotia originating from the primary and secondary infection may differ, and opinions of various authors in this subject are varied (von Békésy, 1956c). For the inoculation of the host plants, conidiospore suspensions are used. The conidia now are obtained by saprophytic cultivation of the particular strain in a submerged condition using sporulation-promoting media. These spores can be rendered suitable by preserving them in a concentrated sugar solution or by lyophilization. Prior to use the most suitable concentration of conidia in the inoculum is prepared by dilution with water. An inoculum containing 3000-5000 spores/mm^3 gives generally satisfactory results. The amount of inoculum in liters per acre is dependent on the inoculation technique. By using inoculation machines the amount of spore suspension required is determined by the length of the needle roller, the needle density, and the roller system (horizontal or vertical). With motor-driven machines 5-10 acres/day can be inoculated. Further agrotechnical details such as the most suitable host plants, stand width of the host plants, soil fertilization, danger of contaminating neighboring fields, and problems of harvest are discussed by Rochelmeyer (1950), W. Hecht (1955), and von Békésy (1956c). Six to seven weeks after inoculation the harvest begins. Hand-picking of the sclerotia is a time-consuming procedure but still gives the best results. Collecting machines are used in a few countries. The best construction for such devices has not yet been described, and a considerable loss of crude ergot by this method has been observed. The yield of ergot per acre depends on different factors. In addition to the inoculation and harvesting technique the weather, the virulence of the inoculum, and the rye species used are important. Under normal conditions the yield should range from 50 to 100 kg/acre. From the point of view of economics, for the large-scale field production of ergot, not only the yield in kilograms per acre but also the alkaloid content of the crop is of considerable importance.

The alkaloid yield of wild ergot is extremely varied. Detailed investigations of German wild ergot has been published by Silber (1952). The first ergot crop, which was produced in 1936 by von Békésy with the newly developed inoculation technique, was alkaloid free! In 1939 von Békésy had already emphasized that the improvement of the alkaloid content is of great importance, economically, and that, by selection, a high-yielding drug can be obtained. He postulated that environmental factors do not exert a great influence because the nature of alkaloids produced by a given strain may be genetically fixed. Many attempts have been made in different countries to obtain a high quality drug (Rochelmeyer, 1954; Silber and Bischoff, 1954; Kybal and Brejcha, 1955; W. Hecht, 1955; Gröger, 1956, 1957; Kybal et al., 1957). First, numerous sclerotia of wild ergot were examined, and the alkaloid-rich sclerotia selected and used as starting material for field cultivation. Later on, screenings were performed using chromatographic techniques, and strains were selected that, in general, were able to produce one main alkaloid or defined alkaloid complex, e.g., ergotoxine. This is important for the extraction and preparation of pure alkaloids on an industrial scale, because the separation of a mixture of lysergic acid derivatives is not easily achieved. For example, Gröger (1956) investigated 15,000 individual sclerotia to obtain suitable strains. A single sclerotium can be a mixed sclerotium, meaning that it is composed of hyphae that originated in different spores. The occurrence of mixed sclerotia was experimentally established by M. Hecht and Hecht (1955) and Silber et al. (1955). The isolation of monospores proved to be a useful method to obtain genetically homogeneous strains. Some countries now use strains for ergot cultivation that are capable of producing 0.4-0.6% selected alkaloids.

The behavior of selected strains and the factors that might influence the alkaloid yield have been repeatedly investigated. Ergot races, which were cultivated under different climatic conditions on the same host plants, generally had the same alkaloid content (Silber and Bischoff, 1954). The importance of the host plant on the determination of the character of the alkaloid was investigated by Meinicke (1956). The total alkaloid content of a selected strain differs considerably after growth on different wild grasses, but the qualitative composition of the fraction of water-insoluble alkaloids is not changed. The chemical character of a given strain is strongly fixed genetically and cannot be modified even under extreme environmental conditions. In mutation experiments leuko-sclerotia were obtained, but the alkaloid character was not altered (Gröger, 1957). A degeneration of ergot strains was described by von Békésy (1956b). In such cases a drastically reduced ergot crop occurred. The reason for this sudden change of a particular strain is still unknown. Thus, the virulence of different ergot strains should be considered. In various strains there was no difference in the amount (kilograms per acre) and the alkaloid yield of the crop when either a vegetative or a generative inoculum was used. Some strains are extremely dependent on the weather, while others are not (von Békésy, 1956b). The results of fertilization experimens with respect to the ergot crop are not clear. A correlation between the total alkaloid content and the total nitrogen content of sclerotia was not observed (Böswart et al., 1957; Gröger, 1958).

C. SAPROPHYTIC PRODUCTION OF ERGOLINE ALKALOIDS

1. GENERAL CONSIDERATIONS

Ergot fungi are easily cultivated in surface culture or under submerged conditions. Preliminary experiments starting with ascospores or conidiospores of the honeydew were performed by Brefeld (1881) and Meyer (1888). Falck (1922) introduced a new technique in order to obtain pure cultures, and this technique is most commonly used today. By this method a piece from the inner part of a sclerotium is transferred to an agar slant under sterile conditions. In his saprophytic cultures Engelke (1902) observed microsclerotia and supposed that in such cultures the same active principle may be formed as in the normal sclerotia. The first patent for the saprophytic production of alkaloids was granted to Weil (1910). Numerous unsuccessful trials were performed by Bonns (1922). In the next decades many attempts were made to produce ergolines in saprophytic culture. Most of the reported results were negative or equivocal. A reliable identification of the active principles without the later chromatographic techniques was practically impossible. Repeatedly low and irregular yields were obtained after a long culture period with emersed cultures. Progress was made and an appreciable improvement of alkaloid yields has been achieved in the past 20 years. Comprehensive reviews on this topic have been presented by Taber (1963), Vining and Taber (1963b), Abe and Yamatodani (1964), Taber (1967), and Voigt (1968). Some aspects of the saprophytic production of ergoline alkaloids will be discussed here.

2. Production of Clavine Alkaloids

In the course of a screening program after World War II, Abe and his co-workers examined ergot fungi parasitic on wild grasses from Japan and other parts of Asia (Abe and Yamatodani, 1964).

The strains that were isolated from ergot parasitic on *Agropyron semi-costatum* Nees and *Agropyron ciliare* Fr. produced agroclavine in surface culture after a period of 30-40 days at 26-27°. The total alkaloid yield amounted to 600 mg/liter. Agroclavine was the first member of a new class of ergot alkaloids that are called now the clavines. The medium used was composed of mannitol (5%), ammonium succinate (0.8%), potassium dihydrogen phosphate (0.1%), magnesium sulfate (0.03%), and tap water adjusted to pH 5.2. This basic medium, first developed by Abe, is especially suited for the saprophytic production of ergoline derivatives. Many authors later adopted this nutrient medium or varied it slightly. Stoll *et al.* (1954) isolated from the culture broth of a strain obtained from the ergot parasitic on *Pennisetum typhoideum* Rich. (Africa) agroclavine and elymoclavine as well as penniclavine. The alkaloids are produced either in surface or submerged culture. At the beginning alkaloid yields between 50 to 100 mg/liter were obtained. The alkaloid yields were markedly increased by strain selection and modification of the medium. Mannitol was replaced by sucrose (10%) and trace elements were added. The culture was allowed to ferment for 30 to 40 days and an alkaloid content of 1.0-1.5 gm/liter was achieved. The alkaloid fraction of this strain consists mainly of elymoclavine and agroclavine, but seto- and isosetoclavine as well as chanoclavine were obtained in crystalline form. Changes of the alkaloid content during growth of *Pennisetum* ergot in surface cultures were investigated by Brack *et al.* (1962). A similar strain from *Pennisetum* was isolated by Tyler (1958), and it is used by other authors to study the physiology of alkaloid formation and biosynthetic problems (strain SD 58 of Gröger; strain PRL 1980 of Taber).

The taxonomic position of *Pennisetum* type ergot fungi seems to be clear. This type does not belong to *C. purpurea* or *Claviceps micro-*

cephala (Abe and Yamatodani, 1964). Loveless (1967a) has described ergot parasitic on *Pennisetum typhoideum* Rich. as a new *Claviceps* species and called it *Claviceps fusiformis*. Formation of conidiospores in saprophytic culture of *C. fusiformis*, which has been described by other researchers, was never seen in the author's laboratory. The effect of varying the medium phosphate concentration on the synthesis of alkaloids of *Pennisetum* and *Agropyron*-type ergot was investigated by Brady and Tyler (1960). In a low phosphate medium an increase of the alkaloid yield with the *Pennisetum* strain was observed, but the reverse results were obtained using the *Agropyron* strain. A semicontinuous process for the production of clavines was developed by Gröger and Erge (1961). Conditions favoring clavine alkaloid accumulation in submerged culture were investigated by Vining and Nair (1966), and a maximum yield of alkaloids, about 1 gm/liter, was reached in 14 days. A mannitol (6.5%), glucose (1%) mixture was the best carbon source. A high C:N ratio was found to be necessary to extend the phase of alkaloid formation. The addition of tryptophan but not mevalonic acid stimulated alkaloid formation. The effect of deuterium oxide on the growth and alkaloid production of a clavines-synthesizing strain was described by Mrtek *et al.* (1965), while a detailed description of the morphology of the sclerotia and the saprophytically grown mycelium of *C. fusiformis* was given by Tonolo and Udvardy-Nagy (1968). In submerged culture the mycelial structure differed from that seen in surface cultures. Optimal alkaloid production was obtained with simple synnemata. The mycelium was reddish brown, with many conidia. After 7 days of cultivation the alkaloid yield reached 1.5 gm/liter on medium containing glycerol (10%) and peptone Difco (2%). With this substrate formation of polysaccharides was avoided, and aeration reached maximum efficiency.

3. PRODUCTION OF SIMPLE LYSERGIC ACID DERIVATIVES

A milestone in ergot research was the announcement of the results of an Italian research team in Rome under Professor E. B. Chain (Arcamone *et al.*, 1960, 1961a). Ergot of a wild grass was again a good starting material. In the course of the investigation of *Claviceps paspali* Stevens & Hall, a long-neglected species, a strain was selected that produced 10 mg/liter alkaloid in submerged culture. The morphology and physiology of this strain differed considerably from other *Claviceps* species. The formation of conidiospores or arthrospores was never observed. Under submerged conditions this strain formed pellets made up of single synnemata. The inner parts of the synnemata revealed the typical structure of a sclerotium. For selection purposes hyphae fragments were spread on agar plates, and the developing colonies were isolated and tested in submerged culture. The isolates tested showed enormous differences with regard to pigmentation and alkaloid production. Finally by a repeated selection process strains were obtained that produced 1-2 gm/liter alkaloid in submerged culture. A modified Abe medium gave the best results. For the propagation of the inoculum the medium is supplemented with chick-pea meal. Alkaloid formation is dependent on the age of the inoculum, the amount of the inoculum used, the pH of the medium, and the temperature. Stimulation of alkaloid production was observed following intense aeration and after addition of tryptophan to the medium. The principal alkaloids of this *C. paspali* strain are lysergic acid methylcarbinolamide and its isomer, as well as ergine and isoergine.

Gröger and Tyler (1963) studied alkaloid formation in Australian *Paspalum* strains. These authors used a fermentation technique that differed from that of Arcamone *et al.* (1961a). Alkaloid formation in these strains was favored by using a corn steep solids-containing medium for the preculture and a fermentation medium supplemented with 1,2-propanediol. Tryptophan had no effect on alkaloid formation in these particular strains. Gröger *et al.* (1961) detected in the sclerotia of *Paspalum* ergot simple lysergic acid derivatives in addition to chanoclavine. The total alkaloid yield in this material was surprisingly low and amounted to 0.001-0.005%.

Extensive studies on *C. paspali* fermentation were made by Schwarting's group (Pacifici *et al.*, 1962, 1963; Mary *et al.*, 1965). For strain selection the Tonolos method was used. Maximum alkaloid yields were obtained after addition of 50 mg tryptophan and 1 gm acetamide/100 ml basal medium. The importance of tap water and minerals as well as trace elements in alkaloid production was investigated. An eight-salt synthetic medium with mannitol and ammonium succinate was found to be optimal. Phosphate depletion was most rapid in cultivation of high-yield strains. Kobel *et al.* (1964) showed that different chemical races also exist in *C. paspali*. In the course of a screening program several hundred strains from around the world were tested (see Section III, B). All strains were able to produce traces of elymoclavine and chanoclavine as well as the major alkaloids. For the first time Kobel *et al.* (1964) achieved sporulation of *C. paspali* strains. For this purpose agar slants were used that contained a medium composed of 250 ml corn steep extract and 500 ml beer-wort per liter. In their fermentation medium Kobel *et al.* (1964) replaced mannitol by sorbitol. In the experiments of Arcamone *et al.* (1961a) sorbitol as a carbon source gave inferior results. A combination of sorbitol and DL-malic acid with various mineral salts were especially suitable for alkaloid production in the strain used by Taber (1967).

In a comparison of the alkaloid production with the total biomass formation under saprophytic conditions two different fermentation patterns
were distinguished (Taber, 1964, 1967). In the first pattern alkaloid
production started after total biomass formation ceased and usually
continued while dry weight of mycelium decreased. Apparently the
decrease in weight is due to the loss in mycelial polyol, lipid, trehalose
and water-soluble nitrogenous compounds. Mostly this pattern is observed when complex organic media are used, and these favor a luxurious
growth. In the second pattern alkaloid formation is observed along with
a more or less constant but reduced rate of total biomass synthesis. The
cultures grow very slowly and never build up much mycelia. Brar *et al.*
(1968) were able to demonstrate that a particular *C. paspali* strain could
produce alkaloid by both patterns depending on the medium used. On a
glucose-yeast extract medium the alkaloid formation began after the
rapid growth was over (pattern No. 1). The other situation was observed
upon culturing the strain on polyol and ammonium succinate media.
Alkaloid production occurred simultaneously with a slow but steady
increase in biomass formation (pattern No. 2). Increased yields were
obtained after the addition of 1-2 gm tryptophan/liter to sorbitol-
succinate medium, but mevalonate had no effect on alkaloid production.

The influence of different aliphatic and cyclic glycols in a basal medium
on alkaloid production in *C. paspali* was investigated by Mizrahi and
Miller (1968a). The same authors reported on different procedures for
long-term preservation of *C. paspali* (Mizrahi and Miller, 1968b).

A selected strain from the ergot parasitic on *Calamagrostis* in China
was used for the production of ergometrine in surface cultures. This
strain, termed *Claviceps microcephala,* grown on a wheat grain-con-
taining medium was found to be optimal for alkaloid formation (Teh-chao
*et al., * 1962).

4. PRODUCTION OF PEPTIDE ALKALOIDS

The production of peptide alkaloids in crystalline form from saprophytic culture was achieved for the first time by Abe *et al.* (1952). In
surface culture the given strain produced 400 mg/liter alkaloid after a
culture period of 30-40 days. This strain, isolated from ergot parasitic
on *Elymus mollis* Trin., was termed *Claviceps litoralis* Kawatani. Ergo-
kryptinine and its isomer were the main peptide alkaloids along with
agroclavine and elymoclavine. A mannitol-ammonium succinate medium
was used for fermentation. This strain produced only a few conidiospores and gave inferior alkaloid yields in submerged culture. Abe's
group (Abe and Yamatodani, 1964) also later employed a *C. purpurea*
strain from Spanish rye ergot and obtained alkaloid yields of 40 mg/

liter. The composition of the peptide type alkaloid mixture was not given. The yields were not always satisfactory and fluctuated from batch to batch due to the instability of the given strain. Low and irregular yields in saprophytic peptide alkaloid production have been also recorded by many other authors. These results are fully discussed in the above-mentioned reviews.

Kobel et al. (1962) grew monospore strains from ergotamine- and ergotoxine-rich ergot in surface cultures for 60 days. In the culture filtrate 100 mg/liter alkaloid and in the dried mycelium 0.16-0.55% alkaloid were detected. The alkaloid spectrum of the saprophytic culture showed only slight differences compared with that of the sclerotia. However within the spectrum the amounts of the individual alkaloids differed considerably. An important advance was made through the investigations of Tonolo (1966) and Amici et al. (1966). These authors isolated C. purpurea strains from ergot parasitic on Triticale (an artificial hybrid between rye and wheat), which grows in Spain. The alkaloid yield reached its maximum after 10-12 days in submerged culture and amounted to 1-1.5 gm/liter, consisting mainly of ergotamine and its isomer. Apparently media with a high osmotic pressure, which contain about 20-30% sucrose or mannitol, favor alkaloid formation. These strains do not form conidiospores, or chlamido- or arthrospores. The mycelium grown in submerged culture showed a typical plectenchymatic structure similar to a sclerotium.

The Hungarian workers Molnar et al. (1965) succeeded in isolating mutants of C. purpurea after irradiation of the conidiospores with a ^{60}Co source. These strains produced mainly ergotoxine and ergometrine in a yield of 400 mg/liter. Minghetti et al. (1969) patented a procedure for the saprophytic production of ergocristine.

The physiology of ergotamine formation of the strain 275 FI was thoroughly studied by Amici et al. (1967a,b). From this strain, which produced large amounts of peptide alkaloids in submerged culture, the strains V, C, and W were spontaneously obtained. The "daughter strains" were unable to produce alkaloids. The strain 275 FI differs markedly from the unproductive strains in its capacity to accumulate lipids and sterols. The alkaloid-producing strain utilizes large amounts of sucrose and citric acid simultaneously. Strain W, which is regarded as the "normal" form, utilizes only an amount of sucrose and citric acid proportional to that required for the growth of the microorganism. The authors concluded that in strain 275 FI regulating mechanisms for the two substrates were lacking or operated in a limited manner. Therefore the utilization rate of the carbon sources is much higher, as in the case of the "normal" form, and the accumulation of lipids and sterols and the formation of alkaloids are favored. The alkaloid-producing strain 275 FI is a heterocaryon and showed segregation in its single components after several generations on agar media. These components (designated V, C, and W) are stable and produce practically no alkaloids. The mycelium of strain FI 275 is composed of hyphae with multinucleate cells and does not produce conidiospores. A new strain (neo-275 FI) was obtained by combining strains V and C, and this strain displayed many features of the original strain 275 FI. According to the Italian authors alkaloid formation is related to heterokaryosis of the alkaloid-producing strains.

A distinct correlation between the morphology of saprophytic mycelium and the capacity to form alkaloids was observed by Mantle and Tonolo (1968). Only plectenchymatic growth forms are able to synthesize alkaloids. Mantle (1969a) isolated a *C. purpurea* strain from ergot parasitic on *Spartina townsendii* in England. After passage on rye plants some isolates could be obtained; they were highly pigmented and exhibited plectenchymatic structures. These isolates were able to produce alkaloids in fairly large amounts. Repeated transfer caused degeneration to pigmentless and nonalkaloid-producing types. The mechanism that triggers the development of alkaloid-producing forms and, also, their degeneration is still unknown. Mantle (1969a) made the interesting observation that the sclerotia used as starting material did contain ergotoxine alkaloids but the resulting saprophytic cultures produced lysergic acid and chanoclavine.

5. REGULATION OF ALKALOID FORMATION

In principle it should be possible that all selected strains that produce large amounts of alkaloids in parasitic culture should, in saprophytic cultures, produce alkaloids. Numerous experiments in many laboratories have demonstrated that a sclerotial form or a sclerotial "phase" of the saprophytic mycelium and a synchronous alkaloid formation are not always achieved. Much information on ergot alkaloid biosynthesis has appeared in the past 15 years. On the other hand our knowledge of the factors involved in the initiation of biosynthesis and their regulation is still fragmentary. One reason might be that our information on the enzymes involved in ergoline biosynthesis is limited. An enzyme or enzyme system specific for ergoline formation has not yet been detected. The effects of stimulators or inhibitors of alkaloid synthesis are in general measured by determining the total alkaloid yield. Investigations in this area have not yet reached the enzyme level.

Alkaloid production in *Claviceps* is, according to Teuscher (1964, 1966), concomitant with an intense tryptophan metabolism. Active tryptophan transport with exogenously administered tryptophan was detected only in alkaloid-producing strains. The uptake of this amino acid is completely inhibited by dinitrophenol. The effect of 4-methyl-, 5-methyl-, and 6-methyltryptophan on the formation of lysergic acid methylcarbinolamide in *C. paspali* was investigated by Arcamone *et al.* (1962). 4-Methyltryptophan was found to be a strong inhibitor, while 6-methyltryptophan exerted a less pronounced effect on alkaloid formation with no inhibition of growth. 5-Methyltryptophan produced a decrease in growth rate but not alkaloid formation. An increase in alkaloid formation in the clavine strain SD 58 was observed when tryptophan was added prior to inoculation or during trophophase. Similar results were obtained with various tryptophan analogs. Mycelium grown in the presence of tryptophan, 7-methyltryptophan, or 1-*N*-methyltryptophan retains the ability to produce more alkaloids than the controls even after transfer to fresh media (Floss and Mothes, 1964). Apparently these indole derivatives act as inductors of alkaloid synthesis. This assumption is supported by the fact that a rapid protein turnover takes place during the idiophase when alkaloids are synthesized (Floss, 1967b; Rothe and Fritsche, 1967). In experiments with a clavine strain, fluctuating rates of alkaloid production were observed by Bu'Lock and Barr (1968). They found that in a particular strain of *Claviceps* mycelial tryptophan induced a metabolically labile enzyme. This "synthetase" is regarded as rate-limiting for ergot alkaloid biosynthesis. Exogenously administered tryptophan inhibits the endogenous tryptophan synthesis. Alkaloid formation stopped within 24 hours after protein synthesis was inhibited by either cycloheximide or *p*-fluorphenylalanine.

The enzymes involved in aromatic biosynthesis in *C. paspali* have been investigated thoroughly (Lingens *et al.*, 1967a,b). DAHP-Synthetase can be separated into three isoenzymes. L-Tryptophan-sensitive isoenzyme accounted for the greater part of the DAHP-synthetase activity. Chorismate mutase can be separated into two isoenzymes. Prephenate dehydrogenase and prephenate dehydratase were inhibited by the corresponding end products L-tyrosine and L-phenylalanine. Anthranilate synthetase the first enzyme in tryptophan biosynthesis showed no feedback inhibition by the end product. This is in contrast to findings in other microorganisms. If there is no control mechanism for anthranilate synthetase operative in *C. paspali* tryptophan might be produced in excess. An internal accumulation of tryptophan could be one of the alkaloid-formation triggering factors. It remains to be seen if, in all alkaloid-producing strains, anthranilate synthetase or other enzymes in aromatic biosynthesis show lack of feedback inhibition. It is interesting to note that in the clavine strain SD 58 tryptophan strongly inhibits anthranilate synthetase (Lingens, 1969). Enzymes of isoprenoid biosynthesis have generally not been investigated in *Claviceps*. Mevalonic acid or related compounds are apparently not limiting factors in ergoline formation, and they cannot induce alkaloid synthesis.

VI. Biological Activity

For a detailed description of the action of the ergot alkaloids and their use as therapeutic agents the reader is referred to textbooks of pharmacology and toxicology and to the chapter on ergot in Hofmann's book (1964). According to Cerletti (1959) six principal effects of all natural ergot alkaloids can be observed; these can be divided into the following groups depending on the site of action: (1) peripheral effects, (2) neurohumoral effects, and (3) effects on the central nervous system. Vasoconstriction and contraction of the uterus are the most important peripheral effects. The classic use of ergot alkaloids in obstetrics depends on its contractile effect on the smooth muscle of the uterus. The neurohumoral effects include antagonism to adrenaline as well as to serotonin. The use of ergot preparations as very potent sympathicolytic agents is based on their adrenolytic action. The central nervous system effects of the ergot alkaloids are various. The alkaloids reduce the activity of the vasomotor center, while the sympathetic structures in the midbrain, especially the hypothalamus, are stimulated by ergot alkaloids.

Different types of ergot alkaloids vary in degree of biological activity in these three areas. For example ergotamine and the alkaloids of the ergotoxine group have similar activity spectra. However the latter are much more toxic. Ergometrine is widely used in obstetrics but exerts practically no adrenolytic action, and central nervous effects are seen only after high doses.

Slight modifications in the chemical structures of the native alkaloids produce pronounced changes in their biological activity. Hydrogenation of the double bond in the 9-10 position of the lysergic acid moiety of the peptide type alkaloids leads to an almost complete change in pharmacological action. Thus these compounds can be used in the treatment of peripheral and cerebral vascular disorders and essential hypertension. Methylergometrine gives a greater effect than ergometrine on the smooth muscle of the uterus and is used on a larger scale as Methergin® as a uterotonic. 1-Methyl-D-lysergic acid butanolamide, a derivative of Methergin®, is the strongest serotonin antagonist known.

Lysergic acid diethylamide (LSD) has been named the "drug of the century" by Bové (1970). More than 2000 scientific papers and numerous reports in other journals have been published on the use and misuse of this semisynthetic compound. It is the prototype of the hallucinogens.

All aspects of this most powerful hallucinogenic agent have been discussed in the excellent review of Hoffer and Osmond (1967).

In the past years other biological activities of ergot alkaloids have been demonstrated. Most surprising are the results of Shelesnyak's group (Shelesnyak, 1954, 1955; Carlsen *et al.*, 1961; Kraicer and Shelesnyak, 1965). Ergotoxine alkaloids, and especially ergocornine, orally administered in female rats and mice inhibit implantation of the ovum. Despite fertilization no pregnancy occurs. The exact mechanism of ergotoxine in interfering with decidualization and nidation is still under investigation. It seems that progesterone metabolism plays a major role. Further experiments will show if this new principle of birth control can be extended to humans. The consequences could be tremendous. The clavine alkaloids have not yet been used as therapeutic agents. Recently Mantle (1968) has shown that agroclavine inhibits mammary hypertrophy and lactation in mice. Furthermore an implantation block by oral administration of agroclavine at a nontoxic dosage in mice has been demonstrated (Mantle, 1969b).

REFERENCES

Abe, M. (1951a). *Ann. Rept. Takeda Res. Lab.* **10**, 73.
Abe, M. (1951b). *Ann. Rept. Takeda Res. Lab.* **10**, 145.
Abe, M. (1963). *Abhandl. Deut. Akad. Wiss. Berlin, Kl. Chem., Geol. Biol.* No. 4, p. 309.
Abe, M. (1966). *Abhandl. Deut. Akad. Wiss. Berlin, Kl. Chem., Geol. Biol.* No. 3, p. 393.
Abe, M. (1969). *4th Intern. Symp. Biochem. Physiol. Alkaloide, Halle, 1969*, Akademic-Verlag, Berlin (in press).
Abe, M., and Yamatodani, S. (1954). *J. Agr. Chem. Soc. Japan* **28**, 501.
Abe, M., and Yamatodani, S. (1955). *Bull. Agr. Chem. Soc. Japan* **19**, 161.
Abe, M., and Yamatodani, S. (1964). *Progr. Ind. Microbiol.* **5**, 205-229.
Abe, M., Yamano, T., Kozu, Y., and Kusumoto, M. (1952). *J. Agr. Chem. Soc. Japan* **25**, 458.
Abe, M., Yamatodani, S., Yamano, T., and Kusumoto, M. (1956). *Bull. Agr. Chem. Soc. Japan* **20**, 59.
Abe, M., Yamano, T., Yamatodani, S., Kozu, Y., Kusumoto, M., Komatsu, H., and Yamada, S. (1959). *Bull. Agr. Chem. Soc. Jap.* **23**, 246.
Abe, M., Yamatodani, S., Yamano, T., and Kusumoto, M. (1961). *Agr. Biol. Chem. (Tokyo)* **25**, 594.
Abe, M., Yamatodani, S., Yamano, T., Kozu, Y., and Yamada, S. (1967). *J. Agr. Chem. Soc. (Tokyo)* **41**, 68.
Abe, M., Ohmomo, S., Ohashi, T., and Tabuchi, T. (1969). *Agr. Biol. Chem. (Tokyo)* **33**, 469.
Acklin, W., Fehr, T., and Arigoni, D. (1966). *Chem. Commun.* p. 799.
Agurell, S. (1964). *Experientia* **20**, 25.
Agurell, S. (1965). *Acta Pharm. Suecica* **2**, 357.
Agurell, S. (1966a). *Acta Pharm. Suecica* **3**, 7.
Agurell, S. (1966b). *Acta Pharm. Suecica* **3**, 71.

Agurell, S., and Lindgren, J. E. (1968). *Tetrahedron Letters*, 5127.
Agurell, S., and Ramstad, E. (1962a). *Lloydia* 25, 67.
Agurell, S., and Ramstad, E. (1962b). *Arch. Biochem. Biophys.* 98, 457.
Agurell, S., and Ramstad, E. (1965). *Acta Pharm. Suecica* 2, 231.
Allport, N. L., and Cocking, T. (1932). *Quart. J. Pharm. Pharmacol.* 5, 341.
Amici, A. M., Minghetti, A., Scotti, T., Spalla, C., and Tognoli, L. (1966). *Experientia* 22, 415.
Amici, A. M., Minghetti, A., Scotti, T., and Tognoli, L. (1967a). *Appl. Microbiol.* 15, 597.
Amici, A. M., Scotti, T., Spalla, C., and Tognoli, L. (1967b). *Appl. Microbiol.* 15, 611.
Apsimon, J. W., Corran, J. A., Creasey, N. G., Sim, K. Y., and Whalley, W. B. (1963). *Proc. Chem. Soc.* p. 209; *J. Chem. Soc.* p, 4130 (1965).
Arcamone, F., Bonino, C., Chain, E. B., Ferretti, A., Pennella, P., Tonolo, A., and Vero, L. (1960). *Nature* 187, 238.
Arcamone, F., Chain, E. B., Ferretti, A., Minghetti, A., Pennella, P., Tonolo, A., and Vero, L. (1961a). *Proc. Roy. Soc.* B155, 26.
Arcamone, F., Chain, E. B., Ferretti, A., and Pennella, A. (1961b). *Nature* 192, 552.
Arcamone, F., Chain, E. B., Ferretti, A., Minghetti, A., Pennella, P., and Tonolo, A. (1962). *Biochim. Biophys. Acta* 57, 174.
Arigoni, D. (1965) "Some Aspects of Mevalonoid Biosynthesis." As quoted in reference (Agurell, 1966b).
Atanasoff, D. (1920). "Ergot of Grain and Grasses" *U.S. Dept. Agr., Bur. Plant Ind.* 1-127.
Ballio, A., Kulaev, I., Moret, V., and Russi, S. (1966). *Biochim. Biophys. Acta* 115, 489.
Barger, G. (1931). "Ergot and Ergotism." Gurney & Jackson, London.
Basmadjian, G., Floss, H. G., Gröger, D., and Erge, D. (1969). *Chem. Commun.* p. 418.
Bauhin, C. (1658). "Theatri botanici liber primus." Basileae, column 434.
Baxter, R. M., Kandel, S. I., and Okany, A. (1961). *Chem. & Ind. (London)* p. 1453.
Birch, A. J., McLoughlin, B. J., and Smith, H. (1960). *Tetrahedron Letters* p. 1.
Bonns, W. W. (1922). *Am. J. Botany* 9, 339.
Böswart, J., Karmazin, M., and Horák, P. (1957). *Abhandl. Deut. Akad. Wiss. Berlin, Kl. Chem., Geol. Biol.* No. 7, p. 236.
Bové, F. J. (1970). "The Story of Ergot." Karger, Basel.
Brack, A., Brunner, R., and Kobel, H. (1962). *Arch. Pharm.* 295, 510.
Brack, A., Brunner, R., and Frey, H. P. (1963). *Lloydia* 26, 75.
Brady, L. R. (1962). *Lloydia* 25, 1.
Brady, L. R., and Tyler, V. E., Jr. (1960). *Lloydia* 23, 8.
Brar, S. S., Giam, C. S., and Taber, W. A. (1968). *Mycologia* 60, 806.
Brefeld, O. (1881). "Untersuchungen aus dem Gesamtgebiet der Mykologie," Part 4. Munster i. Westfalen. (Germany)
Buck, K. W., Chen, A. W., Dickerson, A. G., and Chain, E. B. (1968). *J. Gen. Microbiol.* 51, 337.
Bu'Lock, J. D., and Barr, J. G. (1968). *Lloydia* 31, 342.
Camphell, W. P. (1957). *Can. J. Botany* 35, 315.
Carlsen, R. A., Zeilmaker, G. H., and Shelesnyak, M. C. (1961). *J. Reprod. Fertility* 2, 369.
Castagnoli, N., Jr., and Tonolo, A. (1967). *Proc. 9th Intern. Congr. Microbiol., Moscow, 1966*, Symposia, p. 31. Jvanovski Institute of Virology.
Cerletti, A. (1959). *Proc. 1st Intern. Neuro-Pharmacol., Congr., Rome, 1958* pp. 117-123. Elsevier, Amsterdam.
Der Marderosian, A. (1967). *Lloydia* 30, 23.
Dibbern, H. W., and Rochelmeyer, H. (1963). *Arzneimittel-Forsch.* 13, 7.

Dragendorff, G., and Podwyssotzky, G. (1877). *Arch. Exptl. Pathol. Pharmakol.* **6**, 174.
Dudley, H. W., and Moir, C. (1935). *Brit. Med. J.* **1**, 520.
Engelke, E. (1902). *Hedwigia* **51**, 221.
Falck, R. (1922). *Pharm. Ztg., Ver. Apotheker-Ztg.* **67**, 777.
Fehr, T., and Acklin, W. (1966). *Helv. Chim. Acta* **49**, 1907.
Fehr, T., Acklin, W., and Arigoni, D. (1966). *Chem. Commun.* p. 801.
Feuell, A. J. (1966). *Trop. Sci.* **8**, 61.
Floss, H. G. (1967a). *Chem. Commun.* p. 804.
Floss, H. G. (1967b). *Ber. Deut. Botan. Ges.* **80**, 705.
Floss, H. G. (1969). *4th Intern. Symp. Biochem. Physiol. Alkaloide, Halle 1969.* Akademic-Verlag, Berlin (in press).
Floss, H. G., and Mothes, U. (1964). *Arch. Mikrobiol.* **48**, 213.
Floss, H. G., Mothes, U., and Günther, H. (1964). *Z. Naturforsch.* **19b**, 784.
Floss, H. G., Hornemann, U., Schilling, N., Gröger, D., and Erge, D. (1967a). *Chem. Commun.* p. 105.
Floss, H. G., Günther, H., Gröger, D., and Erge, D. (1967b). *J. Pharm. Sci.* **56**, 1675.
Floss, H. G., Günther, H., Mothes, U., and Becker, I. (1967c). *Z. Naturforsch.* **22b**, 399.
Floss, H. G., Hornemann, U., Schilling, N., Kelley, K., Gröger, D., and Erge, D. (1968). *J. Am. Chem. Soc.* **90**, 6500.
Floss, H. G., Günther, H., Gröger, D., and Erge, D. (1969). *Arch. Biochem. Biophys.* **131**, 319.
Foster, G. E. (1955). *J. Pharm. Pharmacol.* **7**, 1.
Foster, G. E., McDonald, J., and Jones, T. S. G. (1949). *J. Pharm. Pharmacol.* **1**, 802.
Franck, B. (1965). *In* "Beiträge zur Biochemie und Physiologie von Naturstoffen. Festschrift K. Mothes," pp. 153–165. Fischer, Jena.
Franck, B. (1969). *Angew. Chem.* **81**, 269.
Franck, B., and Reschke, T. (1959). *Angew. Chem.* **71**, 407.
Franck, B., and Reschke, T. (1960). *Chem. Ber.* **93**, 347.
Franck, B., and Zimmer, I. (1965). *Chem. Ber.* **98**, 1514.
Franck, B., Baumann, G., and Ohnsorge, U. (1965). *Tetrahedron Letters* p. 2031.
Franck, B., Hüper, F., Gröger, D., and Erge, D. (1966). *Angew. Chem. Intern. Ed. Engl.* **5**, 728.
Franck, B., Hüper, F., Gröger, D., and Erge, D. (1968). *Chem. Ber.* **101**, 1954.
Freeborn, A. (1912). *Pharm. J.* **88**, 586.
Fuentes, S. F., de la Isla, M. L., Ullstrup, A. J., and Rodriguez, A. E. (1964). *Phytopathology* **54**, 379.
Garay, A. S. (1956a). *Physiol. Plantarum* **9**, 344.
Garay, A. S. (1956b). *Physiol. Plantarum* **9**, 350.
Garay, A. S. (1958). *Physiol. Plantarum* **11**, 48.
Garay, A. S., and Kökenyesy, S. (1955). *Phytopathol. Z.* **25**, 109.
Gatenbeck, S. (1958). *Acta Chem. Scand.* **12**, 1211.
Gäumann, E. (1951) "Pflanzliche Infektionslehre." Birkhäuser, Basel.
Gäumann, E. (1964). "Die Pilze." Birkhäuser, Basel.
Genest, K. (1965). *J. Chromatog.* **19**, 531.
Genest, K., and Farmilo, C. G. (1964). *J. Pharm. Pharmacol.* **16**, 250.
Gläz, E. T. (1955). *Acta Microbiol.* **2**, 316.
Grasso, V. (1954). *Ann. Sper. Agrar.* (*Rome*) [N.S.] **8**, 1.
Grasso, V. (1957). *Boll. Staz. Patol. Vegetale* [3] **15**, 317.
Gröger, D. (1956). *Kulturpflanze Suppl.* **1**, p. 226.
Gröger, D. (1957). *Abhandl. Deut. Akad. Wiss. Berlin, Kl. Chem., Geol. Biol.* No. 7, p. 243.

Gröger, D. (1958). *Kulturpflanze* 6, 243.

Gröger, D. (1968). *In* "Das Art- und Rassenproblem bei Pilzen in taxonomischer, morphologischer, cytologischer, physiologischer, biochemischer und genetischer Sicht," (M. Lange-de la Camp, ed.) pp. 169-179. Fischer, Jena.

Gröger, D. (1969). *In* "Biosynthese der Alkaloide" (K. Mothes and H. R. Schütte, eds.), pp. 486-509. Deut. Verlag Wiss., Berlin.

Gröger, D., and Erge, D. (1961). *Planta Med.* 9, 471.

Gröger, D., and Erge, D. (1970). *Z. Naturforsch.* 25b, 196.

Gröger, D., and Tyler, V. E., Jr. (1963). *Lloydia* 26, 174.

Gröger, D., Mothes, K., Simon, H., Floss, H. G., and Weygand, F. (1960). *Z. Naturforsch.* 15b, 141.

Gröger, D., Tyler, V. E., Jr., and Dusenberry, J. E. (1961). *Lloydia* 24, 97.

Gröger, D., Erge, D., and Floss, H. G. (1965). *Z. Naturforsch.* 20b, 856.

Gröger, D., Erge, D., and Floss, H. G. (1968a). *Z. Naturforsch.* 23b, 177.

Gröger, D., Erge, D., Franck, B., Ohnsorge, U., Flasch, H., and Hüper, F. (1968b). *Chem. Ber.* 101, 1970.

Guggisberg, H. (1954). "Mutterkorn, vom Gift zum Heilstoff." Karger, Basel.

Gyenes, I., and Bayer, J. (1961). *Pharmazie* 16, 211.

Hartmann, T. (1965). *Planta* 66, 27 and 191.

Heath, H., and Wildy, J. (1958). *Biochem. J.* 68, 407.

Hecht, M., and Hecht, W. (1955). *Pharmazie* 9, 424.

Hecht, W. (1941). U.S. Patent 2,261,368.

Hecht, W. (1944). *Pharm. Acta Helv.* 19, 112.

Hecht, W. (1955). *Oesterr. Apotheker Ztg.* 9, 75.

Hoffer, A., and Osmond, H. (1967). "The Hallucinogens." Academic Press, New York.

Hofmann, A. (1961). *Planta Med.* 9, 354.

Hofmann, A. (1963). *Harvard Botan. Museum Leaflets* 20, 194.

Hofmann, A. (1964). "Die Mutterkornalkaloide." Enke, Stuttgart.

Hofmann, A. (1965). *Pharm. Weekblad* 100, 1261.

Hofmann, A., and Tscherter, H. (1960). *Experientia* 16, 414.

Hofmann, A., Brunner, R., Kobel, H., and Brack, A. (1957). *Helv. Chim. Acta* 40, 1358.

Hofmann, A., Frey, A. J., and Ott, H. (1961). *Experientia* 17, 206.

Hofmann, A., Ott, H., Griot, R., Stadler, P. A., and Frey, A. J. (1963). *Helv. Chim. Acta* 46, 2306.

Hornemann, U., Speedie, M. K., Kelley, K. M., Hurley, L. H., and Floss, H. G. (1969). *Arch. Biochem. Biophys.* 131, 430.

Jacobs, W. A., and Craig, L. C. (1934). *J. Biol. Chem.* 104, 547.

Jacobs, W. A., and Gould, R. G. (1937). *J. Biol. Chem.* 120, 141.

Jindra, A., Ramstad, E., and Floss, H. G. (1968). *Lloydia* 31, 190.

Julia, M., Le Goffic, F., Igolen, J., and Baillarge, M. (1969). *Tetrahedron Letters* p. 1569.

Jung, M., and Rochelmeyer, H. (1960). *Beitr. Biol. Pflanz.* 35, 343.

Kawatani, T. (1952). *Bull. Natl. Hyg. Lab.* 70, 127.

Kawatani, T. (1953). *Bull. Natl. Hyg. Lab.* 71, 161.

Kharasch, M. S., and Legault, R. R. (1935). *Science* 81, 388 and 614.

Killian, C. (1919). *Bull. Soc. Mycol. France* 25, 182.

Kirchhoff, H. (1929). *Zentr. Bakteriol., Parasitenk., Abt. II* 77, 310.

Klavehn, M., and Rochelmeyer, H. (1961). *Deut. Apotheker-Ztg.* 101, 477.

Klavehn, M., Rochelmeyer, H., and Seyfried, J. (1961). *Deut. Apotheker-Ztg.* 101, 75.

Kobel, H., and Stopp, K. (1967). *Naturwissenschaften* 54, 145.

Kobel, H., Brunner, R., and Brack, A. (1962). *Experientia* 18, 140.

Kobel, H., Schreier, E., and Rutschmann, J. (1964). *Helv. Chim. Acta* 47, 1052.

Kobert, R. (1889). "Zur Geschichte des Mutterkorns. Historische Studien aus dem pharmakologischen Institut der Kaiserlichen Universität Dorpat," Vol. 1, pp. 1-47. Halle, S.

Kordts, D., and Voigt, R. (1961). *Sci. Pharm.* 29, 81.

Kornfeld, E. C., Fornefeld, E. J., Kline, G. B., Mann, M. J., Jones, R. G., and Woodward, R. B. (1954). *J. Am. Chem. Soc.* 76, 5256.

Kraicer, P. F., and Shelesnyak, M. C. (1965). *J. Reprod. Fertility* 10, 221.

Krebs, I. (1936). *Ber. Schweiz. Botan. Ges.* 45, 71.

Kühn, J. (1865). *Vierteljahresschr. Prakt. Pharm.* 14, 8.

Kybal, J. (1964). *Planta Med.* 12, 166.

Kybal, J., and Brejcha, V. (1955). *Pharmazie* 10, 752.

Kybal, J., Horák, P., Brejcha, V., and Kudrnác, S. (1957). *Abhandl. Deut. Akad. Wiss. Berlin, Kl. Chem., Geol. Biol.* No. 7, p. 236.

Langdon, R. F. N. (1954). *Univ. Queensland Papers, Dept. Botany* 3, 61.

Leemann, H. G., and Fabbri, S. (1959). *Helv. Chim. Acta* 42, 2696.

Lewis, R. W. (1959). *Acta Botan. Acad. Sci. Hung.* 5, 71.

Lindquist, J. C., and Carranza, J. M. (1960). *Rev. Fac. Agr. (Argentina)* 36, 151.

Lingens, F. (1969). *4th Intern. Symp. Biochem. Physiol. Alkaloide, Halle, 1969*, Akademic-Verlag, Berlin (in press).

Lingens, F., Goebel, W., and Uesseler, H. (1967a). *Naturwissenschaften* 54, 941.

Lingens, F., Goebel, W., and Uesseler, H. (1967b). *European J. Biochem.* 2, 442.

Lonicer, A. (1582). "Kreuterbuch." Frankfurt am Main.

Loveless, A. R. (1967a). *Brit. Mycol. Soc. Trans.* 50, 15.

Loveless, A. R. (1967b). *Brit. Mycol. Soc. Trans.* 50, 19.

Macek, K., and Vaneček, S. (1955). *Pharmazie* 10, 422.

McCrea, A. (1931). *Am. J. Botany* 18, 50.

McLaughlin, L. J., Goyan, J. E., and Paul, A. G. (1964). *J. Pharm. Sci.* 53, 306.

McPhail, A. T., Sim, G. A., Asher, J. D. M., Robertson, J. M., and Silverton, J. V. (1963). *Proc. Chem. Soc.* p. 210; (1966) *J. Chem. Soc., B* p. 18.

Mantle, P. G. (1968). *Proc. Roy. Soc.* B170, 423.

Mantle, P. G. (1969a). *Brit. Mycol. Soc. Trans.* 52, 381.

Mantle, P. G. (1969b). *J. Reprod. Fertility* 18, 81.

Mantle, P. G., and Tonolo, A. (1968). *Brit. Mycol. Soc. Trans.* 51, 499.

Mary, N. Y., Kelleher, W. J., and Schwarting, A. E. (1965). *Lloydia* 28, 218.

Meinicke, R. (1956). *Flora (Jena)* 143, 395.

Meyer, B. (1888). *Landwirtsch. Jahrb. Schweiz.* 17, 924.

Milovidov, P. (1954). *Preslia* 26, 415.

Milovidov, P. (1957). *Acta Histochem.* 4, 41.

Minghetti, A., and Arcamone, F. (1969). *Experientia* 25, 926.

Minghetti, A., Spalla, C., and Tognoli, L. (1969). German Patent 1,806,984.

Mitchell, D. T., and Cooke, R. C. (1968a). *Brit. Mycol. Soc. Trans.* 51, 721.

Mitchell, D. T., and Cooke, R. C. (1968b). *Brit. Mycol. Soc. Trans.* 51, 731.

Mizrahi, A., and Miller, G. (1968a). *Biotechnol. Bioeng.* 10, 102.

Mizrahi, A., and Miller, G. (1968b). *Appl. Microbiol.* 16, 1100.

Molnar, G., Tétényi, P., Udvardy, E., Weck, G., and Wolf, L. (1965). DDR-Wirtschaftspatent 41,967.

Morris, L. J. (1967). *Lipids* 3, 260.

Morris, L. J., Hall, S. W., and James, A. T. (1966). *Biochem. J.* 100, 29C.

Mothes, K., Winkler, K., Gröger, D., Floss, H. G., Mothes, U., and Weygand, F. (1962). *Tetrahedron Letters* p. 933.

Mrtek, R. G., Crespi, H. L., Blake, M. I., and Katz, J. J. (1965). *J. Pharm. Sci.* **54**, 1450.

Mrtek, R. G., Crespi, H. L., Norman, G., Blake, M. I., and Katz, J. J. (1968). *Phytochemistry* **7**, 1535.

Nelson, U., and Agurell, S. (1969). *Acta Chem. Scand.* **23**, 3393.

Oddo, N., and Tonolo, A. (1967). *Ann. Ist. Super. Sanita* **3**, 16.

Pacifici, L. R., Kelleher, W. J., and Schwarting, A. E. (1962). *Lloydia* **25**, 37.

Pacifici, L. R., Kelleher, W. J., and Schwarting, A. E. (1963). *Lloydia* **26**, 161.

Pantidou, M. E. (1959). *Can. J. Botany* **37**, 1233.

Perlin, A. S., and Taber, W. A. (1963). *Can. J. Chem.* **41**, 2278.

Plieninger, H. (1966). *Abhandl. Deut. Akad. Wiss. Berlin, Kl. Chem., Geol. Biol.* No. 3, p. 387.

Plieninger, H., Fischer, R., Lwowski, W., Brack, A., Kobel, H., and Hofmann, A. (1959). *Angew. Chem.* **71**, 383.

Plieninger, H., Immel, H., and Völk, A. (1967). *Ann. Chem.* **706**, 223.

Pöhm, M. (1953). *Arch. Pharm.* **286**, 509.

Pöhm, M., and Fuchs, L. (1954). *Naturwissenschaften* **41**, 63.

Ramstad, E. (1968). *Lloydia* **31**, 327.

Ramstad, E., Chan Lin, W. N., Shough, R., Goldner, K. J., Parikh, R. P., and Taylor, E. H. (1967). *Lloydia* **30**, 441.

Robbers, J. E., and Floss, H. G. (1969a). *Arch. Biochem. Biophys.* **126**, 967.

Robbers, J. E., and Floss, H. G. (1969b). *Tetrahedron Letters* p. 1857.

Rochelmeyer, H. (1950). *Sueddeut. Apotheker-Ztg.* **90**, 39.

Rochelmeyer, H. (1954). *Deut. Apotheker-Ztg.* **94**, 1.

Rochelmeyer, H. (1958). *Pharm. Ztg., Ver. Apotheker-Ztg.* **103**, 1269.

Röder, K., Mutschler, E., and Rochelmeyer, H. (1967). *Pharm. Acta Helv.* **42**, 407.

Rothe, U., and Fritsche, W. (1967). *Arch. Mikrobiol.* **58**, 77.

Rozumek, K.-E. (1963). *Beitr. Biol. Pflanz.* **38**, 238.

Šantavý, F. (1967). *In* "Dünnschichtchromatographie, Ein Laboratoriumshandbuch" (E. Stahl, ed.), pp. 405-449. Springer, Berlin.

Schlientz, W., Brunner, R., Thudium, F., and Hofmann, A. (1961a). *Experientia* **17**, 108.

Schlientz, W., Brunner, R., Hofmann, A., Berde, B., and Stürmer, E. (1961b). *Pharm. Acta Helv.* **36**, 472.

Schlientz, W., Brunner, R., and Hofmann, A. (1963). *Experientia* **19**, 397.

Schlientz, W., Brunner, R., Stadler, P. A., Frey, A. J., Ott, H., and Hofmann, A. (1964). *Helv. Chim. Acta* **47**, 1921.

Schlientz, W., Brunner, R., Rüegger, A., Berde, B., Stürmer, E., and Hofmann, A. (1967). *Experientia* **23**, 991.

Schreier, E. (1958). *Helv. Chim. Acta* **41**, 1984.

Shelesnyak, M. C. (1954). *Proc. Soc. Exptl. Biol. Med.* **87**, 337.

Shelesnyak, M. C. (1955). *Am. J. Physiol.* **180**, 47.

Shemyakin, M. M., Tchaman, E. S., Denisova, L. I., Ravdel, G. A., and Rodionow, W. J. (1959). *Bull. Soc. Chim. France* p. 530.

Silber, A. (1952). *Pharmazie* **7**, 854.

Silber, A., and Bischoff, W. (1954). *Pharmazie* **9**, 46.

Silber, A., Mothes, K., and Gröger, D. (1955). *Kulturpflanze* **3**, 90.

Skalický, V., and Starý, F. (1962). *Preslia* **34**, 229.

Smith, M. J. (1930). *Public Health Rept. (U.S.)* **45**, 1466.

Smith, S., and Timmis, G. M. (1937). *J. Chem. Soc.* p. 396.

Spilsbury, J. F., and Wilkinson, S. (1961). *J. Chem. Soc.* p. 2085.

Stadler, P. A., and Hofmann, A. (1962). *Helv. Chim. Acta* **45**, 2005.

Stadler, P. A., Guttmann, S., Hauth, H., Huguenin, R. L., Sandrin, E. D., Wersin, G., and Hofmann, A. (1969). *Helv. Chim. Acta* 52, 1549.

Stäger, R. (1922). *Centr. Bakteriol., Parasitenk., Abt. II* 56, 329 (and earlier papers).

Stauffacher, D., and Tscherter, H. (1964). *Helv. Chim. Acta* 47, 2186.

Stauffacher, D., Tscherter, H., and Hofmann, A. (1965). *Helv. Chim. Acta* 48, 1379.

Stauffacher, D., Niklaus, P., Tscherter, H., Werner, H. P., and Hofmann, A. (1970). *Tetrahedron* 25, 5879.

Stearns, J. (1808). *Med. Repository N.Y.* 5, 308.

Steiner, M., and Hartmann, T. (1958). *Flora (Jena)* 146, 472.

Steiner, M., and Hartmann, T. (1964). *Biochem. Z.* 340, 436.

Stein von Kamienski, E. (1958). *Planta* 50, 331.

Stoll, A. (1918). Swiss Patent 79,879.

Stoll, A. (1952). *Fortschr. Chem. Org. Naturstoffe* 9, 114-174.

Stoll, A., and Brack, A. (1944a). *Pharm. Acta Helv.* 19, 118.

Stoll, A., and Brack, A. (1944b). *Ber. Schweiz. Botan. Ges.* 54, 252.

Stoll, A., and Burckhardt, E. (1935). *Compt. Rend.* 200, 1680.

Stoll, A., and Burckhardt, E. (1937). *Z. Physiol. Chem.* 250, 1.

Stoll, A., and Hofmann, A. (1943a). *Helv. Chim. Acta* 26, 1570.

Stoll, A., and Hofmann, A. (1943b). *Helv. Chim. Acta* 26, 944.

Stoll, A., and Hofmann, A. (1943c). *Helv. Chim. Acta* 26, 922.

Stoll, A., and Hofmann, A. (1965). *In* "The Alkaloids" (R. H. F. Manske and H. C. Holmes, eds.) Vol. 8, pp. 725-783. Academic Press, New York.

Stoll, A., and Rüegger, A. (1954). *Helv. Chim. Acta* 37, 1725.

Stoll, A., and Rutschmann, J. (1950). *Helv. Chim. Acta* 33, 67.

Stoll, A., Rutschmann, J., and Schlientz, W. (1950). *Helv. Chim. Acta* 33, 375.

Stoll, A., Brack, A., Kobel, H., Hofmann, A., and Brunner, R. (1954). *Helv. Chim. Acta* 37, 1815.

Taber, W. A. (1963). *Develop. Ind. Microbiol.* 4, 295.

Taber, W. A. (1964). *Appl. Microbiol.* 12, 321.

Taber, W. A. (1967). *Lloydia* 30, 39.

Tanret, C. (1875). *Compt. Rend.* 81, 896.

Tanret, C. (1909). *Compt. Rend.* 149, 222.

Teh-chao, Y., Yung-peng, Y., Jing-ru, L., Shih-yi, L., Li-yi, H., Pin, L., Tung-whei, C., Liang-niang, L., and Qui-cheng, F. (1962). *Sci. Sinica* 11, 917.

Teichert, K., Mutschler, E., and Rochelmeyer, H. (1960a). *Deut. Apotheker-Ztg.* 100, 283.

Teichert, K., Mutschler, E., and Rochelmeyer, H. (1960b). *Deut. Apotheker-Ztg.* 100, 477.

Teuscher, E. (1964). *Flora (Jena)* 155, 80.

Teuscher, E. (1966). *Abhandl. Deut. Akad. Wiss. Berlin, Kl. Chem., Geol. Biol.* No. 3, p. 429.

Thiele, O. W. (1964). *Biochim. Biophys. Acta* 84, 483.

Thielmann, E. (1955). Ph.D. Thesis, T. H. Stuttgart.

Thompson, M. R. (1935). *Science* 81, 636.

Tondo, A. (1966). *Nature* 209, 1134.

Tonolo, A. (1967). *Ann. Ist. Super. Sanita* 3, 613.

Tonolo, A., and Udvardy-Nagy, E. (1968). *Acta Microbiol. Acad. Sci. Hung.* 15, 29.

Tschirch, A. (1923). "Handbuch der Pharmakognosie." Vol. 3, pp. 139-165. Tauchnitz, Leipzig.

Tulasne, L. R. (1851). *Compt. Rend.* 33, 645.

Tulasne, L. R. (1853). *Ann. Sci. Nat. Botan. Biol. Vegetale* [iii] 20, 5.

Tyler, V. E., Jr. (1958). *J. Am. Pharm. Assoc.* 47, 787.

Tyler, V. E., Jr., Mothes, K., Gröger, D., and Floss, H. G. (1964). *Tetrahedron Letters* p. 593.

Uhle, F. C., and Jacobs, W. A. (1945). *J. Org. Chem.* **10**, 76.

van Urk, H. W. (1929). *Pharm. Weekblad* **66**, 473.

Vauquelin, M. (1816). *Ann. Chim. Phys.* **3**, 337.

Vining, L. C., and Nair, P. M. (1966). *Can. J. Microbiol.* **12**, 915.

Vining, L. C., and Taber, W. A. (1963a). *Can. J. Microbiol.* **9**, 292.

Vining, L. C., and Taber, W. A. (1963b). *In* "Biochemistry of Industrial Microorganisms" (C. Rainbow and A. H. Rose, eds.), pp. 341-378. Academic Press, New York.

Voigt, R. (1959). *Pharmazie* **14**, 607.

Voigt, R. (1968). *Pharmazie* **23**, 285, 353, and 419.

Voigt, R., Bornschein, M., and Rabitzsch, G. (1967). *Pharmazie* **22**, 326.

von Békésy, N. (1935). Hungarian Patent 112,711.

von Békésy, N. (1938). *Zentr. Bakteriol., Parasitenk., Abt. II* **99**, 321.

von Békésy, N. (1939). *Biochem. Z.* **303**, 368.

von Békésy, N. (1956a). *Phytopathol. Z.* **26**, 49.

von Békésy, N. (1956b). *Z. Pflanzenzuecht.* **35**, 461.

von Békésy, N. (1956c). *Pharmazie* **11**, 339.

von Münchhausen, O. (1764). "Der Hausvater," Part 1, p. 332. Hannover.

Walker, J. (1957). *Proc. Linnean Soc. N.S. Wales* **82**, 322.

Weil, R. (1910). D. R. Patent 267,560.

Wildy, J., and Heath, H. (1957). *Biochem. J.* **65**, 220.

Yamano, T. (1961). *J. Agr. Chem. Soc. Japan* **35**, 1284.

Yamano, T., Kishino, K., Yamatodani, S., and Abe, M. (1962). *Ann. Rept. Takeda Res. Lab.* **21**, 95.

Yamatodani, S. (1960). *Ann. Rept. Takeda Res. Lab.* **19**, 1.

Yamatodani, S., and Abe, M. (1955). *Bull. Agr. Chem. Soc. Japan* **19**, 94.

Yamatodani, S., and Abe, M. (1956). *Bull. Agr. Chem. Soc. Japan* **20**, 95.

Yamatodani, S., and Abe, M. (1957). *Bull. Agr. Chem. Soc. Japan* **21**, 200.

Author Index

Subject Index

A

Acencoumarol, as anticoagulant drug, 81
Aesculetin
  in plant tissue, 53, 62, 68, 86
  use of, 85
Aesculin
  biological action of, 51, 52
  uses of, 70
Aflatoxins, toxicity of, 4, 86, 87
Agaricic acid
  as mushroom toxin, 309-310
  structure of, 309
Agaricus sp., toxins from, 308
Agroclavine, 339, 345
  structure of, 332, 340
Alfalfa plants, stress metabolite formation
    in, 240
Alkaloids, in ergot, 329-348
Altenin
  isolation and properties of, 173-175, 190
  structure of, 174
Altenuic acids, properties of, 190
Altenusin, properties of, 190
Alternaria kikuchiana toxins, *see* Phytoal-
    ternarins; Altenin
Alternaria solani toxin, isolation and prop-
    erties of, 175-178
Alternaria tenuis toxin, isolation and prop-
    erties of, 178-187
Alternaria toxins, 169-192
  summary of, 190
Alternaria zinniae toxin, isolation and
    properties of, 187-189
Alternaric acid
  biological action of, 51
  isolation and properties of, 175-178, 190
  structure of, 176
Alternariol
  isolation of, 51
  properties of, 190
Altertenuol, properties of, 190
Amanin, 250
  chemistry of, 259
  structure of, 262

Amanita bisporigea, description of, 249
Amanita citrina, description of, 249
  toxins in, 304
Amanita muscaria, muscarine from, 289
Amanita phalloides, description of, 249
Amanita solitaria, toxin from, 299
Amanita tenuifolia, description of, 249
Amanita toxins, 249-280
  chemistry of, 252-265
  in *Galerina*, 249, 284
  lysosomes and, 276
  narcotic-intoxicant type, 296
  occurrence, analysis, and characteri-
    zation of, 249-252
  paper chromatography of, 251-252
  poisoning by, 252
  species sensitivity to, 270
  therapy for, 277-278
  toxicology of, 265-278
    clinical symptoms and pathology of,
      265
    experimental studies on, 266-277
  tryptamine derivatives, 304
Amanita virosa, description of, 249
Amanitins, 250
  chemistry of, 260
  mechanism of action of, 271-274
  pathological changes caused by, 269
  structure of, 262
  toxicity of, 251, 266
Amanullin, 250
  chemistry of, 259, 261
  structure of, 262
Amatoxins
  analysis of, 250-251
  chemistry of, 252, 259-264
  comparison with phallotoxins, 253
  mechanism of action of, 271-274
  toxicity of, 251-252
Angelicin, biological action of, 61
Antamanide
  as antitoxin, 264-265, 277-278
  structure of, 264
Anthoxanthum odoratum, toxic coumarins
    from, 21

392